# PHYSICS for
# ENGINEERS and SCIENTISTS

# PHYSICS for ENGINEERS and SCIENTISTS

**Gebhard von Oppen**
**and**
**Frank Melchert**

**INFINITY SCIENCE PRESS LLC**
Hingham, Massachusetts

INFINITY SCIENCE PRESS LLC
11 Leavitt Street
Hingham, MA 02043
Tel. 877-266-5796 (toll free)
Fax 781-740-1677
info@infinitysciencepress.com
www.infinitysciencepress.com

This book is printed on acid-free paper.

Gebhard von Oppen and Frank Melchert. *Physics for Engineers and Scientists.*
ISBN: 0-9778582-1-9

Library of Congress Cataloging-in-Publication Data

Oppen, G. von (Gebhard von)
    Physics for engineers and scientists / G. von Oppen and F. Melchert.
       p. cm.
    Includes index.
    ISBN 0-9778582-1-9 (hardcover : alk. paper)
    1. Physics. I. Melchert, F. (Frank) II. Title.
QC21.3.O67 2007
530—dc22
                                        2006028115
Printed in the United States of America
6 7 8 9 5 4 3 2 1

# Contents

# Foreword

Physics is the foundation of all the engineering sciences. That is why a set of lectures on physics shows up at the beginning of most engineering curricula. These lectures are usually heard in the first year of study by young people who are setting out on a new and exciting part of their lives with great expectations. A course in such a fascinating area as physics should, therefore, not only impart basic physical knowledge in a compact form, but should also arouse and satisfy the curiosity and interest of the students. Naturally, neither classical nor modern physics should be shortchanged. Without classical physics, modern physics cannot be understood, and without modern physics, an introductory course loses value for many engineering curricula.

Unfortunately, there are limits on the wish list for an introductory course in physics. An academic year has 30 weeks and in most courses there are no more than two lecture hours with at best two more hours in lab or recital. It is obviously impossible to present all of physics in its variety in these 30 weeks or for a student to learn and comprehend it.

Physics is exciting, not only because of its variety, but even more so because of its internal consistency. With a few fundamental concepts and laws an astonishingly large amount of natural phenomena can be described and logically comprehended. It is worthwhile to make the associated concepts of physics comprehensible, establishing the relationship to daily experience and to experiment. It is this interplay between theory and experiment that has stimulated the historical development of physics. Obviously, internally self-consistent concepts such as Newton's mechanics or Maxwell's electromagnetic theory have had to be extended and replaced by new theoretical concepts based on new insights and observations.

In this textbook we have tried to present the fundamental concepts of the physical description of nature within the framework of a one-year introductory course. In order to emphasize the structural connection to a one-year course, we have divided the contents of this book into 33 lectures. Each lecture is intended to be covered in two hours. On the scale of an academic year this is at least three lectures too many. We believe that this excess is justifiable in terms of rounding out the subject matter without overtaxing the students or the teacher as to the choice of class material.

The 33 lectures are grouped into six chapters. The first three chapters, with a total of 17 lectures, deal with classical physics, ranging from mechanics, through thermodynamics, and on to electromagnetic theory. In presenting these disciplines we have endeavored to clarify the differences among the fundamental concepts of these branches of physics, so that students will be prepared for the far-reaching transition to modern quantum physics. In the last three chapters, with a total of 16 lectures, we present those branches of physics which can only be understood comprehensively in terms of quantum mechanics.

This textbook is based on many years of experience in teaching elementary physics at the Technical University of Berlin. It is based on the lecture notes written over many years by one of the authors (G. v. O.), together with his former coworkers W.-D. Perschmann and D. Kaiser, for students in the electrical engineering curriculum. Matching the subject matter to the electrical engineering curriculum and its modern disciplines of semiconductor technology and optoelectronics made it necessary from the beginning to grant an equal amount of space to quantum physics and to classical physics.

In presenting the theoretical foundations, we have tried to use the most elementary mathematics possible. We hope that, in this way, first-year students will have a chance to grasp the essential intellectual concepts of physics. In order to focus the attention of students on the essentials, each lecture is divided into four sections and a brief summary is provided at the end of each section. Problems are scattered throughout the text, so students are invited to go beyond passive learning of the material to active involvement with it.

The authors thank all those who, over many years as teachers or students, have collaborated in the "Physics for Electrical Engineering" course at the Technical University of Berlin and contributed to the concept of this textbook. We especially thank Dr. Thorsten Ludwig for his untiring work in formatting and Dr. Uwe Brinkmann for reviewing the manuscript. We hope that many students will use this book to gain their first insight to the fundamental concepts of physics and hope they will find that working through the material is both a pleasure and fun.

Gebhard von Oppen
Frank Melchert
Berlin, June 2006

# Introduction

*"The leaves fall, fall as though from afar,*
*as though withered in distant gardens of the sky;*
*their fall a tremor of denial.*
*And by night the heavy earth falls*
*from all the stars into solitude."*

The falling of autumn leaves is wonderfully described by Rainer Maria Rilke, but falling leaves can also interest a physicist, as well as a poet. A physicist, however, will not just leave them to the play of the autumn winds, but observe their fall and measure it under well-controlled conditions. The goal of a physical experiment is to create conditions under which measurements are as exactly reproducible as possible. Only when every kind of uncontrollable external influence on the object being studied has been eliminated as completely as possible, can we hope to discover the fundamental regularities in natural events. The *tremor of denial* of those falling leaves that so engages the poet and the readers of poetry, disturbs the experimental physicist because of its randomness. In order to avoid it, one can, in a lecture demonstration, for example, perform experiments on falling objects in a vacuum or using objects with little air friction, such as heavy spheres.

In their efforts to obtain perfect experimental conditions, however, it soon becomes clear to every experimenter that, despite all their pains, measurements are never exactly reproducible. There is always a bit of poetry left. In repeated measurements, the measured values are usually spread about an average value. The mean (square) deviation of a measured value from the average is a measure of the uncertainty associated with a measurement result, so the deviation must, by all means, be indicated along with the measured value.

The greatest desire of any experimenter is to be able to explain measurements that are subject to uncertainties with a new natural law or even one which was already widely suspected. If it works, it delights them and others, much as the poetry of Rilke. But it should not be forgotten that a law of nature, such as the law of gravitation in the form

$$s = \frac{1}{2} g t^2,$$

is only a description of idealized measurements. The mathematical equation represents an exact relationship between the measurements of time and place. A natural law can never be proven so exactly by experiments! Theory not only reproduces the experimental facts, but goes substantially beyond the experimental results to formulate a mathematically precise relationship.

In the course of these lectures we shall see how important it is always to be aware of this exaggerated accuracy of theory. This is because, with the increasing refinement and precision of experimental techniques, we shall keep discovering that natural laws which are considered to be fundamental are only applicable within a limited range of experience and that outside this range they lose their validity. Then the problem becomes one of seeking a better formulation of these laws.

The inability of theory to square with the "poetry" of experiments often has quite radical consequences, since laws of limited validity have to be replaced by more precise formulations. Every theory has an intellectual and conceptual foundation which is based on a world view. A world view of this sort is, like the poetry of Rilke, a work of art of the human spirit and not an objectively preset reality. Evolved as a result of unexpected, new experimental results, the world view of physics has undergone many decisive changes over the centuries. Besides the desire to provide a compact survey of the concepts of physics, especially for engineers, one of the concerns of these lectures is to make these changes in the world view of physics both clear and convincing.

We begin with the mechanics of Newton. Newtonian mechanics is based on the assumption that material objects move in space and time under the influence of external forces. These motions are continuous. *Natura non facit saltus*—nature makes no leaps—is a fundamental assumption of classical physics. In fact, daily experience appears to teach us that all objects change their positions in space steadily and not discontinuously. This continuum hypothesis of classical physics is, however, brought into question by the atomic and quantum hypotheses of modern physics.

In order to introduce the basic ideas of Newtonian mechanics, in Chapter 1 we limit the discussion to the simplest objects, such as point masses and rigid objects. Both are extreme idealizations of the things studied in experiments. This is because all material objects have a finite size and are more or less deformable. A civil engineer, for example, will obviously be interested in the load-carrying capacity and the deformation under load of structural materials. Then why, in a textbook for engineers, are idealizations discussed, rather than realistic objects? We can say two things in response: 1. The physics of the extended and deformable objects of interest to a civil engineer also relies on the elementary laws of physics, which will be explained here with the aid of simple idealizations. Rather than provide specialist knowledge, an introductory course in physics should enable budding engineers to appreciate the elementary concepts behind the often difficult and complex practical and theoretical procedures of engineering science and, thereby, enable them to acquire a deeper understanding. 2. The fundamental laws of nature are most clearly revealed through studies of simple objects and processes, not only to researchers, but also to students. The elementary laws of nature are, themselves, always related to idealized limiting cases, which must be seen in terms of the underlying world view.

In the following chapters, it will be shown how the world view of physics has changed with experimental advances. In Chapter 2 a discrete structure shows up for the first time in the physical world view with the atomic hypothesis. It is closely linked to the idea of chance, which conflicts sharply with the determinism of classical mechanics. Chapter 3 proceeds from the phenomena of wave propagation and equilibration processes in material objects to Faraday's idea of the electromagnetic field. At first the field seems to be a perfect continuum and appears to be entirely consistent with the classical continuum hypothesis. Hence it was surprising when Max Planck concluded from measurements of the electromagnetic radiation from black bodies that the electromagnetic field also has a discrete structure. We shall present the arguments for the quantum hypothesis in Chapter 4.

On one hand, the quantum hypothesis represents a giant step toward a unified physical world view, in which the discrete structure of matter made evident by the atomic hypothesis and the associated randomness in natural phenomena are brought into consistency with the physics of the electromagnetic field as a result of the quantum hypothesis. On the other hand, the quantum hypothesis implies a dualism between the wave and particle pictures that is hard to grasp in terms of the world view imprinted by classical physics. Sometimes it is advantageous to treat the propagation of electromagnetic waves as a wave motion and at other times, as a particle motion. In Chapter 5 we show that this wave-particle dualism is not a special feature of the electromagnetic field, but also applies to material objects such as electrons. In combination with the wave nature of electrons, the quantum hypothesis offers the key to understanding the physics of atoms and molecules. The quantum mechanical idea of a wave-particle dualism also opens up the path to solid-state physics and quantum optics, which are the basis of modern electronics. The ideal gas model of classical physics has a quantum physical parallel: the quantum gas. This will be discussed in Chapter 6. Because of the wave-particle dualism, such variegated processes as lattice vibrations of crystals and electromagnetic oscillations in resonators or the motion of electrons in metals, insulators, and semiconductors can be described to a good approximation using a simple model of a quantum gas.

At the end of Chapter 6 we return again to the fundamental problem of the so-called exact natural sciences: the experimental uncertainty of all measurements. This seems to be an elementary trait of all natural sciences and points the way to a deeper understanding of the apparent internal contradictions in the physical world views.

Chapter

# 1 Mechanics of Idealized Bodies

## Summary

- Motion in space and time
- Dynamics of point masses
- Conservation of energy and momentum
- Angular momentum
- Dynamics of rigid bodies
- Oscillations

The mechanics founded by Newton stands at the beginning of physics in the modern era. Until around 1900, phenomena observed in nature or in experiments were regarded as physically understood if they could be explained by the laws of mechanics. With the development of the theory of relativity and quantum mechanics, mechanics lost its central significance in physics. The range of validity of the laws of mechanics is bounded by the natural constants $c$ (the speed of light) and $h$ (Planck's constant). Nevertheless, mechanics has made a persistent impression on our ideas of nature. Many concepts, such as mass, energy, momentum, and angular momentum, which were initially introduced in the framework of mechanics, still retain their original significance in modern physics. For this reason alone, today a basic understanding of mechanics continues to be a prerequisite for the understanding of physical relationships in general.

## 1.1  MOTION IN SPACE AND TIME

The perception of motion is among the first experiences in human life. In order to understand and represent motions, we use both measurement devices, such as rulers and clocks, and abstract concepts of space and time. In this lesson the basic concepts and laws of kinematics will be discussed with the aid of some simple examples, such as free fall, uniform motion along straight lines and circles, and planetary motion.

### 1.1.1  Space and time

The concepts of space and time are fundamentally connected to our perceptions of distances and the propagation of light. For example, we consider an edge to be straight if it coincides with the path of a ray of light. Lengths and times have to be measured in order to characterize geometric structures and describe motions. For this purpose, units of measurement are necessary. These units of measurement must be realized through suitable techniques. In the following we shall use the internationally agreed *SI units* (Système International d'Unités). In realizing these units, it is important to develop techniques that are as precise as possible, and which yield good, reproducible results at all times and all places. It should be no surprise that the modern techniques for realization of SI units cannot be understood physically right from the beginning of a short lecture course in physics, but only at the end. Hence, here we shall limit ourselves to just a few explanatory comments regarding the SI units of length and time. Both units will be defined in terms of the quantum structure of the atom (Section 5.1.3) and the wave properties of light or other electromagnetic waves (Section 3.6). The *SI unit of time* is the second and has been defined since 1960 as the duration of a fixed number of oscillations of the Cs atom (transitions between hyperfine levels of the ground state of $^{133}$Cs):

$$1\,\text{s} = 9{,}192{,}631{,}770 \text{ ground state oscillations of } ^{133}\text{Cs}.$$

The so-called cesium atomic clock makes it possible to measure intervals of time, $\Delta t$, with a relative accuracy of $\delta t/\Delta t = 10^{-12}$ s; that is, the time according to two cesium atomic clocks will differ by at most $\delta t \sim 1\,\text{s}$ after $10^{12}$ s ($\approx 10^{7}$ days $\approx 30{,}000$ years).

Since 1983 the SI unit of length has been defined using the fact that light propagates at a fixed velocity in space (Section 1.2.4). Today the *speed of light* is set arbitrarily at

$$c = 299792458 \text{ m/s}.$$

The *SI unit of length*, the meter, is thereby defined. 1 m is the distance that light will travel in a vacuum in $(1/299792458)$ s $\approx 3.3 \times 10^{-9}$ s.

Once the value of $c$ is established, of course, the unit of length is defined; however, suitable measurement procedures will be needed in order to measure precisely the distances between specified points in space. One possibility is to measure the specified distance by comparison with the wavelength $\lambda = c/\nu$ of light in a spectrum line of known frequency $\nu$. A comparison of this sort can be carried out using the Michelson interferometer, described in Section 4.2.2.

Since the wavelength of visible light is on the order of $\lambda \sim 10^{-6}$ m, the classical Michelson interferometer can be used to measure lengths of order ranging from $10^{-6}$ m to $10^{-2}$ m. But physics is concerned with objects that extend over 40 orders of magnitude, rather than just four (Table 1.1). Depending on the order of magnitude of the size of the object to be measured, very different measurement procedures will be needed. The same holds for time measurements, for the durations of physical processes likewise range over about 40 orders of magnitude.

**Table 1.1** Orders of magnitude of lengths and distances (in m).

| | |
|---|---|
| Radius of the proton | $10^{-15}$ |
| Radius of an atom | $10^{-10}$ |
| Size of a virus | $10^{-7}$ |
| Wavelength of visible light | $10^{-6}$ |
| Height of a person | $10^{0}$ |
| Diameter of the earth | $10^{7}$ |
| Distance from the earth to the sun | $10^{11}$ |
| 1 light year | $10^{16}$ |
| Diameter of the Milky Way | $10^{21}$ |
| Distance to the furthest galaxies | $10^{26}$ |

Measurements of length are the basis of *geometry*. Physical space is generally known to have 3 dimensions (length, width, and height) and is usually assumed to be Euclidean. Whether this assumption is valid or not is to be decided by experimental studies. For example, the sum of the three angles in a triangle is 180° in Euclidean geometry. Whether this proposition applies to physical space was first proved with great accuracy by the famous mathematician **C.F. Gauss**

**(1777–1855)**. In 1821–1823 he measured the sum of the angles in the triangle formed by the mountain peaks of Brocken, Hohe Hagen, and Inselsberg (of which the longest side is about 100 km in length).

In order to determine the angle between two intersecting straight lines, one makes a precise measurement of the ratio $b/r$ of two lengths, specifically of a circular arc $b$ to its radius $r$. This (dimensionless) ratio of the two lengths is the *angle in radians* (*rad*, for short). A full revolution corresponds to turning by an angle of $2\pi$. Correspondingly, $180° = \pi$.

To within the experimental error, Gauss' measurements did yield a value of $\pi$ for the sum of the three angles and thereby confirmed Euclidean geometry. At cosmic distances, however, spatial curvature is quite detectable. This was predicted by Einstein in his general theory of relativity. In the course of these lectures, we shall be concerned only with the physics of earthbound processes, so the entire discussion will be based on Euclidean geometry.

In order to label the position of an object in space, one must rely on *reference frames*. As an example, in the lecture hall we could use a table to specify a *reference system*. The three edges emerging from a corner of the tabletop could, for example, be chosen as the $x$-, $y$-, and $z$-axes of a (righthanded, rectangular) *Cartesian coordinate system*. This makes it clear that physical space is, indeed, three dimensional. The position of an arbitrary point in pace can then be described by specifying three coordinates, which are combined into a *position vector* $\mathbf{r} = (x, y, z)$. Instead of the three Cartesian coordinates $x$, $y$, $z$, the position vector can be specified in terms of three *polar coordinates* $(r, \theta, \varphi)$. Figure 1.1 shows that $r = \sqrt{x^2 + y^2 + z^2}$ is the length of the vector, $\theta$ is the angle between the vector and the $z$-axis ($0 \leq \theta \leq \pi$), and $\varphi$ is the angle between the projection of the vector on the $x$-$y$ plane and the $x$-axis ($0 \leq \varphi \leq 2\pi$). Thus,

$$x = r \sin\theta \cos\varphi$$
$$y = r \sin\theta \sin\varphi$$
$$z = r \cos\theta$$

The values of the coordinates $x$, $y$, $z$ and the angles $\theta$ and $\varphi$ depend on the choice of coordinate system. On the other hand, the length $r$ of the vector corresponding to the distance between two given points in space is independent of the choice of coordinate system; that is, the length $r$ is a scalar. For two given vectors $\mathbf{r}_1$ and $\mathbf{r}_2$, the lengths of the two vectors, as well as the angle $\theta_{1,2}$ between them, and the scalar product

$$\mathbf{r}_1 \cdot \mathbf{r}_2 = x_1 x_2 + y_1 y_2 + z_1 z_2 = r_1 r_2 \cos\theta_{1,2}$$

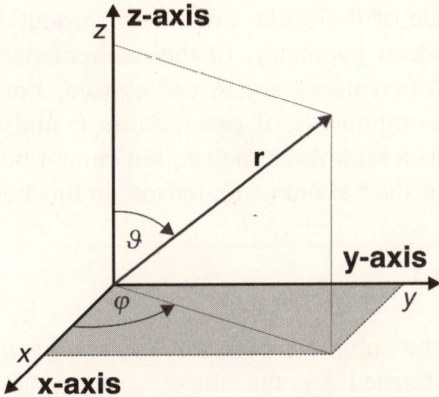

**FIGURE 1.1** Cartesian coordinates and polar coordinates of a vector **r**.

are scalars. Besides the scalar product, the area of the parallelogram formed by the two vectors is of interest. This area is described by the vector **A**, which specifies both the magnitude and the orientation in space of the area. The vector **A** is perpendicular to $\mathbf{r}_1$ and $\mathbf{r}_2$, and the triplet $(\mathbf{A}, \mathbf{r}_1, \mathbf{r}_2)$ obeys the righthand rule (Figure 1.2). The length $A = |\mathbf{A}| = r_1 r_2 |\sin \theta_{1,2}|$ of **A** is the magnitude of the area of the parallelogram.

$$\mathbf{A} = \mathbf{r}_1 \times \mathbf{r}_2$$

is referred to as the *vector product* of $\mathbf{r}_1$ and $\mathbf{r}_2$. Its (Cartesian) coordinates are

$$A_x = y_1 z_2 - y_2 z_1$$
$$A_y = z_1 x_2 - z_2 x_1$$
$$A_z = x_1 y_2 - x_2 y_1.$$

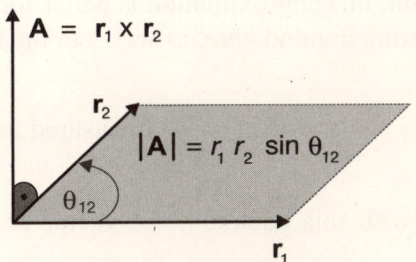

**FIGURE 1.2** Illustrating the vector product **A** of two vectors $\mathbf{r}_1$ and $\mathbf{r}_2$.

The significance of the scalar and vector products discussed here is intuitive in terms of Euclidean geometry. In the mathematical description of physical relationships it is often necessary to use *tensors*. For example, the $3 \times 3$ product of the Cartesian components of two vectors $\mathbf{r}_1$ and $\mathbf{r}_2$ form a tensor. This can be written down as a rank three matrix, but cannot be visualized as easily as vectors and scalars. We shall avoid using tensors in this book.

---

### Problem 1.1

Calculate the volume $V = (\mathbf{r}_1 \times \mathbf{r}_2) \cdot \mathbf{r}_3$ (*scalar triple product*) of the parallelepiped formed by the three vectors $\mathbf{r}_1 = (1,0,0)$, $\mathbf{r}_2 = (1,1,0)$, and $\mathbf{r}_3 = (1,1,1)$, as well as the angles between these vectors.

---

> **Notes**
>
> The properties of physical space are to be determined experimentally. For example, in order to measure the sum of the angles of a triangle, three points in space must first be designated, the straight lines which join them must be distinguished physically, a unit of length and a suitable measurement procedure must be established, and, finally, the lengths of the line segments and circular arcs must be measured. In each step the experimental errors must be taken into account.

## 1.1.2  Kinematics

A moving object changes its position in space with time. The object is idealized as a point mass, so its motion can be described by specifying a *trajectory* $\mathbf{r}(t)$. Besides the position $\mathbf{r}(t)$ of the object at time $t$, we are also interested in its *velocity* $\mathbf{v}(t)$ and its *acceleration* $\mathbf{a}(t)$ at time $t$. The velocity $\mathbf{v}(t)$ is measured by determining the position $\mathbf{r}$ of the object at time $t$ and at a somewhat later time $t + \Delta t$. Then we have approximately $\mathbf{v}(t) \approx [\mathbf{r}(t + \Delta t) - \mathbf{r}(t)]/\Delta t$. From a theoretical standpoint, this approximation is better for a smaller chosen $\Delta t$. (Keep the experimental errors in mind here, as well.) In the limit $\Delta t \to 0$, we have

$$\mathbf{v}(t) = \frac{d\mathbf{r}}{dt} \quad \text{(measured in m/s)}.$$

In accordance with this equation, the vector components of $\mathbf{v} = (v_x, v_y, v_z)$ are given by

$$v_x = \frac{dx}{dt}, \, v_y = \frac{dy}{dt}, \, v_z = \frac{dz}{dt}.$$

The acceleration $\mathbf{a}(t)$ of the object at time $t$ is correspondingly measured by determining the velocities $\mathbf{v}(t)$ and $\mathbf{v}(t + \Delta t)$. Then we have

$$\mathbf{a} = \frac{d\mathbf{v}}{dt} = \frac{d^2\mathbf{r}}{dt^2} \quad \text{(measured in m/s}^2\text{)}.$$

A simple example is provided by uniform linear motion, for which $\mathbf{r}(t) = \mathbf{r}_0 + \mathbf{v}_0 t$, $\mathbf{v}(t) = \mathbf{v}_0$, and $\mathbf{a}(t) = 0$. Another example is free fall, for which $\mathbf{a}(t) = \mathbf{g}$ is constant in time. The *acceleration of gravity* $\mathbf{g}$ is a vector directed toward the center of the earth, with a magnitude of $g = 9.81 \, \text{m/s}^2$ in Central Europe. The velocity $\mathbf{v}(t)$ and distance travelled $\mathbf{r}(t)$ for free fall are found by integrating with respect to time:

$$\mathbf{v}(t) = \mathbf{v}_0 + \mathbf{g}t$$

$$\mathbf{r}(t) = \mathbf{r}_0 + \mathbf{v}_0 t + \frac{1}{2}\mathbf{g}t^2.$$

Here $\mathbf{r}_0$ and $\mathbf{v}_0$ are the position and velocity at time $t = 0$. If $\mathbf{r}_0 = \mathbf{v}_0 = 0$, then $s = |\mathbf{r}(t)|$ is the distance an object initially at rest falls in time $t$ and is given by

$$s = \frac{1}{2}gt^2.$$

---

**Experiment 1.1   Free fall**

---

Experiment confirms that, to within the experimental error, all objects in free fall experience the same acceleration $g = 9.81 \, \text{m/s}^2$. In order to avoid air friction, let us have the objects fall in an evacuated glass tube (Figure 1.3). The vacuum pump reduces the air pressure to about one thousandth of the normal pressure. The time of fall is measured from the starting position to arrival at a light barrier using an electronic timer. The timer is actuated

*continued*

when the electromagnet is turned off to start the fall and it is switched off when the falling object passes the light barrier. A thumbtack with feather attached to it requires about (within the measurement accuracy) the same time to fall as a steel ball.

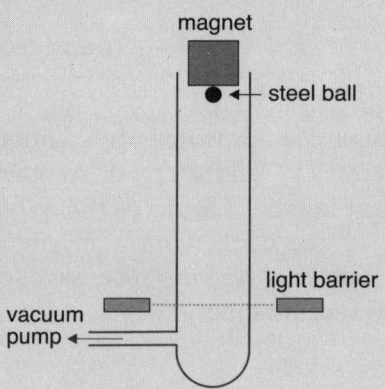

**FIGURE 1.3** Drop tube with a light barrier.

The description of *uniform circular motion* is somewhat more difficult. If a trajectory of radius $r$ lies in the $x$-$y$ plane and the object is on the $x$-axis at time $t = 0$, then a uniform counterclockwise circular motion will be described by the equations:

$$\mathbf{r}(t) = r(\cos \omega t, \sin \omega t, 0)$$
$$\mathbf{v}(t) = \omega r(-\sin \omega t, \cos \omega t, 0)$$
$$\mathbf{a}(t) = -\omega^2 r(\cos \omega t, \sin \omega t, 0).$$

Here $\omega t = \varphi$ is the angle $\varphi$ between the radius vector $\mathbf{r}$ and the $x$-axis, which increases linearly with time. The constant $\omega$ (measured in $\text{s}^{-1}$) is referred to as the *angular velocity* of the circular motion.

The relationships between $\mathbf{r}$, $\mathbf{v}$, and $\mathbf{a}$ can be formulated in a more elegant way that is independent of the choice of coordinate system by using vector products. For this, we treat the angular velocity $\boldsymbol{\omega}$ as a vector that is perpendicular to the circular trajectory in a righthand sense. $\boldsymbol{\omega} = (0,0,\omega)$ in the above coordinate system. Then, at each time $t$, we have

$$\mathbf{v} - \boldsymbol{\omega} \times \mathbf{r}$$

$$\mathbf{a} = \boldsymbol{\omega} \times \mathbf{v} = -\omega^2 \mathbf{r}.$$

Since $\mathbf{a} = d^2\mathbf{r}/dt^2$, the second equation shows that uniform circular motion obeys the simple differential equation

$$\frac{d^2\mathbf{r}}{dt^2} = -\omega^2 \mathbf{r}.$$

### Experiment 1.2    Velocity measurements

Let us use these properties of uniform motion to measure the velocity of a rifle bullet (Figure 1.4). To do this, the uniform rectilinear motion of the bullet will be compared with the uniform circular motion of two disks which are separated by a distance l and mounted on an axis that rotates with angular velocity $\omega$. The bullet, moving at velocity $v$, perforates the two disks at times separated by $\Delta t$ and leaves two holes separated by an angle $\Delta\varphi = \omega\Delta t$. The rotation frequency $\nu = \omega/2\pi$ (number of rotations per s) of the disks is measured (by comparison, for example, with the frequency of a suitable stroboscope, i.e., a periodically flashing lamp). Then, we have $v = l/\Delta t = 2\pi\nu l/\Delta\varphi$. The velocity of an air rifle pellet is somewhat higher than 100 m/s.

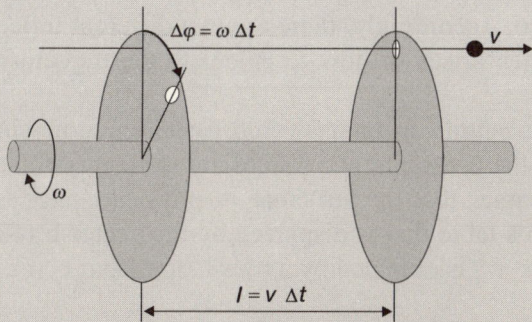

**FIGURE 1.4** Experimental setup for measuring the velocity of a bullet.

**Problem 1.2**

Design a practical realization of this experiment. How high a rotation frequency would you choose? Discuss the precision of the velocity measurement.

---

| Notes | All objects experience the same acceleration $g = 9.81\,\mathrm{m/s^2}$ during free fall (at a given location on the earth's surface). |
|---|---|

In uniform circular motion with angular velocity $\omega$, the velocity vector $\mathbf{v}$ and acceleration vector $\mathbf{a}$ of the object are given by the vector products of the angular velocity $\omega$ with the radius vector $\mathbf{r}$ and linear velocity $\mathbf{v}$, respectively:

$$\mathbf{v} = \omega \times \mathbf{r}$$
$$\mathbf{a} = \omega \times \mathbf{v}.$$

These two equations imply that the acceleration vector $\mathbf{a}$ is directed toward the center of the circular motion, i.e.,

$$\mathbf{a} = -\omega^2 \mathbf{r}.$$

---

### 1.1.3 Changing the reference system

As noted in Section 1.1.2, in order to be able to specify the trajectory $\mathbf{r}(t)$ of a moving object, we must refer to a reference frame. Two observers $A$ and $B$ can describe the motion of a given object in this way, but choosing different frames of reference. Accordingly, there are two different trajectories $\mathbf{r}_A(t)$ and $\mathbf{r}_B(t)$. The question then arises of how to calculate the trajectory $\mathbf{r}_B(t)$ from the trajectory $\mathbf{r}_A(t)$.

This question can be answered based on some simple mathematical considerations if the two reference frames are stationary relative to one another, insofar as one assumes that the structure of physical space is Euclidean. Suppose $B$ is attached to a table that is displaced by the vector $\mathbf{b}$ relative to the reference table of $A$; then $\mathbf{r}_B(t)$ is obtained by a translation from $\mathbf{r}_A(t)$; i.e.,

$$\mathbf{r}_B(t) = \mathbf{r}_A(t) - \mathbf{b}.$$

Here it is implicitly assumed that both observers rely on the same clock time.

If $A$ and $B$ choose reference frames that move relative to one another, the question of converting the trajectories cannot be answered by mathematical considerations alone. It was long assumed that in this case, as well, $A$ and $B$ can rely on the same clock time. For the treatment of rectilinear motions, this assumption yields a law for the addition of velocities. Let us assume then that the reference frame of $B$ moves at velocity $\mathbf{v}_0$ relative to the reference frame for $A$. Then the *Galilean transformation* (Figure 1.5) gives $\mathbf{r}_B(t) = \mathbf{r}_A(t) - \mathbf{b}(t)$, with $\mathbf{b}(t) = \mathbf{v}_0 t$. Differentiating with respect to time gives the following addition rule for velocities:

$$\mathbf{v}_A = \mathbf{v}_B + \mathbf{v}_0. \qquad \text{(Galilean addition rule for velocities)}$$

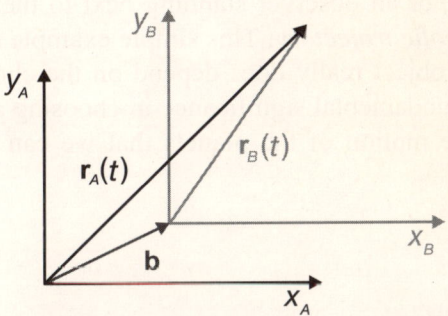

**FIGURE 1.5** The Galilean transformation.

This addition rule was brought into question when experiments with the Michelson interferometer (Section 4.2.2) showed that the propagation velocity $c$ of light in a vacuum does not vary when the reference frame is changed. This discovery was obviously inconsistent with the Galilean transformation. As Einstein showed in 1905, the contradiction lies in an erroneous conception of space and time. Only when $v_0 \ll c$ can $A$ and $B$ rely on essentially the same clock time. But they cannot do this if $v_0$ is not very much less than the velocity of light. The familiar laws then have to be changed in accordance with the special theory of relativity. Collinear (in the same direction) velocities add according to the equation

$$v_A = \frac{v_B + v_0}{1 + v_B v_0 / c^2}. \qquad \text{(relativistic addition rule for velocities)}$$

In the limit of $v_0 \ll c$ or $v_B \ll c$, the two addition rules give the same result to within experimental error.

---

### Problem 1.3

How accurately do you have to measure the velocities of two cars each travelling at 100 km/h in opposite directions on a superhighway in order to be able to detect a deviation in their relative velocity from a value of $v_{rel} = 200$ km/h?

---

Finally, a simple example (Figure 1.6) illustrates how a change in the reference system leads to a transformation in the trajectory of an object, as well as in its velocity. A ball player on a railway car that is moving at constant velocity throws a ball vertically upward in order to be able to catch it again. (The effect of the airstream can be neglected.) For the ball player, the ball moves along a straight line. For an observer standing next to the car, however, the ball moves along a *parabolic trajectory*. This simple example shows that the form of the trajectory of an object really does depend on the choice of reference system. This insight is of fundamental significance in choosing a suitable reference system for describing the motion of the planets that we can see in the night sky (Section 1.1.4).

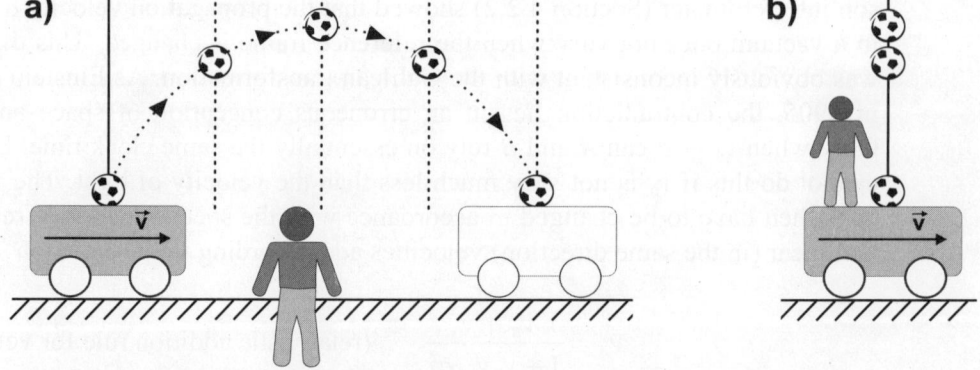

**FIGURE 1.6** A ball player *B* on a moving railway car (b) and an observer *A* standing on the railway embankment (a) see different trajectories. For *A* the ball moves along a projectile parabola and for *B*, along a straight line.

| Notes | In order to describe the motion of an object, one must rely on a reference frame. Every observer is free to choose a reference frame. The trajectories of a given object as described by two observers relying on different reference frames can be very different. In order to be able to compare the results of the two observers, transformation equations are required. If the reference frames move slowly relative to one another in uniform rectilinear motion ($v \ll c$), the Galilean transformation can be used. |

### 1.1.4 Planetary motion

Up to now we have only considered the motion of objects on the earth. For this, the earth is naturally chosen to be the reference frame. It is more difficult to choose suitable reference frames for describing the motion of heavenly bodies. On one hand, we see how they move once around the earth during the course of a day. On the other, we can also see that most heavenly bodies do not move relative to one another; that is, over many centuries their relative distances (more precisely, the observed angular separations between them) remain essentially unchanged. We call these heavenly bodies *fixed stars*. In addition, there are some heavenly bodies which move relative to the fixed stars. They are known as *planets*. When we observe the locations of the planets night after night and measure their positions against the background of the fixed stars, the motion of the planets relative to the fixed stars appears to be very complicated. Recognizing that the planets move (approximately) along circular orbits if one chooses the sun instead of the earth as the coordinate origin and fixes the directions of the coordinate axes relative to the fixed stars was the great achievement of **Nicholas Copernicus (1473–1543)**. This change from a geocentric to a heliocentric reference frame marks the historical beginning of physics in the modern era.

The astronomer **Tycho Brahe (1546–1601)**, who worked at the Danish court, measured the motion of the planets with a high accuracy for that time. Even without a telescope, he was able to determine the positions of the planets with an uncertainty of about 2 angular minutes ($2' \approx 0.6 \times 10^{-3}$ rad). Based on these measurements, **Johannes Kepler (1571–1630)** could calculate the trajectories of the planets in a heliocentric reference frame. The results are summarized in Kepler's three laws (Figure 1.7):

## Kepler's Laws

1. The planets move along elliptical orbits $\mathbf{r}(t)$ with a common focus located at the sun.

2. The radius vector of a planet sweeps out equal areas $dA$ in equal time intervals $dt$; that is, the areal velocity $dA/dt$ of the planetary motion is constant. With $\mathbf{v} = d\mathbf{r}/dt$ we also have

$$\frac{d\mathbf{A}}{dt} = (\mathbf{r} \times \mathbf{v})/2 = const.$$

3. The squares of the periods of revolution $T$ of the planets are proportional to the cubes of the major semiaxes $a$ of their elliptical orbits:

$$T^2 \propto a^3.$$

Proceeding from these laws, a full half century later (1687) Isaac Newton formulated the law of gravity for the mutual attraction of masses and, thereby, *explained* the motion of the planets. He also gave the concepts of *force* and *mass* a precise physical meaning and lay the foundation of classical mechanics.

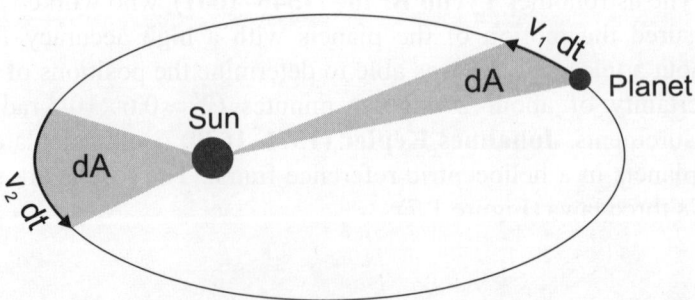

**FIGURE 1.7** The trajectory of a planet in a heliocentric reference frame.

## Problem 1.4

Calculate the angular velocity, linear velocity, and areal velocity of the moon's motion about the earth and the earth's motion about the sun. For simplicity, assume that both objects move in circular orbits (with radii $R_M \approx 1 \, \text{s} \cdot c$ and $R_S \approx 8 \, \text{min} \cdot c$, respectively) in geo- and heliocentric reference frames, respectively.

**Notes**  In a heliocentric reference frame, where the sun and the fixed stars are at rest, the planets move in elliptical orbits around the sun. The sun is a common focus for all the elliptical orbits. Each planet moves at a constant areal velocity and the squares of the orbital periods of the planets are proportional to the cubes of the major axes of their orbits.

## 1.2 DYNAMICS OF POINT MASSES

Kinematics deals with the description of the motions of objects in space and time. Dynamics examines the causes and seeks to explain the motions. Contrary to the earlier assumption that the motion of an object was to be explained in relation to the earth considered as absolutely at rest, **Isaac Newton (1643–1727)** proceeded from the idea that an object which is in uniform rectilinear motion requires no further influence, but that for every change in its state of motion, an external force acting from outside the object is necessary. From this point of view, the concepts of force and mass acquire a central significance. We shall introduce these concepts in the first two sections of this lecture and then use them to account for accelerated motion. The constancy of the speed of light referred to in Section 1.1.1 imposes limits on the domain of applicability of Newtonian mechanics. These limits will be outlined in the last section of this lecture.

### 1.2.1 Inertial and gravitational mass

The masses of objects can be compared by two fundamentally different measurement methods. Accordingly, a distinction is made between the inertial and gravitational mass of an object.

The measurement technique for *inertial* masses (dynamic mass comparison) is based on velocity measurements. In order to avoid the effect of friction and gravitation on the velocities of moving objects, we study the motion of objects on, for example, horizontal air tables. The air flowing out of numerous small pores on the tabletop produces an air pillow between the objects and the tabletop, on which the objects can move nearly friction free in straight lines. Only in a collision of two objects will their velocities change. In particular, let us study the velocities of two objects $A$ and $B$, which are initially at rest and then are pushed apart by releasing a spring under tension (Figure 1.8). These measurements show that the oppositely directed velocities $\mathbf{v}_A$ and $\mathbf{v}_B$ imparted after launch can be very different, depending on the tension of the spring. But, as long as the experiment is carried out with the same objects, the ratio of the magnitudes $v_A/v_B$ is always the same. Because of this regular behavior, it is possible to assign an inertial mass to every object.

**FIGURE 1.8** Dynamic mass comparison.

In addition, a mass of $m_A = 1\,\text{kg}$ is assigned to a particular object, the standard kilogram located in Sèvres near Paris, and this is used to define the *SI unit of mass*. The mass of an arbitrary object $B$ can then be compared with the mass $m_A$ of the standard kilogram using the above repulsion technique. The inertial mass of $B$ is then given by

$$m_B = m_A \frac{v_A}{v_B}.$$

The measurement technique for *gravitational* masses (static mass comparison) relies on weighing by means of position measurements. The masses of two objects $A$ and $B$ are compared using a beam balance (Figure 1.9). Here the objects are hung so that the beam is balanced and the lengths $l_A$ and $l_B$ from the suspension points to the pivot point are measured. In fact, one of the suspension points can be chosen arbitrarily, but the other is then determined and the ratio of the lengths always has same value here, as well. Consequently, a gravitational mass $M_B$ can be assigned to object $B$ using the following equation relating it to the gravitational mass $M_A = m_A = 1\,\text{kg}$ of the standard kilogram, which is again chosen as the unit of mass:

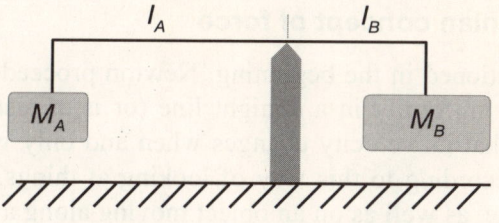

**FIGURE 1.9** Static mass comparison.

$$M_B = M_A \frac{l_A}{l_B}.$$

It might seem surprising that for two given objects the velocity ratio $v_A/v_B$ obtained from the repulsion measurement is always the same, as is the length ratio $l_A/l_B$ obtained by weighing. Even more surprising, however, is the fact that these ratios are equal; that is, $v_A/v_B = l_A/l_B$. The immediate consequence of this experimental result is the equivalence of inertial and gravitational mass. Thus, in the following we shall merely speak of the mass of an object.

The equivalence of inertial and gravitational mass is, moreover, of fundamental significance. The general theory of relativity formulated by Einstein in 1916 is essentially based on this equivalence. This theory revolutionized our concept of space and time. The description of cosmic processes is based on a curved space-time metric, rather than the flat metric of Euclidean space.

### Problem 1.5

Construct a beam balance and try to determine the ratio of the gravitational masses of two objects using the procedure described above. How does the mass of the beam impair the measurement process? How would you compare masses more accurately using an apothecary's (tare) balance?

| Notes | The masses of two objects can be compared dynamically and statically. Both procedures yield the same values for all objects, as long as the same mass unit is adhered to. |
| --- | --- |

### 1.2.2 The Newtonian concept of force

As we mentioned in the beginning, Newton proceeded from the idea that an object moves uniformly in a straight line (or is at rest) without an external influence, and that its velocity changes when and only when a force acts on it from outside. According to this way of looking at things, an external force acts on a falling stone, as well as on an object moving along a trajectory. From this insight and making use of our daily experience with the effect of forces, Newton formulated the three basic axioms of mechanics:

---

**Newton's Axioms**

---

- ■  **Axiom 1 (inertia principle):** An object remains in a state of rest or in uniform rectilinear motion if no external forces act on it.
- ■  **Axiom 2 (action principle):** If a force $F$ acts on an object with mass m, the object will be accelerated in accordance with the formula Force = Mass × Acceleration: $\mathbf{F} = m\mathbf{a}$.
- ■  **Axiom 3 (reaction principle):** The forces which two objects exert on one another are equal in magnitude but in opposite directions: action = reaction.

---

The second axiom, in particular, shows that force is a directed quantity, or vector. This determines the *SI unit of force*, the *Newton*:

$$1\,\text{N} = 1\,\text{kg} \cdot \text{m} \cdot \text{s}^{-2}.$$

If a force acts on an object $A$, then there is another object $B$ from which the force originates. Conversely, the object $A$ acts on object $B$, as we have seen in the spring-driven repulsion experiment. This experimental result is the basis of the third axiom.

It is said that these three axioms form the intellectual foundation of classical mechanics. But before we use them to explain motions such as free fall or planetary motion, we must mention the problem of choosing a reference system. An object that is at rest in the lecture hall, or moves uniformly in a straight line there, is undergoing an accelerated motion in a heliocentric reference system. Just taking the daily rotation of the earth about its north-south axis into account, an object at the equator experiences an acceleration $a = 0.034\,\text{m/s}^2$ according to Section 1.1.2.

## Problem 1.6

Estimate the value of $a$ on your own. (The length of the day = almost $10^5$ s and the earth's circumference $= 2\pi R = 40{,}000$ km.)

Compared to the acceleration of gravity, we can neglect this small acceleration in many earthbound experiments. But in large scale motions on the earth, as during the formation of cyclones in the atmosphere (Figure 1.10), or in precision experiments, as with the Foucault pendulums in the Deutsches Museum in Munich or the German Technical Museum in Berlin, it does play a role.

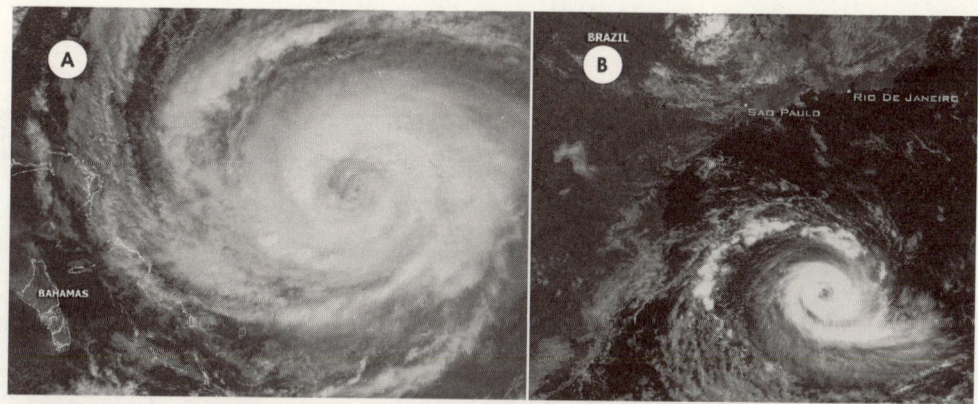

**FIGURE 1.10** Cyclones over the Atlantic Ocean: (a) northern hemisphere and (b) southern hemisphere.

This discussion leads to a fundamental question: which reference system should we rely on in order to determine a force from the acceleration of an object? Newton chose a heliocentric reference system attached to the fixed stars as a so-called *inertial system* and, therefore, as a reference system in which the inertia principle holds. From the Kepler laws of planetary motion, he could then determine the gravitational force acting on the planets (Section 2.2.3).

Reference to the fixed stars for establishing a reference system presents no problem as long as we are considering motions in our solar system. However, motions also occur in outer space, which appreciably change the relative positions of the fixed stars and entire galaxies over millions or billions of years. The spiral structures of galaxies (Figure 1.11) illustrate this. Relative to a reference system defined in terms of the positions of distant galaxies (such as the Andromeda galaxy, which is about $2.2 \times 10^6$ light years $= 2.1 \times 10^{22}$ m from us), the stars in our galaxy (including the sun) are also orbiting the center of the Milky Way. The sun, indeed, takes about 250 million years to do this. It moves roughly in an orbit with a radius of $3 \times 10^4$ light years. The concept of Newtonian mechanics is ultimately inadequate for describing cosmic motions of this type, since, first of all, the choice of an inertial system would be dubious. Here Einstein's general theory of relativity offers a way out of the conceptual difficulties of Newtonian mechanics. In the context of a lecture, however, we can confidently rely on Newtonian mechanics and use a heliocentric reference system as an inertial system in which the inertia principle is essentially exact. This is because in all earthbound processes, the discrepancies to be expected owing to the inadequacy of this inertial system are far smaller than the measurement errors encountered in experimental research.

**FIGURE 1.11** The spiral structure of the galaxy M51 in the constellation Canes Venatici (Hunting Dogs).

To measure some of the forces which play a role in many processes of daily life, we can even use the lecture hall for some examples without making too great an error. In the following we shall list three of these forces and describe them briefly:

**Gravity.** The force of gravity, $\mathbf{F}_G$, accelerates objects in free fall. Hence,

$$\mathbf{F}_G = m\mathbf{g}.$$

Here $m$ is the mass of the object and $\mathbf{g}$ is the acceleration of gravity, directed toward the center of the earth and with a magnitude of $g = 9.81 \, \text{m/s}^2$. Gravity also acts between objects on the earth's surface and the earth, itself, which is accelerated in accordance with the reaction principle.

---

**Problem 1.7**

Estimate the acceleration of the earth as an apple falls. Is this measurable?

---

**Springs.** If a weight is suspended on a spring balance (Figure 1.12), then the force of gravity $F_S$ acting on the weight stretches the spring along its length $z$. According to the third axiom, a force $F_F$ owing to the stretching of the spring acts against gravity. For stretching that is not too great, Hooke's law applies, i.e.,

$$F_F = -F_G = -Dz. \hspace{3cm} \text{(Hooke's law)}$$

The spring constant $D$ is a measure of the elasticity of the spring.

**Friction.** On earth every object eventually comes to rest (even those moving on a "frictionless" table). In terms of the concepts of Newtonian mechanics, therefore, a braking force acts on all earthbound objects. It is referred to as *friction*. Friction plays an extraordinarily large role in daily life and in all engineering equipment with moving parts. Unlike for gravity, however, there is no simple law of force for friction. As everyone knows from their own experience, sliding friction and static friction depend sensitively on the properties of the surfaces rubbing against each other. Likewise, the air friction of, for example, a moving car, depends in a complicated way on the shape and properties of the surface and increases with the velocity of the car. Since the effect of friction on the motion of objects is not only hard to monitor and, therefore, hard to reproduce, we initially treat friction as a troublesome nuisance and try, as in the case of free fall, to avoid it as much as possible. Many phenomena of classical mechanics are more clearly recognized in (idealized) frictionless processes. Finally, friction is closely related to the thermal motion of atoms (Section 2.1.3) and, therefore, belongs not to mechanics, but to a different branch of physics. This relationship can be represented quantitatively for some simple model systems. Thus, for example, the internal friction of gases can be interpreted quantitatively in terms of simple models (Section 3.2.1).

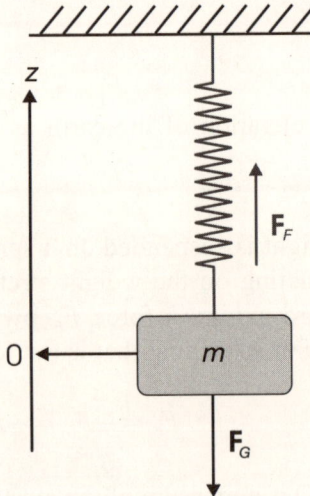

**FIGURE 1.12**  A spring balance.

**Notes**    In inertial systems the acceleration **a** of an object with mass $m$ is given by the force **F** acting on it, i.e.,

$$\mathbf{F} = m\mathbf{a}.$$

For an object not subject to any force, therefore, $\mathbf{a} = 0$. It will thus move in an inertial system with a velocity **v** that is constant in time.

### 1.2.3  Gravitation

The planets move in space essentially without friction. According to Newtonian mechanics, the elliptical orbits of the planets in a heliocentric inertial system that Kepler calculated imply an attractive force acting between the sun and the planets that falls off as the square of the distance $r$, i.e.,

$$F_G \propto \frac{1}{r^2}.$$

Newton recognized that this force, known as *gravity*, acts not only between the sun and the planets, but is also a universal force which acts between all pairs of objects with masses $m$ and $M$ and is proportional to these masses. That is,

$$F_G = G\frac{mM}{r^2}. \qquad \text{(the law of gravity)}$$

The proportionality constant

$$G = 0.672 \times 10^{-10} \text{ m}^3 \cdot \text{kg}^{-1} \cdot \text{s}^{-2} \qquad \text{(the gravitational constant)}$$

is a universal constant of nature. It is known as the gravitational constant. Gravitation works between earthbound objects, as well as between objects in the heavens, so it is also ultimately responsible for the free fall of objects on the earth's surface. In that case, the force of gravity at the earth's surface is a special case of the law of gravity. Accordingly, the acceleration of gravity at the earth's surface is given in terms of the mass $M_E$ of the earth and the earth's radius $R_E$ by

$$g = G\frac{M_E}{R_E^2}.$$

In order to determine the gravitational constant $G$, one must measure the force of gravity between two objects with known masses. Usually, we do not perceive the gravitational force between the objects around us, because their masses are much less than $M_E$. Two 1 kg masses separated by a distance $r = 0.1$ m attract each other with a force of only $F = 0.67 \times 10^{-8}$ N. This corresponds to the weight of a mass of 0.68 µg. With an extremely sensitive balance, a so-called gravitational balance, this force can be measured in the laboratory and the gravitational constant can be determined.

## Problem 1.8

Use the law of gravity to determine the masses of the earth and the sun. What are the average densities (mass/volume) of these objects?

Newton used Kepler's laws to find the law of gravity. Conversely, the Kepler laws can be derived from the law of gravity. To do this we begin with the action principle and set **F** equal to the force of gravity. This yields the following *equation of motion* for the trajectories $\mathbf{r}(t)$ of the planets:

$$m_P \frac{d^2\mathbf{r}}{dt^2} = -G \frac{m_P m_S}{r^2} \cdot \frac{\mathbf{r}}{r}.$$

Kepler's elliptical orbits are solutions of this differential equation. We shall omit proving this statement in general. The proof is, however, very simple if we consider the special case of (circular) orbits. In this case the planets move with a constant angular velocity $\omega$ and a constant distance $r$ from the center. The sun is only approximately at the center, for, according to the reaction principle, the sun also moves in an orbit about this center, although, of course, with a very much smaller radius.

This adjustment also shows that the first and second Kepler laws are evidently satisfied by the orbits. The third Kepler law also follows from the gravitational force of the sun, which balances the *centrifugal* acceleration $\mathbf{a} = -\omega^2\mathbf{r}$ of the planet,

$$G \frac{m_P m_S}{r^2} = m_P \omega^2 r.$$

Since the period is $T = 2\pi/\omega$, this equation yields $T^2 = (2\pi)^2/Gm_S r^3$; that is, the ratio $T^2/r^3$ has the same value for all the planets. That value is determined by the gravitational constant $G$ and the sun's mass $m_S$,

$$\frac{T^2}{r^3} = \frac{4\pi^2}{Gm_S}.$$

> **Notes**
>
> The gravitational force operates between all pairs of objects with masses $M$ and $m$:
>
> $$F_G = G\frac{mM}{r^2}. \qquad \text{(the law of gravity)}$$
>
> All objects attract each other mutually. The attractive gravitational force falls off with the square of the distance $r$.
>
> The action principle and the law of force yield a differential equation for the motion of objects on which this force acts. By integrating this *equation of motion* the possible trajectories $\mathbf{r}(t)$ of the objects can be calculated.

## 1.2.4  Constancy of the speed of light

There are several assumptions at the root of Newtonian mechanics that are not at all self-evident. It is worth contemplating the basic assumptions of Newtonian mechanics in order to identify the limits of its validity. In Section 1.2.2 we pointed out that the choice of a suitable inertial system would be problematic when describing the motions of galaxies that can be detected by modern measurement techniques. The experimental finding that light propagates at the same velocity $c$ in all reference systems is further indication that the laws of Newtonian mechanics have only a limited range of validity.

Since the speed of light, $c$, is independent of the choice of reference system, $c$ is, like the gravitational constant $G$, a constant of nature. We shall encounter other constants of nature later, specifically the Boltzmann constant $k$ in Section 2.1.3, the elementary charge $e$ in Section 3.4.1, and Planck's quantum of action $h$ in Section 4.3.2. These latter constants are all beyond the scope of mechanics. Hence, the discovery of these constants led to decisive changes in the world view of physics. Here we shall call attention to a few of the consequences of the constancy of the speed of light.

Newtonian mechanics is based, among other things, on the assumption that the simultaneity of two events in different places is independent of the state of motion of the observer. The simultaneity of two events thus came to have an absolute significance. The constancy of the speed of the light, on the contrary, implies that, under certain conditions, two observers moving relative to one another will have different perceptions of the temporal ordering of two events that occur at different locations (Figure 1.13). In 1905 Einstein explained this contradiction and showed clearly that Newtonian mechanics is only applicable for velocities $v \ll c$. His analysis led to the *special* and *general theories of relativity*.

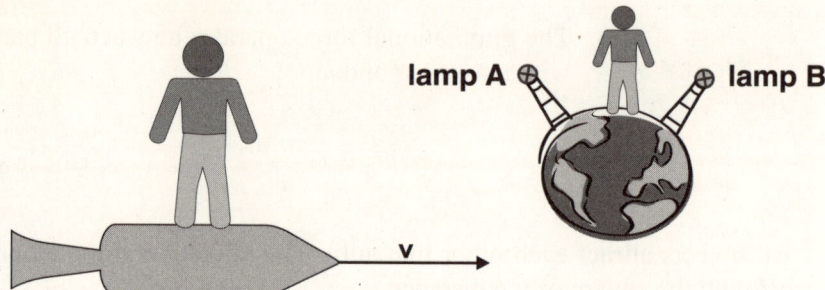

**FIGURE 1.13** The relativity of simultaneity: the flashes of two signal lamps $A$ and $B$ separated by a distance $d = 10,000$ km which flash at the same time for us, the inhabitants of earth, seem to be separated in time by $\Delta t = vd/c^2 = 1\,\mu s$ for a space traveller moving the direction $A \rightarrow B$ at a velocity of $v = 9$ km/s. (The duration of the signal transmission is taken into account in the time specification.)

According to the special theory of relativity, events which appear simultaneous to one observer can appear sequentially to another observer. For distances $d$ on the earth ($d < 10^7$ m), these time shifts $\Delta t$ are usually negligible (but not always; consider, for example, the Global Positioning System, GPS). At high relative velocities, $v \sim c$, this still yields a time difference $\Delta t \le d/c$ much smaller than 1 s. Nevertheless, the relativity of simultaneity is of fundamental interest, since it raises doubts about the Newtonian concept of action at a distance and, in particular, the law of gravitation at large distances. This force is, on one hand, proportional to the square of the *instantaneous* distance $r$ of two objects, as well as to the distance at which the two objects are in simultaneity. Because of the relativity of simultaneity, this assumption makes no sense in terms of the theory of relativity. Therefore, the idea of *action at a distance* had to be replaced subsequently by the idea of the *field* (Section 3.4.2).

These remarks are sufficient to make it clear why the discovery of the constancy of the speed of light led to a revolutionary reorientation of the world view of physics. Unfortunately, in this textbook, we cannot discuss the theoretical foundations of the theory of relativity in detail and thus limit ourselves to a few comments:

## A Look at Relativistic Mechanics

- Only if $v \ll c$ can velocities simply be added vectorially. In general, the relativistic addition law for velocities must be used (Section 1.1.3).
- The speeds of material objects are always less than the speed of light.
- Newtonian mechanics contains a law of mass conservation: the mass $m$ of an object made up of two objects with masses $m_1$ and $m_2$ is given by $m = m_1 + m_2$. By contrast, according to the special theory of relativity, mass and energy are equivalent. Mass can be converted into energy and energy into mass. The famous equation $E = mc^2$ holds.

Thus, for example, the mass $m_d$ of a deuteron, which is made up of a proton and a neutron, is less than the sum of the masses $m_p$ and $m_n$ of a proton and a neutron. The quantity $\Delta m = m_p + m_n - m_d$ is known as the mass defect. Breaking up a deuteron into a proton and a neutron requires an energy $E = (m_p + m_n - m_d)c^2$.

### Problem 1.9

Calculate the error in the position reading of a Global Positioning System whose time base is accurate to only about $\pm 1\,\mu s$.

**Notes** Light propagates at the same speed $c = 3 \times 10^8$ m/s in all inertial systems. Since the speed of light is independent of the choice of reference system, the time sequence of two events that take place at great distances from one another can appear to be different to two observers who are moving relative to one another.

## 1.3 CONSERVATION OF ENERGY AND MOMENTUM

In all branches of physics the concept of energy plays a central and fundamental role. It is familiar (or at least seems to be) from the daily political discussions about "energy supply." Its physical significance derives from the principle of energy conservation. The experimental foundation of the principle of energy conservation becomes clear in a discussion of cyclic processes. In the framework of mechanics, it is based on the concept of *work*. Initially the conservation of energy is valid solely under the restrictive condition that only conservative forces are acting (Section 1.3.1). It only acquires its universal significance when thermodynamic processes are introduced (Section 2.3.2).

By contrast, the principle of the conservation of momentum is of unrestricted validity in mechanics. The experimental foundation is formed by collision experiments, to which we have already referred in our definition of inertial mass (Section 1.2.1).

## 1.3.1 Work and cyclic processes

Everyone who has made a round trip on a bicycle knows that you cannot go downhill all the time. Every downhill run is followed by an uphill stretch. Only in the fantasy of an artist such as M. C. Escher (Figure 1.14) can a closed water circulation exist in which the water is always flowing downhill. The elementary experience of downhill and uphill stretches in the course of a round trip is the basis for assigning a *potential* to every point on the earth's surface, specifically the height above sea level. A generalization of this elementary perception can be expressed in the statement that *there can be no perpetual motion machine* and the *principle of energy conservation*. The following considerations proceed from that.

When an object (initially we mean a point mass) moves from a position $\mathbf{r}_1$ rectilinearly to a position $\mathbf{r}_2$ under the action of a force $\mathbf{F}$, *work* is performed. Physically, work is defined as the scalar product of force and path length:

$$W = \mathbf{F} \cdot (\mathbf{r}_2 - \mathbf{r}_1).$$

The *SI unit of work* is defined as $1\,\mathrm{N} \cdot \mathrm{m} = 1\,\mathrm{kg} \cdot \mathrm{m}^2 \cdot \mathrm{s}^{-2} = 1\,\mathrm{J}$. It is named after the physicist **James Prescott Joule (1818–1889)**. If the force depends on position and the path is curved, then the work is calculated by summing over many small rectilinear path segments $d\mathbf{r}$ as the integral along the path from a starting point $A$ to the end point $Z$:

$$W = \int_{A}^{Z} \mathbf{F}(\mathbf{r}) \cdot d\mathbf{r}.$$

In general, after the motion the object has a different position and velocity than it had at the start. But, if the object returns to its starting point and to its initial velocity, then we say that it has undergone a *cyclic process*. The resulting work is then given by a path integral along a closed path:

**FIGURE 1.14** Water circulating in a constantly downhill flow (M. C. Escher).

$$W = \oint \mathbf{F}(\mathbf{r}) \cdot d\mathbf{r} = 0.$$

Experimentally, it turns out that the work performed in a cyclic process is always greater than or equal to zero, but never less than zero.

A corresponding result is obtained for complex mechanical machines. A machine (for example, the flywheel reservoir of a Stirling motor (Figure 1.15), which will be discussed in Section 2.4.4) is said to run through a cyclic process when all the point masses in the machine return to their initial positions and velocities at the end of the process.

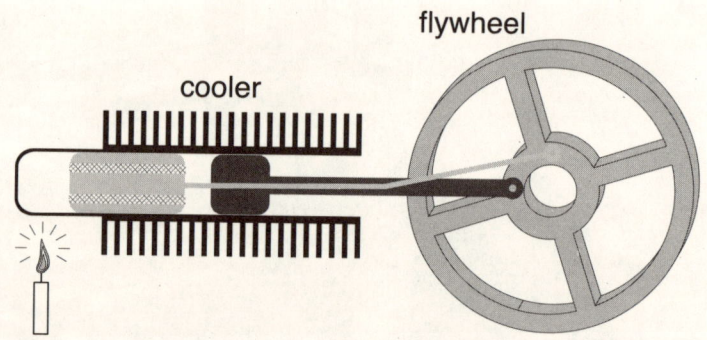

**FIGURE 1.15** The flywheel reservoir of a Stirling motor (Section 2.4.4).

In this sort of cyclic process, in one part work $W_{in}$ will be performed (as the flywheel is driven) and in the other work $W_{out}$ will be delivered (when the flywheel compresses the gas in the Stirling motor). Figure 1.16 is a schematic representation of the operation of such a cyclical process. For centuries, many inventors have tried to build a so-called *perpetual motion machine*, for which $W_{out} > W_{in}$. All these attempts have ultimately only shown that, no matter how artful the machine, the work output of a cyclical process is always less than the work input; that is,

$$W_{out} \leq W_{in}$$

(the experimental basis of the conservation of energy in mechanics). Only in an idealized limiting case, when the machine runs without friction, is $W_{out} = W_{in}$.

Studies of cyclic processes have rendered possible the following fundamental classification of the forces found in nature:

**FIGURE 1.16** Work diagram for mechanical cyclical processes.

- **Conservative forces**, or forces which lead to no loss of work, and
- **Dissipative forces**, or forces which cause a loss of work during a cyclical process.

Conservative forces include weight, gravitation, and the force of a spring (as long as the spring is completely elastic). Dissipative forces appear during friction and plastic deformations.

## Problem 1.10

Calculate the work performed in lifting a 10 kg weight. How much work will be done if the weight is lifted onto a 1 m high table and how much if it is then put back on the floor? Is it positive or negative?

**Notes**

At the end of a cyclical mechanical process all the point masses in the system under consideration have the same position and the same velocity as at the beginning. If only conservative forces act in the system during the cyclic process, the system performs exactly the same amount of work as has been applied to it.

In the work integral, applied work is given a positive value and work output is negative. With this sign convention, the work performed in a cyclic process (with conservative forces) is zero.

## 1.3.2 Potential and kinetic energy

In the following we shall consider mechanical systems in which no dissipative forces act. In this case, $W_{in} = W_{out}$ in a cyclic process. Otherwise, in general, when the system goes from an initial state $A$ to a final state $Z$ (Figure 1.17), $W_{in} \neq W_{out}$. But in cyclic processes the applied and output energy are similar in magnitude, and for these open processes it follows that the difference $W_{in} - W_{out}$ is independent of the path followed by the system in going from state $A$ to state $Z$. Then different paths would yield different values of $W_{in} - W_{out}$, so it would be possible to take the system along one path and bring it back to its final state along another and thus obtain a cyclic process with $W_{in} \neq W_{out}$, contradicting our initial assumption that only conservative forces are acting.

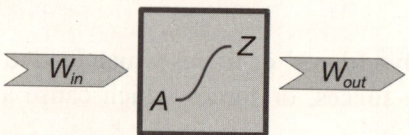

**FIGURE 1.17** An energy diagram for a conservative process that takes a mechanical system from an initial state $A$ to a final state $Z$.

The *path independence* of the difference $W_{in} - W_{out}$, or, generally speaking, of the work integral $\int \mathbf{F} \cdot d\mathbf{r}$ allows us to define a new quantity, the *energy*, that characterizes the state of a mechanical system. To do this, one chooses an arbitrary initial state $A$ of the system as the zero point for the energy scale. The energy $E$ of any other state $Z$ of the system is then equal to the net work ($W_{in}$ is positive and $W_{out}$ negative) that must be applied to bring the system from state $A$ to state $Z$:

$$E = W(A \rightarrow Z) = \int_A^Z \mathbf{F}(\mathbf{r}) \cdot d\mathbf{r}.$$

The *SI unit of energy* is 1 J, the same as the unit of work. In processes where work is neither done on a mechanical system nor extracted from it, the energy of the system obviously remains constant. Thus, an energy conservation law holds for conservative and dynamically closed systems.

For many analyses, there is a great advantage in knowing the energy as a function of the position and velocity coordinates of a system. We shall consider the energy of a point mass as an example. In general, the total energy $E_{tot}$ can be broken up into a potential part (dependent only on the spatial coordinate) and a kinetic (dependent only on the velocity coordinate) part; i.e.,

$$E_{tot} = E_{kin} + E_{pot}.$$

In order to calculate the *potential energy* $E_{pot}$ of a point mass acted on only by gravity $\mathbf{F} = m\mathbf{g}$, one calculates the work required to lift the point mass from height 0 to height $h$. In this case, $E_{pot} = mgh$. In general, however, the position dependence of the forces acting on a particle must be taken into account and it is necessary to evaluate the *work integral*

$$E_{pot} = \int\limits_0^z \mathbf{F}(\mathbf{r}) \cdot d\mathbf{r}.$$

For a mass point suspended on a spring, the potential energy is obtained from Hooke's law (Section 1.2.2), $F(z) = -Dz$. During a displacement from the position of rest at $z = 0$ to position $z$ the potential energy rises to

$$E_{pot}(z) = \frac{1}{2}Dz^2.$$

The potential energy of a planet (with mass $m_P$) acted on by the gravitational force of the sun (with mass $m_S$) follows from the law of gravity (Section 1.2.3):

$$E_{pot}(r) = -G\frac{m_S m_P}{r}.$$

Here the energy at $r \rightarrow \infty$ is taken to be the zero point for the energy scale.

---

### Problem 1.11

Prove the above formulas for the potential energy of a mass $m$ on a spring and a planet of mass $m_P$ orbiting the sun with mass $m_S$ and plot the functions $E_{pot}(z)$ and $E_{pot}(r)$ graphically.

---

The kinetic energy is a simple function of the velocity $v$. It equals the work required to accelerate a mass $m$ from velocity $v = 0$ to a velocity $v > 0$,

$$E_{kin} = \frac{1}{2}mv^2.$$

---

### Problem 1.12

Prove the formula for the kinetic energy by calculating the work expended in uniform acceleration.

---

Many questions about the motion of objects can be answered without having to examine their trajectories in detail. Important results can be obtained immediately when just the energies of the initial and final states are known.

For example, in order to determine the velocity that a rocket must gain near the earth's surface in order to escape the gravitational attraction of the earth, there is no need to calculate the rocket's trajectory. It is enough to calculate the energy $E_{kin} = mv_0^2/2$ and the potential energy $E_{pot} = -GmM_E/R_E$ of the rocket at the earth's surface ($r = R_E$) and in space ($r \to \infty$) and then to note that for $r \to \infty$ the kinetic energy must have a positive value. Since both the kinetic energy and the potential energy of the rocket are proportional to the mass $m$ of the rocket, the *escape velocity*, as this minimum velocity is known, is independent of the rocket's mass. Of course, the effect of friction has been neglected in this calculation.

---

### Problem 1.13

Calculate the escape velocity $v_0$ for rockets launched from the earth's surface. The result is about $v_0 \approx 11 \, \text{km/s}$.

---

**Notes**

The kinetic energy of an object of mass $m$ moving at velocity $v$ is

$$E_{kin} = \frac{1}{2}mv^2.$$

The potential energy of an object subject to the gravitational force of a central star at distance $r$ is

$$E_{pot}(r) = -G\frac{m_S m_P}{r}.$$

In general, the potential energy $E_{pot}(Z)$ of a system in state $Z$ is given by the work integral taken from the initial state $A$ to the final state $Z$.

### 1.3.3 Conservation of momentum

The *momentum* of an object is defined as the product of its mass and its velocity:

$$\mathbf{p} = m\mathbf{v}.$$

Momentum is also a vector. Since the mass of an object does not usually change with time, $d\mathbf{p}/dt = m\mathbf{a}$. Therefore, according to the action principle, the momentum of an object changes only when an external force $\mathbf{F}$ acts on the object; that is,

$$\mathbf{F} = \frac{d\mathbf{p}}{dt}.$$

If two objects 1 and 2 collide with each other, forces that are equal in magnitude, but have opposite directions, act on both of the objects (the reaction principle). Thus, the *momentum is conserved* in collisions:

$$m_1\mathbf{v}_1 + m_2\mathbf{v}_2 = m_1\mathbf{u}_1 + m_2\mathbf{u}_2. \qquad \text{(momentum conservation)}$$

The sum of the momenta before the collision is equal to the sum of the momenta after the collision. Here the $\mathbf{v}_i$ are the velocities of the masses $m_i$ before the collision and the $\mathbf{u}_i$ are corresponding velocities after the collision. The conservation of momentum is a universal law, as is the reaction principle and does not depend on whether conservative forces act alone during a collision or dissipative forces are also involved.

The momentum of an object can only change if an external force acts on it. Since forces always act between two objects, another object experiences a counterforce so that its momentum is changed in the opposite direction. The sum of the momenta of the two objects does not change. An abrupt impulse $\mathbf{p} = \int \mathbf{F}\, dt$ is transferred, as when a bullet is fired, the shooter experiences an equally large *recoil*.

*Rocket propulsion* also depends on the conservation of momentum (Figure 1.18). A rocket continuously ejects mass at a velocity $\mathbf{u}$ (which is as high as possible relative to the rocket) and is accelerated by the recoil. If mass $dm$ is ejected at velocity $\mathbf{u}$ during a time interval $dt$, then owing to the conservation of momentum, a rocket with instantaneous mass $m$, will undergo a velocity change $d\mathbf{v}$ according to

$$m\, d\mathbf{v} = -dm\, \mathbf{u}.$$

Integration over the burn time yields the final velocity $\mathbf{v}_f$ of the rocket:

$$\mathbf{v}_f = \mathbf{u} \ln \frac{m_i}{m_f}.$$

The mass of the rocket decreases during the burn from $m_i$ at launch to the final mass $m_f$. In order to attain the highest possible final velocity, the mass loss $m_i/m_f$ must be as great as possible.

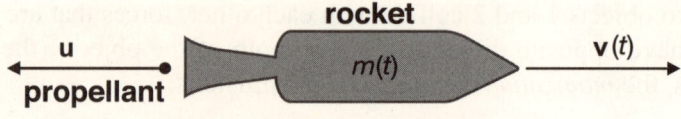

**FIGURE 1.18** Rocket propulsion.

## Experiment 1.3    Rockets

A rocket-shaped plastic vessel is filled with compressed air. When the rocket orifice is opened, the air streams out because of the overpressure and the vertically mounted rocket lifts off. Naturally, it only climbs about 1 m high. The mass loss is too small. The experiment works much better if the rocket is first filled roughly half way with water and then filled with compressed air. This is because the mass of the expelled water is a much larger fraction of the mass of the rocket casing.

## Problem 1.14

Compare the kinetic energies of two objects with masses $m_1 < m_2$ which have momenta of comparable magnitudes. Which of the two bodies has the higher kinetic energy and by what factor do the kinetic energies differ?

| **Notes** | When two objects collide, the sum $m_1\mathbf{v}_1 + m_2\mathbf{v}_2$ of the momenta of the two objects is unchanged. |

### 1.3.4 Elastic and inelastic collisions

The conservation of momentum is universal, while the conservation of energy is valid in mechanics only when no dissipative forces are acting. When we speak of collisions, we distinguish between elastic and inelastic collisions. In an *elastic collision* the energy is conserved, as well as the momentum, while in an *inelastic collision*, only the momentum is conserved. If two balls collide *elastically*, then both

$$m_1\mathbf{v}_1 + m_2\mathbf{v}_2 = m_1\mathbf{u}_1 + m_2\mathbf{u}_2$$

and

$$m_1 v_1^2 + m_2 v_2^2 = m_1 u_1^2 + m_2 u_2^2.$$

In particular, if ball 1 collides with a ball 2 that is at rest and has the same mass, then (Pythagorean theorem) the two equations $\mathbf{v}_1 = \mathbf{u}_1 + \mathbf{u}_2$ and $v_1^2 = u_1^2 + u_2^2$ imply that the two balls fly off in perpendicular directions after the collision. In the special case of a head-on collision, $\mathbf{u}_1 = 0$. Ball 2 then acquires the momentum and energy of ball 1 during the collision.

### Experiment 1.4   Newton's pendulum

The behavior of (almost) elastic collisions of balls of equal mass can be well illustrated experimentally using the apparatus consisting of steel balls shown in Figure 1.19. If the first ball in the series strikes the next ball head-on, and so on, then the momentum and energy propagate from ball to ball, so that ultimately the ball at the other end swings up with the original energy of the first ball.

*continued*

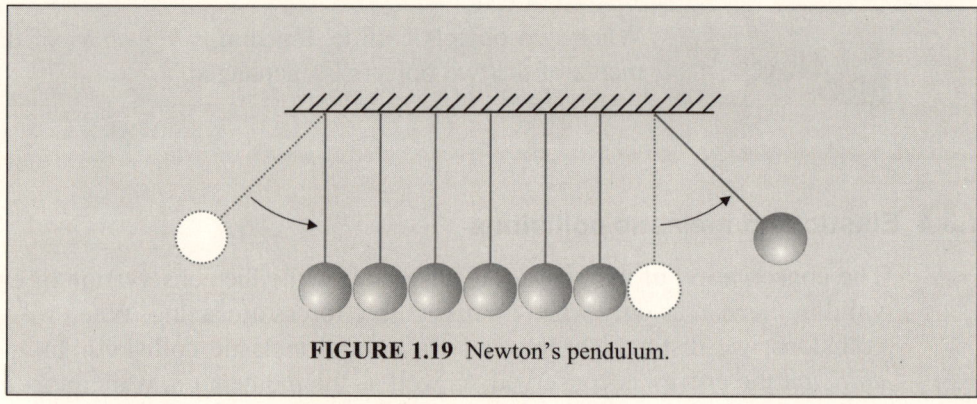

**FIGURE 1.19** Newton's pendulum.

But, following an extremely inelastic collision, the two balls adhere and fly off with the same velocity. The combined momentum of the two balls, in fact, does not change, but the kinetic energy of both balls is lower after the collision than before.

## Experiment 1.5   Ballistic pendulum

This kind of collision is used to measure the velocity of a rifle bullet (Figure 1.20). The rifle bullet impinges on a plasticine ball which is suspended on a thread. Since the bullet remains stuck in the plasticine, the ball and bullet move together after the collision. The pendulum consisting of the thread and plasticine rises up to a height $h$. The height $h$ and the masses $m_G$ and $m_S$ of the bullet and plasticine ball are measured.

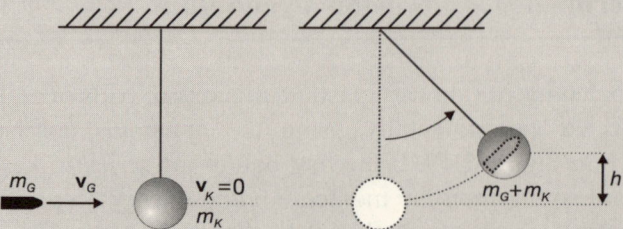

**FIGURE 1.20** A ballistic pendulum. ($v_B$ is the velocity of the bullet, $v_S$ is the velocity of the plasticine sphere with the bullet stuck in it immediately after impact.)

The conservation of momentum in the collision and the conservation of energy in the pendulum's motion imply that the velocity $v_B$ of the bullet before the collision is

$$v_B = \sqrt{2gh}\,\frac{m_B + m_S}{m_B}.$$

### Problem 1.15

Prove this result. Calculate $v_S$ in the first step and then $v_B$ in the second. For which part of the motion does energy conservation alone hold and for which does momentum conservation alone hold? Why? What fraction of the kinetic energy of the rifle bullet is lost as mechanical energy?

**Notes**    In elastic collisions the momentum and kinetic energy of the combined collision system remain fixed and in inelastic collisions, only the momentum.

## 1.4 ANGULAR MOMENTUM

Besides the conservation of energy and momentum, in mechanics the conservation of angular momentum is of central importance. In particular, angular momentum is one of the quantities that determines the evolution of rotational motion. Like energy and momentum, angular momentum plays an important role in other branches of physics besides mechanics. Quantum mechanics shows that there is a smallest unit of angular momentum in nature, the *Planck quantum of action* (Planck's constant) $\hbar$. (See Section 5.1.3 on the quantization of angular momentum.) Planck's constant has the value $\hbar = 1.05 \times 10^{-34}$ J·s. Compared to the angular momentum of the rotational motions we see in daily life, this value is infinitesimally small. Thus, in the following discussion, we can safely treat angular momentum as a continuously variable quantity.

### 1.4.1 The angular momentum of moving point masses

In daily life we make use of the conservation of angular momentum for all rotational motions, such as in sports and in dancing. Examples include somersaults or the pirouette. In order to become familiar with the basic quantities used for physically describing rotational motion, we first consider the motion of a point mass around a center of force (Section 1.1.4).

A point mass, such as the planets surrounding the sun, moving along a trajectory $\mathbf{r}(t)$ around the center of force does indeed have a momentum $\mathbf{p}(t) = m\mathbf{v}$ that changes owing to the action of the force, while the areal velocity $d\mathbf{A}/dt$ is constant according to Kepler's second law. This special circumstance should be considered in a broader context. For this purpose, we define the *angular momentum* of a point mass moving along a trajectory $\mathbf{r}(t)$ around a center of force at $\mathbf{r} = 0$ as

$$\mathbf{L} = \mathbf{r} \times \mathbf{p} = m(\mathbf{r} \times \mathbf{v}). \qquad \text{(definition of angular momentum)}$$

The SI unit of angular momentum is $\text{kg} \cdot \text{m}^2 \cdot \text{s}^{-1}$ and, therefore has the same dimensions as the products *momentum × length* and *energy × time*. Since $\mathbf{F} \propto -\mathbf{r}$ for a force directed toward the center $\mathbf{r} = 0$ and, therefore, according to the action principle $d\mathbf{p}/dt \propto -\mathbf{r}$, as well, so both vectors have the same direction as the radius vector, for the time derivative of the angular momentum we obtain

$$\frac{d\mathbf{L}}{dt} = m(\mathbf{v} \times \mathbf{v}) + \mathbf{r} \times \frac{d\mathbf{p}}{dt} = 0.$$

Since $d\mathbf{L}/dt = 0$, the angular momentum $\mathbf{L}$ of a point mass moving around a central object is constant. In accordance with this, the angular momentum of the planets orbiting the sun is also conserved (Kepler's second law, Section 1.1.4).

---

### Experiment 1.6    Rotation with dumbbells

In order to illustrate the significance of the conservation of angular momentum for rotational motion in sport or in technology, let us have a test subject stand on a rotating table with two dumbbells (Figure 1.21). If the dumbbells are held on opposite sides, the forces acting on this person during rotation will cancel out. There is no danger of falling off the table, if the person stands in the middle of the table! The subject can, therefore, safely lift the dumbbells away from their body or draw them nearer and, thereby, see how their rotational velocity changes.

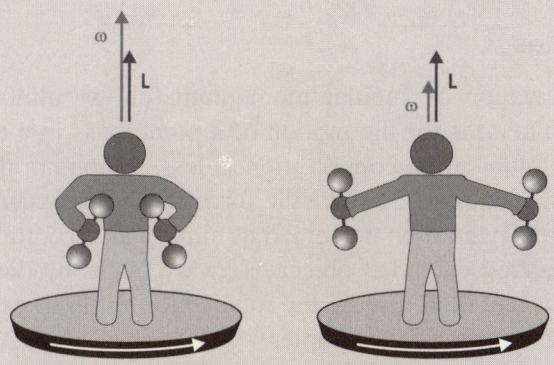

**FIGURE 1.21**  A rotating table experiment with dumbbells.

Under free rotation the angular velocity $\omega$ increases if the dumbbells are drawn nearer to the body and it falls off if the distance of the dumbbells from the body is again increased.

For simplicity we proceed from the highly simplified assumptions that the contribution of the test subject and the rotating table to the angular momentum of the combined system can be neglected and that the dumbbells can be treated as point masses with masses $m$. Then the changes in the angular velocity $\omega$ with varying distance $r$ of the dumbbells from the axis of rotation can be calculated. Since the angular momentum is conserved during rotation, in this case we have $L = 2mr^2\omega = const$. If the distance is halved, then the angular velocity will be a factor of 4 greater.

## Problem 1.16

Calculate the increase in kinetic energy as $r$ is reduced. By what factor does the kinetic energy of the dumbbells increase when their distance from the axis of rotation is halved? Is this result consistent with the conservation of energy?

> **Notes**
>
> The angular momentum **L** of an object remains constant during free rotation; that is,
>
> $$\mathbf{L} = m\mathbf{r} \times \mathbf{v} = const.$$

## 1.4.2  Rigid bodies

The conservation of angular momentum follows almost immediately from the Newtonian axioms for the motion of a point mass in a central field. The validity of angular momentum conservation is less obvious in the case of rotational motion of extended objects, since the forces acting between different parts of an extended object cannot be described as simply as a central force. A *rigid body* is a highly idealized extended object. Here we begin by defining and characterizing rigid bodies and then discuss the rotational motion of extended objects. The rigid body serves as a helpful example.

An extended body with mass *m* can be represented as a set of many small mass elements *dm*, which behave as point masses. In general, each of these point masses can move more or less independently of the others. For example, consider gases, liquids, or plastic and elastically deformable media. Here we consider objects whose mass elements cannot be displaced relative to one another, i.e., a *rigid* body. An ideal rigid body is as rare in nature as is a point mass. Like a point mass, however, the idea of a rigid body is well suited for elaborating the fundamental principles of natural laws.

The position of a point mass in space is characterized by specifying three position coordinates. Correspondingly, a point mass has three *degrees of freedom*. In order to characterize the position of a *rigid body* in space, one stipulates the position of one fixed reference point in the body and the orientation of the body in space (Figure 1.22). This requires three position and three angular coordinates. (Try this with a practical example, by uniquely describing the position in space of a rigid body such as a table using six coordinates.) Thus, a rigid body has six degrees of freedom.

Besides the six position coordinates, six velocity parameters corresponding to the six degrees of freedom must also be specified in order to characterize the motion of a rigid body at time *t*. The velocity parameters are the three components of the velocity vector $\mathbf{v}(t)$ of a mass element and the three components of the angular velocity $\boldsymbol{\omega}(t)$ at which the spatial orientation of the object changes.

At first it seems unimportant which reference point is chosen within an object for describing its position and state of motion. In analyzing the dynamics

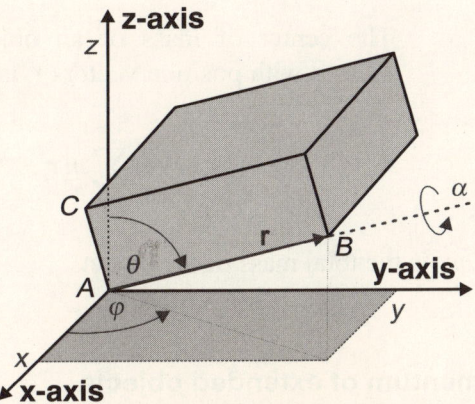

**FIGURE 1.22** Characterizing the position of a rigid body in space using three position coordinates and three direction angles. In this example, the three position coordinates determine the position of the corner $A$ of the block and the two direction angles $\theta$ and $\varphi$ determine the direction of one edge $A \rightarrow B$. The third angle $\alpha$ specifies the direction of the edge $A \rightarrow C$.

(Sections 1.4.3 and 1.5.2) and statics (Section 1.5.1) of rigid bodies, however, it turns out that there is some advantage in choosing the center of mass of the object as a reference point.

The *center of mass* (or *of gravity* or *of inertia*) of an object with mass $m$ has the position vector (Figure 1.22)

$$\mathbf{r}_C = \frac{1}{m} \int \mathbf{r} \, dm. \qquad \text{(definition of the center of mass)}$$

This formula implies, for example, that the center of mass of a dumbbell consisting of two point masses $m_1$ and $m_2$ is located at $\mathbf{r}_C = \left( m_1 + m_2 \right)^{-1} \left( m_1 \mathbf{r}_1 + m_2 \mathbf{r}_2 \right)$.

Referring to the center of mass, from here on we shall describe the motion of rigid bodies as the superposition of the translational motion $\mathbf{r}_C(t)$ of the center of mass with a velocity vector $\mathbf{v}_C = d\mathbf{r}_C / dt$ and a rotational motion of the object about an axis passing through the center of mass at an angular velocity $\boldsymbol{\omega}(t)$.

### Problem 1.17

Prove that the center of mass of a rigid body consisting of three points with equal masses lies at the intersection of the bisectors of the triangle formed by the point masses.

> **Notes**
>
> The center of mass of an object consisting of $N$ point masses with position vectors $\mathbf{r}_i$ is located at
>
> $$\mathbf{r}_C = \frac{1}{m} \sum_{i=1}^{N} m_i \mathbf{r}_i.$$
>
> Here $m = \sum m_i$ is the total mass of the object.

### 1.4.3  Angular momentum of extended objects

By definition the angular momentum $\mathbf{L}$ of a point mass which moves in a plane around a center is perpendicular to $\mathbf{r}$ and $\mathbf{p}$; thus, it is perpendicular to the plane of the trajectory. It is harder to determine the direction of the angular momentum of a rotating rigid body. This is because, in general, the angular momentum is not parallel to the axis of rotation of the body. The simple example of a rotating dumbbell consisting of two rigidly coupled point masses shows that the angular momentum and the axis of rotation can have different directions (Figure 1.23).

Let us, therefore, consider such a dumbbell made up of two point masses of equal mass $m$ rotating with angular velocity $\omega$ about a fixed axis (in the direction of $\boldsymbol{\omega}$). The masses can have a separation $2d$ and the axis of rotation can intersect

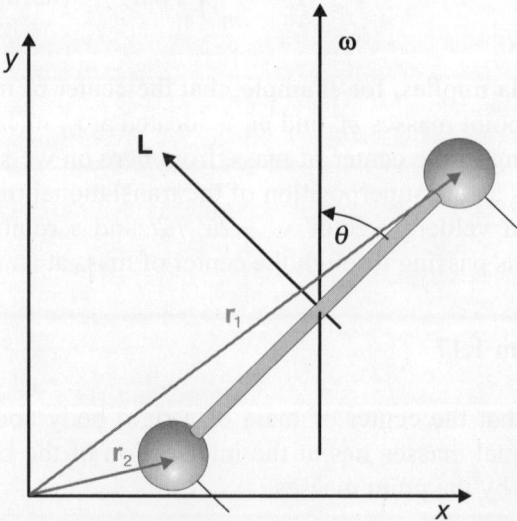

**FIGURE 1.23**  The angular momentum of a dumbbell.

the axis of the dumbbell at the center of mass located at the midpoint of the axis at an angle $\theta$. In order to calculate the angular momentum of the dumbbell, we take the sum of the angular momenta of the two point masses,

$$\mathbf{L} = \mathbf{r}_1 \times \mathbf{p}_1 + \mathbf{r}_2 \times \mathbf{p}_2.$$

Since the axis of rotation intersects the dumbbell at its midpoint, $\mathbf{p}_1 = -\mathbf{p}_2$, so that the angular momentum of the dumbbell,

$$\mathbf{L} = (\mathbf{r}_1 - \mathbf{r}_2) \times \mathbf{p}_1,$$

depends only on the relative position $\mathbf{r}_1 - \mathbf{r}_2$ of the two point masses and not on the choice of coordinate origin. Since $\mathbf{p}_1$ is perpendicular to the axis of the dumbbell, $\mathbf{L}$ lies in the plane formed by the dumbbell and the axis of rotation and is perpendicular to the axis of the dumbbell. Thus, $\mathbf{L}$ is not generally parallel to $\boldsymbol{\omega}$. If $\theta \neq 0$ and $\theta \neq \pi/2$, then the direction of $\mathbf{L}$ changes continuously as the dumbbell rotates. Only the projection $L_z$ of $\mathbf{L}$ on the axis of rotation remains constant.

An arbitrary extended object can, as explained in Section 1.4.2, be thought of as made up of many small elements of mass $dm$ (atoms). The angular momentum of the entire object is then the sum of the angular momenta of all the elements of mass:

$$\mathbf{L} = \int dm (\mathbf{r} \times \mathbf{v}).$$

Thus, the angular momentum depends on the motion of the position vectors $\mathbf{r}$ of the mass elements $dm$ about the origin. Let us assume at first that this is the center of mass of the object and that it lies on the axis of rotation about which the object rotates with angular velocity $\boldsymbol{\omega}$. In this case, the center of mass is motionless and (Section 1.1.2) $\mathbf{v} = \boldsymbol{\omega} \times \mathbf{r}$, so that

$$\mathbf{L} = \int dm (\mathbf{r} \times (\boldsymbol{\omega} \times \mathbf{r})).$$

In general, the center of mass of a rotating object also moves along a trajectory $\mathbf{r}_C(t)$ (Figure 1.24). The total angular momentum of the object is then obtained from the motion of the center of mass and the rotation about a centroid axis:

$$\mathbf{L} = \int dm (\mathbf{r}_C + \mathbf{r}) \times (\mathbf{v}_C + \boldsymbol{\omega} \times \mathbf{r}).$$

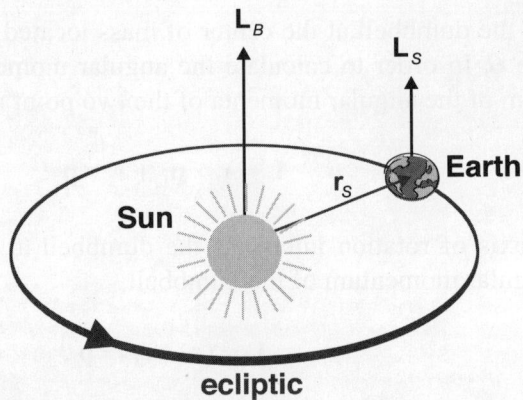

**FIGURE 1.24** The orbital angular momentum and the intrinsic angular momentum of the earth.

Since the origin of the vectors **r** is assumed to be the center of mass, this formula simplifies to

$$\mathbf{L} = m(\mathbf{r}_C \times \mathbf{v}_C) + \int dm(\mathbf{r} \times (\boldsymbol{\omega} \times \mathbf{r})).$$

**Problem 1.18**

Prove this statement. Decompose the integrand into a sum of four products and use the facts that $\int \mathbf{r}\, dm = 0$ and $\int dm = m$ in integrating the individual terms.

The total angular momentum of the object is, therefore, the sum of the orbital angular momentum $\mathbf{L}_B$ associated with the trajectory of the center of mass and the intrinsic angular momentum (spin) $\mathbf{L}_S$ associated with the rotation about a centroidal axis; that is,

$$\mathbf{L} = \mathbf{L}_B + \mathbf{L}_S.$$

**Problem 1.19**

Estimate the orbital angular momentum and spin of the earth. How large are they compared to $\hbar$?

**Notes** The angular momentum **L** of a rigid body rotating at angular velocity ω is usually not parallel to ω. The angular momentum **L** of a dumbbell is not parallel to ω, if the axis of the dumbbell is neither parallel nor perpendicular to the axis of rotation (ω).

## 1.4.4  Conservation of angular momentum

The orbital angular momentum of a point mass moving around a center of force is constant in time (Section 1.4.1). Likewise, the intrinsic angular momentum of the earth (its spin), or the daily rotation of the earth about its north-south axis, remains constant, as we know from experience. Precise measurements do, of course, reveal small deviations and variations in the earth's rotation, which we shall ignore here. They originate, for example, in a weak coupling of the earth's rotation to the motion of the moon (tidal ebb and flow) and to the orbits of the planets. Laboratory experiments show, however, that the angular momentum of many earthbound objects is conserved if they can rotate undisturbed. Some *experiments* should substantiate the universal validity of the conservation law for angular momentum.

Let us first consider the rotation of objects about the fixed axis (pointing in the $z$ direction) of a rotating table (Figures 1.21 and 1.25). The experiment on *rotation with dumbbells*, discussed in Section 1.4.1, has already shown that when the dumbbells are moved, the angular velocity of the rotation does indeed change, but not the $z$-component $L_z$ of the angular momentum. For the conservation of angular momentum it is only important that the rotation not be disturbed by a force acting from outside. The conservation law is valid, regardless of how the rotating object is distorted and what forces operate between the elements of mass of which the body is composed.

### Experiment 1.7   Rotating table

The universal validity of the conservation of angular momentum can be still more impressively demonstrated with experiments in which a test subject on a rotating table moves a rotating wheel (Figure 1.25).

If, for example, a rotating wheel with its axis vertical is taken by a test subject initially at rest and then turned so the wheel's axis is horizontal, the test subject will then rotate on the turntable together with the wheel with the original angular momentum of the wheel. This happens because, when the axis of the wheel is rotated, the condition $L_z = const$ holds for the combined system as it rotates about the $z$-axis. Since the wheel has angular momentum $L_z(\text{wheel}) = L_R$ before its axis is turned, and $L_z(\text{wheel}) = 0$ after the axis is turned, the combined system rotates on the turntable with angular momentum $L_z = L_R$.

If the wheel's axis is turned a further 90° to a vertical orientation opposite its original direction, the test subject will rotate at twice the angular velocity because $L_z = const$, while now $L_z(\text{wheel}) = L_R$.

In a third step, the still rotating wheel is brought to a stop by the test subject. After that, we again have $L_z(\text{wheel}) = 0$. The test subject thus again has the same angular velocity as after the first step.

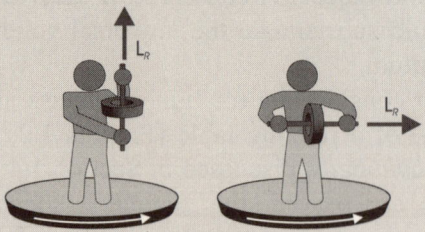

**FIGURE 1.25** A rotating table experiment with a wheel.

In these experiments only the $z$-component of the angular momentum is conserved, since the axis of rotation of the rotating table cannot move with respect to the lecture hall. Things are different for the earth. Its axis of rotation can move freely in space. Nevertheless, its direction in space (more precisely, in an inertial system specified in terms of the fixed stars) and, therefore, the inclination of the earth's axis relative to the plane of the earth's orbit (of the ecliptic) remains approximately constant. For, in this case, the conservation law holds for all three components of the angular momentum. In order to observe an object freely rotating in the laboratory, one can throw it into the air with a spin or let it rotate in a *gimbal* (Cardin) *mount* (Figure 1.26).

If you move a rotating object held in a gimbal (about its axis of symmetry), thereby changing the position of the mount, the direction of the angular momentum (here, the direction of the axis of rotation) will remain unchanged in space. Here, in addition, it should be noted that the three axes of rotation permit independent rotations; thus, in particular, no two of the axes are parallel to one another. With the gimbal mount shown in Figure 1.26, the outer hoop rotates about a vertical axis, while the inner hoop rotates about a horizontal (always perpendicular) axis. But the rotor axis can have an arbitrary direction and could, in particular, be parallel to the axis of rotation of the outer hoop.

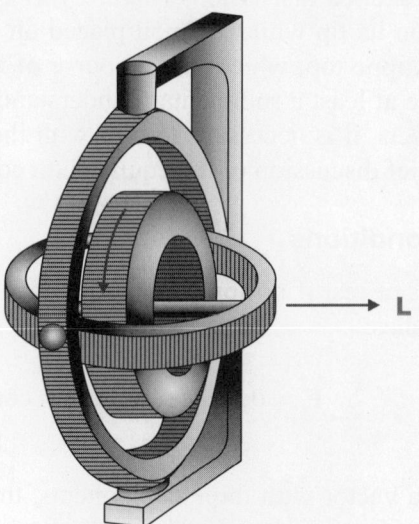

**FIGURE 1.26** A gimbal mount.

---

### Problem 1.20

Do the experiment with the rotating wheel on the turntable yourself and try to find out which forces act between the wheel and your body if, for example, you turn the axis of the wheel from vertical to a horizontal direction. The rotation about the axis of the turntable changes because of these forces.

---

> | **Notes** | During free rotation about a fixed axis ($z$-axis), only the $z$-component of the angular momentum is conserved: $L_z = const$. |
>
> During free rotation of an object in space, all three components of the angular momentum, i.e., the angular momentum vector, are conserved: $\mathbf{L} = const$.

## 1.5 DYNAMICS OF RIGID BODIES

A top spinning on its tip not only delights children, but can stir famous physicists to wonder and excited discussions (Figure 1.27). Why does a spinning top continue to stand on its tip while a pencil placed on its point falls over? Even more amazing is the tippe top, which in the course of rotating first stands on its stem. In order to have at least a rudimentary understanding of the astonishing behavior of rotating objects, it is necessary to deal with the dynamics of rigid bodies. We begin with a brief discussion of the equilibrium conditions for rigid bodies.

### 1.5.1 Equilibrium conditions

A point mass is at rest if no force is applied to it or the sum of the forces $\mathbf{F}_i$ applied to it vanishes:

$$\sum_i \mathbf{F}_i = 0. \qquad \text{(equilibrium condition for translational motion)}$$

Since force is a vector with three components, this vector equation encompasses three conditions. In accordance with the three translational degrees of freedom of a mass point, three equilibrium conditions have to be satisfied in order for a point mass to be at rest. Besides the three translational degrees of freedom, a rigid body has three rotational degrees of freedom. Therefore, in order for a rigid body to be at rest, three additional equilibrium conditions must be satisfied. Let us find these now.

In order to set a wheel mounted on a fixed axis to rotating, it is best to apply a tangential force to the rim. This well known example makes it clear that accelerating a rigid body depends not only on the magnitude and direction of the force but also where the force is applied. Thus, the position vector $\mathbf{r}$ of the point where the force is applied should be indicated. Then the force $\mathbf{F}$ also delivers a *torque* $\mathbf{T}$ to the object, where

**FIGURE 1.27** A tippe top fascinates two Nobel prize winners, Niels Bohr and Wolfgang Pauli.

$$\mathbf{T} = \mathbf{r} \times \mathbf{F}. \qquad \text{(definition of torque)}$$

In order for a rigid body to be at rest and, in particular, for it not to tip over, the forces applied to it must balance out mutually, and so must the torques $\mathbf{T}_i$ acting on it, i.e.,

$$\sum_i \mathbf{T}_i = 0. \qquad \text{(equilibrium condition for rotation)}$$

Applied to a beam balance, this vector equation yields the well-known equilibrium condition *force × force arm = load × work arm*.

For a rigid body which, like the balance, is supported at only one point, we also have the condition that the sum of the torques vanishes: this object is in equilibrium if and only if the support point and the center of mass lie on a vertical line. The equilibrium is referred to as *stable* if the center of mass lies below the support point. In this case, the object swings about the equilibrium position following any excursion from that position. The equilibrium is referred to as *unstable* (unsteady) if the center of mass lies above the support point and *indifferent* if the center of mass and support point are coincident.

---

### Problem 1.21

A three-legged stool must stand stably on a sloping plane with a specified slope. Formulate an equilibrium condition that is to be taken into account in making the stool.

---

**Notes**   Six equilibrium conditions must be satisfied for a rigid body to be at rest: the center of mass is at rest if the sum $\sum \mathbf{F}_i = 0$ of the forces acting on the body vanishes and the rigid body does not rotate if the sum $\sum \mathbf{T}_i = 0$ of the torques applied to it vanishes. The total number of equilibrium conditions for the three force components and the three torque components corresponds to the six degrees of freedom of a rigid body.

## 1.5.2 Moment of inertia

If a force $\mathbf{F}$ acts on a point mass, its momentum $\mathbf{p}$ changes:

$$\mathbf{F} = \frac{d\mathbf{p}}{dt}. \qquad \text{(action principle for translational motion)}$$

This equation is equivalent to the Newtonian action principle, $\mathbf{F} = m\mathbf{a}$, since the momentum of a point mass is given by $\mathbf{p} = m\mathbf{v}$.

The angular momentum $\mathbf{L}$ of a rigid body changes correspondingly if a torque $\mathbf{T}$ is applied to it:

$$\mathbf{T} = \frac{d\mathbf{L}}{dt}. \qquad \text{(action principle for rotation)}$$

Thus far, the basic equation for rigid body rotation corresponds to the same equation for translation of point masses. But while $\mathbf{p} = m\mathbf{v}$ and $\mathbf{v}$ are parallel to one another, the corresponding rotational parameters, the angular momentum $\mathbf{L}$ and angular velocity $\boldsymbol{\omega}$, can have different directions, as the example of the rotating dumbbell shows (Section 1.4.3). **L** and $\omega$ are related by a tensor transformation and do not differ merely by a scalar factor, as do **p** and **v**. Because of this difference, the theory of rotational motion is substantially harder, but also more interesting, than the theory of translational motion.

In order to get a preliminary glimpse at the dynamics of rigid bodies, let us first consider the potential energy of a rigid body under the force of gravity and its kinetic energy. As when we calculate the angular momentum of rigid bodies (Section 1.4.3), here there is also some advantage in describing the motion and position of an element of mass $dm$ of the rigid body relative to its center of mass and, thereby, decomposing the motion of a rigid body into its *center of mass motion* and its *relative motion*. The position and velocity of the center of mass are described by $\mathbf{r}_C$ and $\mathbf{v}_C$ and the relative motion of an element of mass $dm$, by $\mathbf{r}$ and $\mathbf{v}$. The potential energy of a mass element under the (constant in space) force of gravity is $dE_{\text{pot}} = dm\, g\,(h_C + h)$, where $h_C$ and $h$ are the vertical components of the position vectors. After integrating over all elements of mass $dm$, we obtain the potential energy of the rigid body,

$$E_{\text{pot}} = mgh_C. \qquad \text{(potential energy of a rigid body)}$$

It can, therefore, be calculated in the same way as the potential energy of a point mass. It is harder to calculate the kinetic energy of a rigid body. The kinetic energy of one of its elements of mass $dm$ is $dE_{\text{kin}} = \frac{1}{2}dm(\mathbf{v}_C + \mathbf{v})^2$, with $\mathbf{v} = \omega \times \mathbf{r}$. Integrating over $dm$ yields the kinetic energy of the rigid body,

$$E_{\text{kin}} = \frac{1}{2}mv_C^2 + E_{\text{rot}}. \qquad \text{(kinetic energy of a rigid body)}$$

The first term in this sum is the *translational energy* $E_{\text{trans}}$ of the rigid body owing to the motion of the center of mass. It corresponds to the kinetic energy of a mass point whose mass equals the total mass $m$ of the rigid body. In addition, there is a term which originates from the rotational motion of the rigid body,

$$E_{\text{rot}} = \frac{1}{2}\int dm(\omega \times \mathbf{r})^2.$$

After integration over $m$, this yields the *energy of rotation* $E_{\text{rot}}$,

$$E_{\text{rot}} = \frac{1}{2}\mathbf{J}_\omega \omega^2. \qquad \text{(energy of rotation of a rigid body)}$$

Here

$$\mathbf{J}_\omega = \int \rho^2 \, dm$$

(definition of the moment of inertia for rotation about the axis of rotation ω)

is the *moment of inertia* of the rigid body with respect to an axis of rotation passing through the center of mass (it has the same direction as ω) and $\rho$ is the distance of the element of mass *dm* from the axis of rotation (Figure 1.28), i.e., the component of **r** perpendicular to ω.

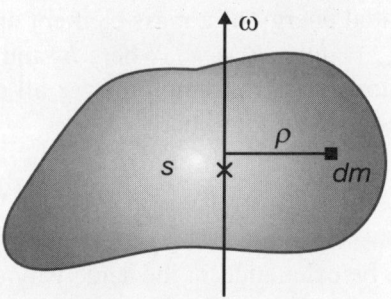

**FIGURE 1.28** Illustrating the definition of the moment of inertia of a rigid body.

The dependence of the moment of inertia on the direction of the axis of rotation is a consequence of the tensor relationship between **L** and ω. This tensor dependence implies further that in every rigid body there are three mutually perpendicular axes of rotation (body axes) attached to the body for which **L** and ω are parallel. These are referred to as the *principal axes of inertia* of the rigid body and the corresponding moments of inertia are referred to as the *principal moments of inertia*. If the directions of the principal axes of inertia and the principal moments of inertia of a rigid body are known, these can be used to calculate the moments of inertia for all other rotation axes of the rigid body.

For rotationally symmetric objects, the axis of symmetry is a principal axis of inertia. All axes through the center of mass perpendicular to it are, likewise, principal axes of inertia. The principal moments of inertia associated with these axes are all equal. The principal moment of inertia $J_{sym}$ associated with the axis of symmetry can often be calculated easily. For example, in a hollow cylinder of mass *m* all the elements of mass have roughly the same distance *R* from the axis of symmetry. Thus, in this case $J_{sym} = mR^2$. For a solid cylinder with a homogeneous distribution of mass, however, a simple integration gives $J_{sym} = \frac{1}{2}mR^2$.

## Experiment 1.8    Cylinder on an inclined plane

Once the moments of inertia $J_{svm}$ are known, we can discuss the rolling of rotationally symmetric objects down an inclined plane (Figure 1.29) or the falling motion of a yo-yo (Figure 1.30) using the conservation of energy. Since the kinetic energy of a rigid body is the sum of the translational energy of the center of mass motion and the rotational energy of the rotational motion relative to the center of mass, we have

$$E_{pot} + E_{trans} + E_{rot} = const.$$

For a rolling cylinder the translational velocity is $v_{trans} = \omega R$. The conservation of energy implies $mgh + \frac{1}{2}m(\omega R)^2 + \frac{1}{2}J_{svm}\omega^2 = const$. This yields the angular velocity $\omega$ of the rolling cylinder as a function of the height $h$.

A question: which cylinder wins if a hollow cylinder and a solid cylinder (with the same radii) roll down the inclined plane in a race?

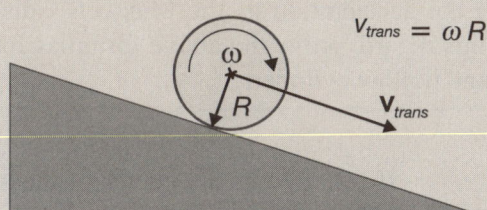

**FIGURE 1.29**  A cylinder on an inclined plane.

## Experiment 1.9    Maxwell's disk

When a yo-yo (Maxwell's disk, Figure 1.30) falls, the rate of fall $v_{trans} = \omega r$ is proportional to the radius $r$ of the axis of the yoyo, while the moment of inertia $J_{svm} = mR^2$ depends on the radius $R$ of the disk. Thus, as the yo-yo falls, the potential energy is primarily converted into rotational energy of the disk.

*continued*

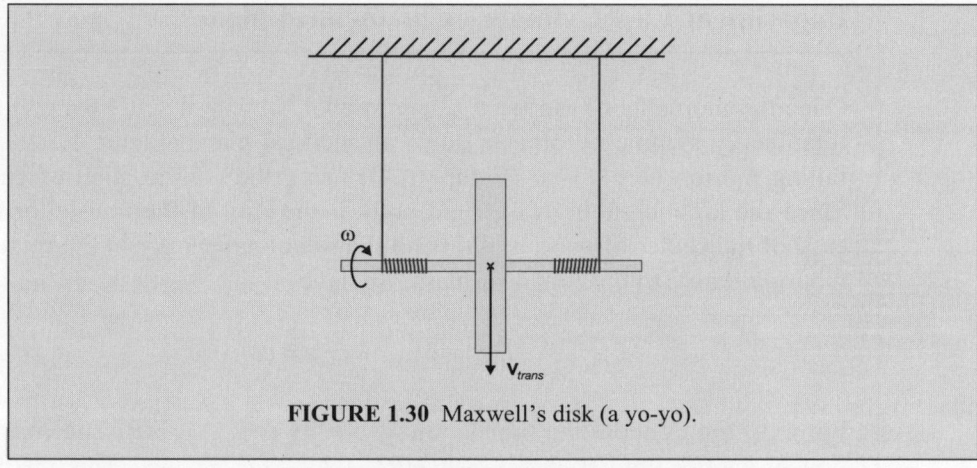

FIGURE 1.30  Maxwell's disk (a yo-yo).

**Problem 1.22**

Calculate the acceleration as the Maxwell's disk shown in Figure 1.30 falls. In this regard, prove the above formulas for the moments of inertia of solid and hollow cylinders.

---

**Notes**

If (relative to the center of mass) a torque $\mathbf{T} = \mathbf{r} \times \mathbf{F}$ is applied to a rigid body, the angular momentum $\mathbf{L}$ of its rotation (relative to the center of mass) changes. The rotational motion satisfies the equation

$$\mathbf{T} = \frac{d\mathbf{L}}{dt}.$$

This equation for the rotational motion of a rigid body corresponds to the Newtonian action principle $\mathbf{F} = d\mathbf{p}/dt$ for the translational motion of point masses.

### 1.5.3  Motion of a top

In order to be able to calculate the motions of rigid bodies, the forces and torques acting on them must be known as functions of time and of the coordinates describing their position. Corresponding to the six degrees of freedom of a rigid body, one then obtains six equations of motion. These can be summarized as two vectorial equations of motion, an equation of motion

$$\mathbf{F} = \frac{d\mathbf{p}_C}{dt}$$

for the translational motion of the center of mass and an equation of motion

$$\mathbf{T} = \frac{d\mathbf{L}}{dt}$$

for the rotational motion about an axis passing through the center of mass.

Usually **F** and **T** are complicated functions of the state of motion of a rigid body. Thus, it is often difficult to derive suitable equations of motion and even harder to solve them. In most toy tops, for instance, **F** and **T** arise not only from the effect of gravity on the top but also, to an important extent, from the adhesion of the top to the floor. Here we limit ourselves to discussions of a few simple examples:

**Precession.** When a top is supported exactly at its center of mass, all the externally acting forces and torques cancel out. In that case, therefore, the direction of the angular momentum remains fixed in space. But only when the axis of rotation coincides with a principal axis of inertia (for example, the axis of symmetry), will the position of the axis of rotation also be fixed in space. On the other hand, if the top is supported at a point other than the center of mass along the axis of symmetry, then precession results (Figure 1.31). Let $\mathbf{r}_C$ be the position vector of the center of mass relative to the point of support and $m$ be the mass of the top. Then a torque $\mathbf{T} = m\mathbf{r}_C \times \mathbf{g}$ acts on the top. The angular momentum **L** of the rapidly rotating top is parallel to $\mathbf{r}_C$. Hence, **T** is perpendicular to **L** and directed horizontally. **L** then precesses about the vertical at a frequency $\omega_P$, but its magnitude does not change. The tip of the angular momentum vector thus describes a circle during precession.

From the equation $\mathbf{T} = d\mathbf{L}/dt$ we obtain the following equation of motion for the precession:

$$\frac{d\mathbf{L}}{dt} = m\mathbf{r}_C \times \mathbf{g}.$$

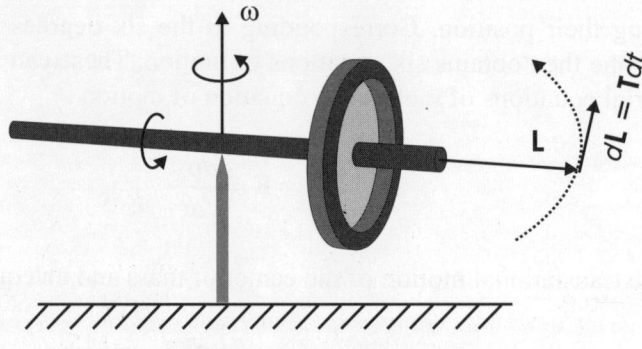

**FIGURE 1.31** Precession of a top.

This yields a simple relationship for calculating the precession frequency $\omega_P$. For circular motion (Section 1.1.2) we then obtain

$$\frac{d\mathbf{L}}{dt} = \boldsymbol{\omega}_P \times \mathbf{L}.$$

(Comment: here we have neglected the vertical component of the angular momentum owing to the precessional motion compared to the angular momentum $\mathbf{L}$ of the rotation about the axis of the top.)

**Stable and unstable rotational motion of rotating blocks.** The principal axes of inertia of a block lie parallel to the edges (and pass through the center of mass). During rotation about one of these axes, therefore, the position of the axis of rotation in the object should not change since $\mathbf{L}$ and $\boldsymbol{\omega}$ are parallel. Experiments show, however, that stable motion occurs only for the principal axes of inertia with the largest and smallest moments of inertia. But the block begins to totter during rotation about the principal axis of inertia with the intermediate value of the moment of inertia. Convince yourself of this by tossing a block with different rotations into the air! In order to explain your observations, the influence of air friction on the rotational motion should also be taken into account.

## Problem 1.23

Experiment with tops and try to understand how a tippe top works. Think about the direction of the angular momentum owing to gravity for a top standing on its tip and the direction of the angular momentum for a tippe top, as it (for a deep lying center of mass) rotates on its conical end. It is assumed that neither top is standing vertically.

| **Notes** | A horizontal angular momentum perpendicular to the axis of symmetry acts on a rotating top that is rotating about its horizontally oriented axis of symmetry and is supported at a |

point on its axis of symmetry other than its center of mass (Figure 1.31). In this way the top precesses with a rapid rotation in the horizontal plane and does not tilt downward.

### 1.5.4 Static and dynamic imbalance

For machines with parts that rotate rapidly about a fixed axis, it is important to make sure that these parts rotate about a principal axis of inertia. If this condition is not met, components of this sort, such as the wheels of a car, have to be *balanced* by installing supplementary weights. Otherwise, an *imbalance* develops if the axis of rotation does not pass through the center of mass. Then the center of mass moves at the rotation frequency $\omega$ about the axis of rotation and is thereby accelerated. The accelerating centrifugal force $F = m\omega^2 \rho_C$ (where $\rho_C$ is the distance of the center of mass from the axis of rotation) is proportional to the square of the rotation frequency $\omega$, so it becomes a heavy load on the axis of rotation at high frequencies.

Therefore, for all rapidly rotating machine parts care must be taken to ensure that the center of mass lies exactly on the axis of rotation. Since deviations can easily be measured by static measurements, we speak of a *static imbalance* when the center of mass of a rotating part does not lie on the axis of rotation.

Besides static imbalance, an equally dangerous *dynamic imbalance* is possible. This occurs when the axis of rotation is not coincident with one of the principal axes of inertia. Then the angular momentum vector does not point in the direction of the axis of rotation. Consequently, during rotation the angular momentum vector moves along the envelope of a cone surrounding the axis of rotation. The time variation in the angular momentum means that a torque $\mathbf{T} = d\mathbf{L}/dt$ must act constantly on the rotating part. This torque also increases with the square of the rotation frequency, imposes a load on the axis, and can cause serious accidents. Note that the rotational energy, which can cause damage if an axle breaks, also increases as the square of the rotation frequency.

In all rapidly rotating parts it is important to ensure that the center of mass lies on the axis of rotation, but also to ensure that the axis of rotation is coincident with one of the principal axes of inertia. When this is not so, we speak of a *dynamic imbalance*. An imbalance of this sort can only be found in rotating objects.

**Problem 1.24**

Calculate the torque with which a rotating dumbbell loads the axis of rotation if the axis of the dumbbell is not exactly perpendicular to the axis of rotation, but deviates by a small angle $\delta\varphi$ from perpendicular. Assume that the center of mass of the dumbbell lies on the axis of rotation.

**Notes**   Parts rotating about fixed axes can have static and dynamic imbalances. In a static imbalance the center of mass of the rotor does not lie on the axis of rotation. In a dynamic imbalance the axis of rotation is not along one of the principal axes of inertia.

## 1.6  OSCILLATIONS

Periodically recurring events, such as the change from day to night or summer to winter, define our sense of time. For thousands of years, the periodic cycle of the stars served as the basis of the definition of the unit of time. The periodic movement of the pendulum in a grandfather clock or of the balance wheel in a (mechanical) pocket watch are still used today for measuring time. Because of this relationship to the time scale, periodic motions are of central importance in physics. Uniform circular motion is an elementary periodic motion. We became acquainted with it in Section 1.1.2. *Harmonic oscillation* is another elementary periodic motion. The two are closely related mathematically. We shall use this relationship in order to solve the equations of motion for oscillating objects.

## 1.6.1  Circular motion and the harmonic oscillator

Circular motions take place in a plane. Thus, it is possible to use the complex plane $Z = x + iy$, where $i^2 = -1$, for describing circular motions in the $x$-$y$ plane mathematically (Figure 1.32). A uniform circular motion with radius $r$ about the origin $Z = 0$ of the complex plane can be described using the following function $Z(t)$:

$$Z(t) = r(\cos \omega t + i \sin \omega t).$$

This expression can be simplified using the Euler formula for complex numbers, $e^{i\varphi} = \cos \varphi + i \sin \varphi$, to give

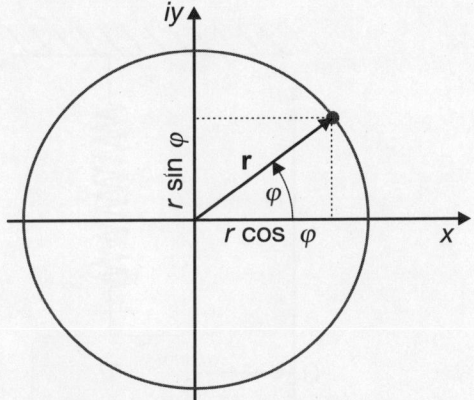

**FIGURE 1.32**  The circular trajectory $Z(t) = r(\cos \varphi + i \sin \varphi)$ in the complex plane with $\varphi = \omega t$.

$$Z(t) = r \exp(i\omega t).$$

The function $Z(t)$ satisfies the same differential equation as in Section 1.1.2 for the vectorial trajectories $\mathbf{r}(t)$:

$$\frac{d^2 Z}{dt^2} = -\omega^2 Z. \qquad \text{(differential equation for uniform circular motion)}$$

This differential equation is linear. Thus, besides the solutions $\exp(i\omega t)$ and $\exp(-i\omega t)$, all linear combinations $Z(t) = A_+ \exp(i\omega t) + A_- \exp(-i\omega t)$ of the two basis solutions with complex *amplitudes* $A_+$ and $A_-$ are solutions of this differential equation.

The equation of motion for a *harmonic oscillator* has an analogous structure. As a simple example of a harmonic oscillator we consider a *spring pendulum* (oscillating in the $z$ direction; Figure 1.33). Once a mass $m$ suspended on a spring is displaced, it oscillates about its equilibrium (rest) position. The equation of motion for this oscillation follows from the action principle and Hooke's law (Section 1.2.2):

$$m\frac{d^2z}{dt^2} + Dz = 0. \quad \text{(equation of motion for a spring pendulum)}$$

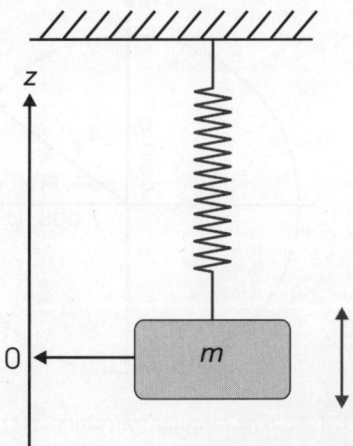

**FIGURE 1.33**  A spring pendulum.

Like the differential equation for circular motion, this equation of motion can be solved if the real trajectory $z(t)$ is supplemented with an appropriate imaginary part to form a complex function $Z(t)$. A comparison of the two equations immediately gives the *natural frequency* or *eigenfrequency* $\omega_0$ of the spring pendulum:

$$\omega_0 = \sqrt{\frac{D}{m}}. \quad \text{(natural frequency of a spring pendulum)}$$

The solutions $z(t)$ are the projections of the complex solutions $Z(t) = A\exp(i\omega_0 t)$ on the real axis of the complex plane: $z(t) = |A|\cos(\omega_0 t + \varphi)$, where

$A = |A| \exp(i\varphi)$. If the displacement of the pendulum at time $t = 0$ is at the maximum $z_0 = |A|$, then the solution is

$$z(t) = z_0 \cos(\omega t).$$

Another realization of the harmonic oscillator is the clock pendulum consisting of a round disk suspended a distance 1 from its center of mass (Figure 1.34). For a pendulum of this sort, the total angular momentum $\mathbf{L} = \mathbf{L}_B + \mathbf{L}_S$, the sum of the angular momentum of the trajectory and relative motion. In the limit where the disk is a point mass $m$ (a mathematical pendulum), $L_S = 0$ and $L_B = mlv$. The torque acting on the pendulum, $\mathbf{T} = \mathbf{r}_C \times \mathbf{F}$ has magnitude $T = mgl \sin \alpha$ that depends on the deflection angle $\alpha$. Thus, the equation of motion for a *mathematical pendulum* has the form

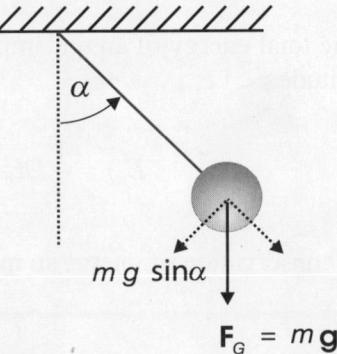

**FIGURE 1.34** The swinging pendulum of a grandfather clock.

$$ml^2 \frac{d^2\alpha}{dt^2} + mgl \sin \alpha = 0. \quad \text{(equation of motion for a mathematical pendulum)}$$

Since the restoring angular momentum depends on $\sin \alpha$ rather than on $\alpha$, the oscillations of the mathematical pendulum are not, in general, harmonic (i.e., not a pure sinusoidal oscillation). For small deflections, however, $\sin \alpha = \alpha$ to a good approximation, so that the oscillations are harmonic; that is, they will be described by a time dependent sinusoidal function $\alpha(t) = \alpha_0 \sin(\omega_0 t + \varphi)$. Since the mass $m$ cancels out in the equation of motion (owing to the equivalence of inertial and gravitational mass), the natural frequency $\omega_0$ of the pendulum is independent of $m$ and is given by

$$\omega_0 = \sqrt{\frac{g}{l}}.$$    (natural frequency of a mathematical pendulum)

As these examples show, an object oscillates harmonically if, when it is displaced from its equilibrium position, a force proportional to the displacement operates to bring it back to its equilibrium position. The equations of motion for these oscillations have a structure analogous to that of the differential equation for circular motion, so they can be solved using the same approach.

We conclude by analyzing the energy of the pendulum. The energy of a harmonic oscillator is the sum of its kinetic and potential energies. For example, for a spring pendulum we have

$$E_{\text{tot}} = \frac{1}{2} m \left( \omega_0 z_0 \right)^2 \sin^2 \left( \omega_0 t \right) + \frac{1}{2} D z_0^2 \cos^2 \left( \omega_0 t \right).$$

Since $m\omega_0^2 = D$, the total energy of an undamped pendulum is proportional to the square of the amplitude $z_0$, i.e.,

$$E_{\text{tot}} = \frac{1}{2} D z_0^2.$$

In accord with the conservation of energy in mechanics, it is constant in time.

---

### Problem 1.25

Calculate the length of a mathematical pendulum that will move back and forth once in a second.

**Notes** The equations of motion for harmonic oscillators can be written in the following standard form:

$$\frac{d^2 Z(t)}{dt^2} = -\omega_0^2 Z(t).$$

The natural frequency of the oscillator is $\omega_0$. The solutions of the oscillator equation are

$$Z(t) = A \exp(\pm i\omega_0 t).$$

Here $A$ is the (generally) complex amplitude of the oscillation.

### 1.6.2 Damped oscillators

Thus far we have dealt with strictly periodic oscillations. But an oscillation is strictly periodic only if no energy is lost during the oscillatory process or any energy loss is compensated by an energy input. Actually, frictional losses do occur with any freely oscillating oscillator. Thus, the amplitude of the oscillations decreases in time, along with the energy of the oscillator. Here we speak of a damped oscillator.

In order to take *damping*, such as that produced by air friction, into account theoretically, a frictional force $F_R$ that retards the motion has to be included in the equation of motion. We shall assume that the frictional force is proportional to the velocity $v$ of the oscillator, i.e., $F_R = -kv$. The proportionality constant $k$ is known as the *friction constant*. The equation of motion of a damped spring pendulum, therefore, takes the form

$$m\frac{d^2 z}{dt^2} + k\frac{dz}{dt} + Dz = 0.$$

With the parameters $\delta = k/2m$ (the damping factor) and $\omega_0 = \sqrt{D/m}$ (the natural frequency), both of which have dimensions $s^{-1}$, we reduce the equation of motion to the normal form

$$\frac{d^2 z}{dt^2} + 2\delta\frac{dz}{dt} + \omega_0^2 z = 0. \quad \text{(equation of motion for a damped oscillator)}$$

In order to solve this homogeneous linear differential equation, we assume a complex trial solution of the form $Z(t) = A\exp(i\Omega t)$. This trial function satisfies the equation of motion if $-\Omega^2 + 2i\delta\Omega + \omega_0^2 = 0$, or

$$\Omega = i\delta \pm \sqrt{\omega_0^2 + \delta^2}.$$

Depending on the sign on the square root, the motion evolves in different ways (Figures 1.35 and 1.36): in the weakly damped case with $\omega_0 > \delta$, the pendulum oscillates back and forth with an exponentially decaying amplitude $A\exp(-\delta t)$ (*oscillatory case*, Figure 1.35). All the solutions $Z(t)$ can be written as superpositions of two basis solutions $Z_+(t)$ and $Z_-(t)$, where

$$Z_\pm(t) = A_\pm \exp\left[\left(-\delta \pm i\sqrt{\omega_0^2 + \delta^2}\right)t\right].$$

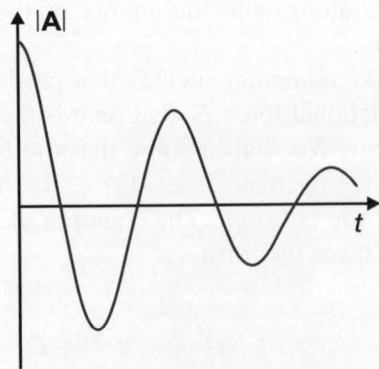

**FIGURE 1.35** The displacement of a weakly damped pendulum as a function of time.

The energy $E$ of the oscillator is proportional to the square of the absolute value of the complex amplitude and, therefore, falls off exponentially with time as

$$E(t) = \frac{1}{2}D|A_\pm|^2 e^{-2\delta t}.$$

In the strongly damped case with $\omega_0 < \delta$, following an initial displacement the pendulum creeps (without oscillating) back to its equilibrium position (*strongly damped case*, Figure 1.36). Here, also, there are two basis solutions,

$$Z_{\pm}(t) = A_{\pm} \exp\left[-\left(\delta \pm \sqrt{\omega_0^2 + \delta^2}\right)t\right].$$

The asymptotic behavior of the function $Z(t)$ as $t \to \infty$ is determined by the solution with the smaller damping factor in the exponent of the exponential function. If $\delta \gg \omega_0$, the smaller damping factor is given by $(\omega_0/\delta)^2 \, \delta/2$. Thus, it can take a very long time for the pendulum to return to its initial position. In the *aperiodic limit* $\omega_0 = \delta$, one solution is

$$Z(t) = A \exp(-\delta t).$$

(In this case, a second basis solution is $Z(t) = t \exp(-\delta t)$.) For a given natural frequency $\omega_0$ of the pendulum, it returns to its equilibrium position in the shortest time possible after an initial displacement. The aperiodic limit is of practical interest, for example, in the design of readout devices and shock absorbers, where a rapid return of the oscillating component to its initial position is desired.

---

## Problem 1.26

Show that the solutions given here satisfy the equation of motion for a damped oscillator.

---

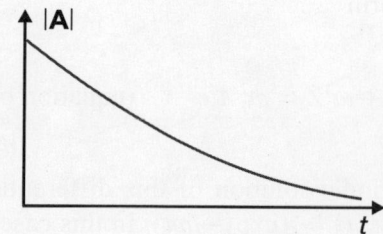

**FIGURE 1.36** The displacement of a strongly damped pendulum as a function of time.

| Notes | The normal form of the equation of motion for a damped oscillator is |
|---|---|

$$\frac{d^2z}{dt^2} + 2\delta\frac{dz}{dt} + \omega_0^2 z = 0.$$

It has different solutions, depending on whether $\delta < \omega_0$ (oscillatory case), $\delta > \omega_0$ (strongly damped case), or $\delta = \omega_0$ (aperiodic limit).

## 1.6.3 Forced oscillations

If a child on a swing wants to fly high, someone must push him synchronously with the oscillations of the swing, or he must himself add to the swinging with appropriately rhythmic movements. In both cases we are speaking of *resonant excitation* of a pendulum. In the first case the pendulum is pushed periodically from outside (*independent* or *external excitation*). In the second case, the child works on the swing and supplies the desired energy of oscillation to the pendulum (*self-excitation*). *Forced oscillations* of these types will be studied here.

Let us consider a spring pendulum whose suspension point moves up and down periodically (Figure 1.37), driven, for example, by a cam disk. If the suspension point oscillates harmonically at frequency $\omega$ with amplitude $A_0$, i.e., its position changes sinusoidally with time with $z_0(t) = A_0 \sin(\omega t)$, then the (complex) equation of motion for the position $z(t)$ of the pendulum is

$$\frac{d^2Z}{dt^2} + 2\delta\frac{dZ}{dt} + \omega_0^2\left(Z - A_0 e^{-i\omega t}\right) = 0.$$

Here it has been kept in mind that the tension of the spring depends on the distance $z_0(t) - z(t)$ of the pendulum object from the suspension point. This is an inhomogeneous linear differential equation. In order to emphasize this, we write it in the normal form

$$\frac{d^2Z}{dt^2} + 2\delta\frac{dZ}{dt} + \omega_0^2 Z = \omega_0^2 A_0 e^{-i\omega t}. \quad \text{(equation of motion for forced oscillations)}$$

A strictly periodic solution of this differential equation can be found using the trial solution $Z(t) = A \exp(-i\omega t)$. In this case the pendulum moves exactly at

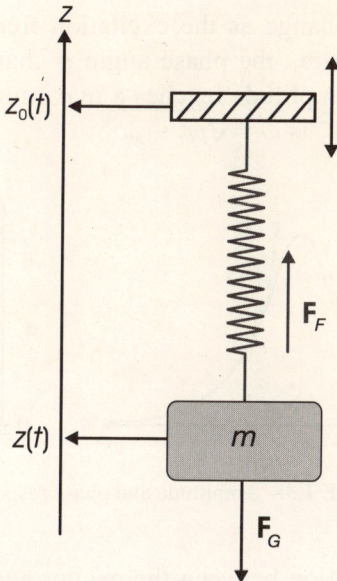

**FIGURE 1.37** Forced oscillations of a spring pendulum.

the excitation frequency $\omega$. Besides these periodic solutions, there are transient solutions which describe the transient oscillatory phenomena. The duration $\tau \sim \delta^{-1}$ of these transient effects is determined by the damping factor $\delta$. We shall not discuss them further here. The periodic trial solution is a solution if $A(-\omega^2 + 2i\delta\omega + \omega_0^2) = \omega_0^2 A_0$. For the complex amplitudes of the forced oscillations this gives

$$A = \frac{\omega_0^2}{\omega_0^2 - \omega^2 + 2i\delta\omega} A_0. \quad \text{(amplitude of forced oscillations)}$$

This varies resonantly if the excitation frequency $\omega$ is tuned across frequencies $\omega \approx \omega_0$. For $\omega = \omega_0$ the complex amplitude is given by

$$A = -i\frac{\omega_0}{2\delta} A_0. \qquad \text{(resonance amplitude)}$$

For very weak damping ($\delta \ll \omega_0$), at the resonance ($\omega = \omega_0$) the amplitude of the oscillator is very large in magnitude (*resonance peaking*) compared to the excitation amplitude (by a factor of $\omega_0/2\delta$) and the phase of the oscillator lags the phase of the exciter by an angle $\varphi = \pi/2$ (Figure 1.38). The magnitude and angle of the amplitude change as the excitation frequency is varied. On crossing the resonant frequency $\omega_0$, the phase angle $\varphi$ changes from $\varphi = 0$ to $\varphi = \pi$ and the magnitude of the amplitude reaches a maximum at $\omega = \omega_0$. The exact position of the maximum of $|A|^2$ is $\omega = \sqrt{\omega_0^2 - 2\delta^2}$.

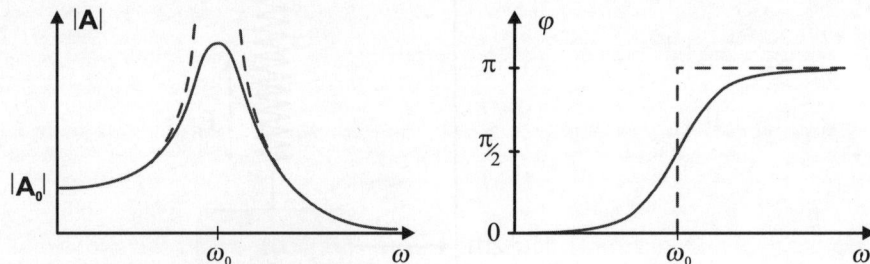

**FIGURE 1.38**  Amplitude and phase responses of forced oscillations.

The phase relation between the exciter and oscillator must be taken into account in the generation of *feedback* of an undamped oscillation. After a bit of practice, any child can bring the pendulum motion of a swing into the correct phasing. Technically, self-excitation through feedback of this type was first developed by Huygens in 1656 for a clock pendulum (Figure 1.39). When the clock is wound (the clock weights are pulled up), the weights drive the toothed escapement wheel, which, in turn, drives the clock pendulum at the correct rate. Precise time measurements became possible because of this discovery. Today the feedback principle is used in many areas of technology.

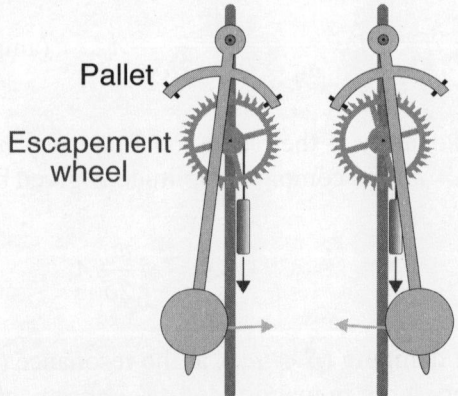

**FIGURE 1.39**  Feedback for the movement of a grandfather clock with pallet and escapement wheel.

For a very weakly damped pendulum with $\delta \ll \omega$, near the resonance we have approximately

$$|A|^2 = \frac{1}{4} \cdot \frac{\omega_0^2}{(\omega_0 - \omega)^2 + \delta^2} |A_0|^2.$$

The oscillator energy has a sharp resonance peak in this case, as well, with a width determined by the damping factor $\delta$. The separation between the two frequencies at which the oscillator energy is half the maximum value is the so-called *half width* (Figure 1.40). It equals $\Delta\omega = 2\delta$.

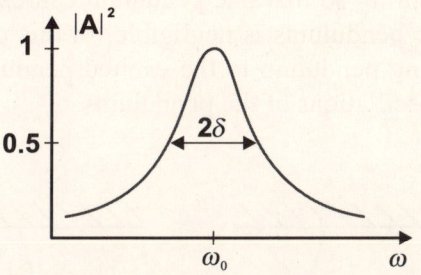

**FIGURE 1.40**  Amplitude response of a very weakly damped ($\delta \ll \omega_0$) pendulum.

For $\omega = \omega_0$, $|A|^2$ is a factor of $(\omega_0/2\delta)^2$ greater than $|A_0|^2$. The same holds for the relationship between the energies of the oscillator and exciter. This sort of resonance peaking can be useful in engineering or, when it shows up unexpectedly, it can have a destructive effect. The latter is referred to as a *resonance catastrophe*. Especially dangerous situations occur when, for example, a small imbalance of a rotating component excites resonant vibrations in a piece of equipment.

---

**Problem 1.27**

Prove that an exciter delivers a time averaged power $P = 2\delta E_{\mathrm{osc}}$ when it drives a sufficiently weakly damped oscillator.

---

### 1.6.4 Coupled pendulums

The pendulum oscillations discussed in Section 1.6.3 were driven by an exciter with a constant oscillation amplitude $A_0$. There energy is transferred from the exciter to the pendulum so that the energy lost through damping is restored to the pendulum. We now consider two coupled pendulums (Figure 1.41) which are joined with a spring so that one pendulum can excite the other. Assume that the damping of the pendulums is negligible. In this case, energy will be transferred from the exciting pendulum to the excited pendulum, thereby changing the amplitude of the oscillations of the pendulums.

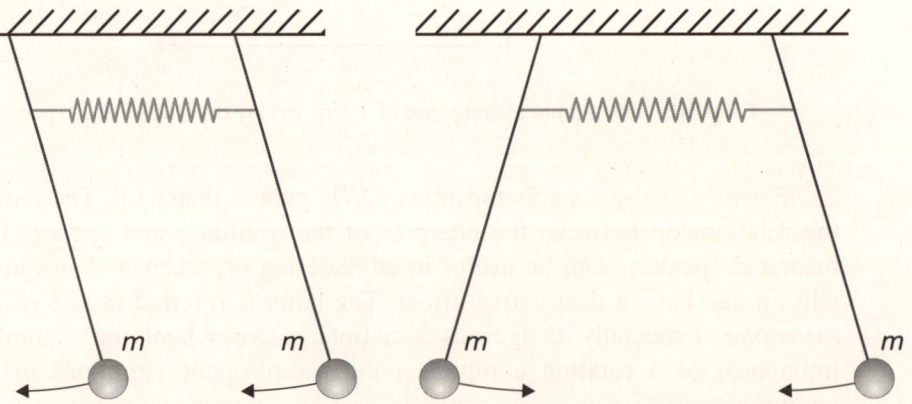

**FIGURE 1.41** The natural (or characteristic) oscillations of coupled pendulums.

In particular, consider two mathematical pendulums of equal length with the same natural frequency $\omega_0$ when they are uncoupled. When the coupling is actuated, if one of the pendulums is struck, it excites oscillations in the other until it has given up all its oscillation energy up and has, itself, come to rest. Then

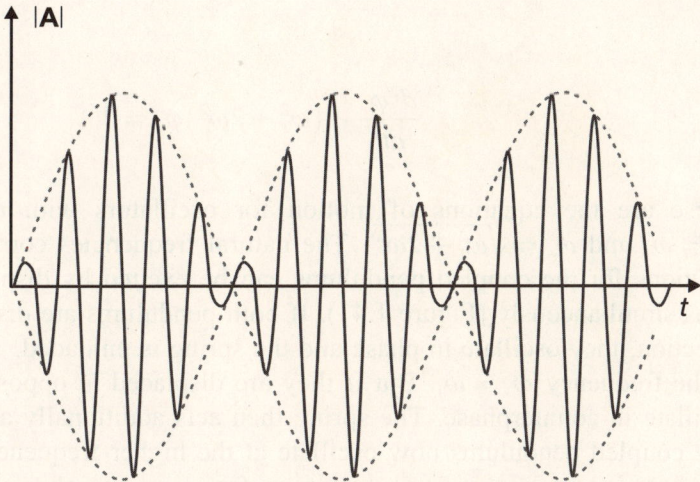

**FIGURE 1.42**  Beating of the amplitude of one of the coupled pendulums.

energy transfer takes place in the reverse direction (Figure 1.42). The oscillation energy also shifts back and forth periodically between the two pendulums at a *beat frequency* $\omega_S$.

In order to analyze this beating quantitatively, the equations of motion of the coupled pendulums have to be solved. When the deflection angles of the pendulums, denoted by $\varphi_1$ and $\varphi_2$, are assumed small and the effect of the spring on the pendulum motion is taken to be proportional to $\Delta\varphi = \varphi_1 - \varphi_2$ (the angular acceleration induced by the spring is $\omega_1^2 \Delta\varphi$), these equations take the form

$$\frac{d^2\varphi_1}{dt^2} + \omega_0^2\varphi_1 + \omega_1^2(\varphi_1 - \varphi_2) = 0$$

and

$$\frac{d^2\varphi_2}{dt^2} + \omega_0^2\varphi_2 + \omega_1^2(\varphi_2 - \varphi_1) = 0.$$

By subtracting and adding these two equations, in which the coupling frequency $\omega_1 \ll \omega_0$ is a measure of the strength of the coupling, we obtain two uncoupled linear differential equations for the functions $\varphi_+ = \varphi_1 + \varphi_2$ and $\varphi_- = \varphi_1 - \varphi_2$:

$$\frac{d^2\varphi_+}{dt^2} + \omega_0^2\varphi_+ = 0$$

and

$$\frac{d^2\varphi_-}{dt^2} + \left(\omega_0^2 + 2\omega_1^2\right)\varphi_- = 0.$$

These are the equations of motion for oscillators with natural frequencies $\omega_+ = \omega_0$ and $\omega_- = \sqrt{\omega_0^2 + 2\omega_1^2}$. The natural frequencies corresponding to these solutions for the coupled pendulums can be excited by displacing both pendulums simultaneously (Figure 1.41). If both pendulums are displaced in the same direction, they oscillate in phase and the spring is unloaded. Thus, they oscillate at the frequency $\omega_+ = \omega_0$. But if they are displaced in opposite directions, they oscillate in counterphase. The spring then acts additionally as a restoring force. The coupled pendulums now oscillate at the higher frequency $\omega_-$. Here we are encountering an important phenomenon for many branches of physics and technology, the *resonant splitting* of coupled oscillators (Figure 1.43). The splitting, i.e., the difference frequency $\omega_- - \omega_+$, is higher the more strongly coupled the pairs of pendulums are.

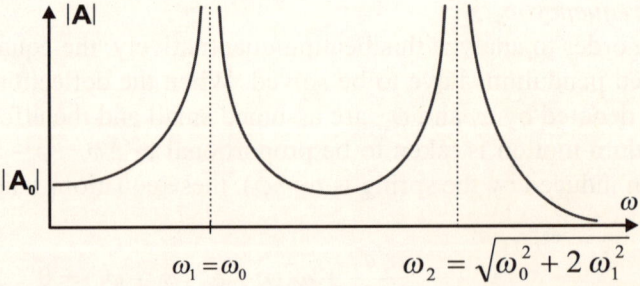

**FIGURE 1.43** The amplitude response of the forced oscillations of coupled pendulums.

The beating described previously can now be understood as a superposition of the two natural modes of the coupled pendulums. This is because, with the two natural modes, every linear combination of the two solution functions (of the uncoupled equations) is a solution of the (linear) equations of motion for the coupled pendulums. The beat frequency $\omega_S = \omega_- - \omega_+$ is equal to the difference in frequency between the two natural frequencies.

## Problem 1.28

Show that the superposed oscillation $Z(t) = 2[\cos(\omega_1 t) + \cos(\omega_2 t)]$ corresponds to beating of the sort shown in Figure 1.42 and prove that the beat frequency $\omega_S$ for the energy of the oscillations is equal to the difference frequency $\Delta\omega = |\omega_1 - \omega_2|$.

**Notes**    When two harmonic oscillators whose natural frequencies are the same are weakly coupled to one another, the system of coupled oscillators has two different resonance frequencies $\omega_1$ and $\omega_2$. The difference frequency serves as an indicator of the strength of the coupling. At the corresponding natural frequencies the oscillators either oscillate in phase or in counterphase.

If only one pendulum is initially excited to oscillations, then beating occurs in which the oscillation energy is transferred back and forth from one pendulum to the other at the beat frequency $\omega_S = |\omega_1 - \omega_2|$. The beating can be represented as a superposition of the two natural modes of oscillation.

Chapter **2**

# Macrophysics of Matter

## Summary

- Gases
- States of aggregation
- Energy principle
- Entropy principle
- Low temperatures

According to mechanics, all natural processes seem to obey strictly deterministic laws. A demon who knows the state of the world, that is, the position and velocity of all point masses, at a time $t = 0$, along with the forces acting between them, should be able to calculate the state of the world at any other time (*the Laplace demon*). But, as we have warned repeatedly, the laws of mechanics do not have unrestricted validity. They can be confirmed or refuted experimentally only to within the currently available measurement precision. One should, therefore, be careful not to conclude that the mechanical description of nature is an exact picture of nature.

As in Chapter 1, here we shall be concerned with macroscopic objects that, on the whole, obey the laws of mechanics, but we now include the fact that these objects can be hot or cold. The state of these objects cannot be described merely in terms of the position and velocity of their point masses, but the temperatures of these objects must also be specified. On one hand, it seems that the temperature of an object can be understood in terms of the mechanical world view as the average kinetic energy of its building blocks, namely the atoms and molecules of which it consists. On the other, we must also assume that the movements of these atoms and molecules are subject to the laws of chance, as well as the deterministic laws of mechanics. In fact, the development of quantum mechanics over the last century has shown that, in nature, the laws of chance play an important role alongside the deterministic laws.

## 2.1 GASES

Material objects can be gaseous, liquid, or solid. Apart from its chemical properties, a gas contained in a vessel of volume $V$ is characterized physically by its *density* $\rho$ (mass per volume), the *pressure P* it exerts on the walls of the vessel, and its *temperature T*. Here it is assumed that the gas is in *thermal equilibrium*; that is, there are no pressure or temperature gradients. The gas in the vessel, with a total mass $M$, also has a constant density $\rho = M/V$ over the entire volume $V$. Macroscopically, no motion can be discerned in the gas.

The pressure of the gas is measured with a *barometer* or a *manometer* and represents the force **F** acting on a surface (of area) **A** (Figure 2.1):

$$\mathbf{F} = P\mathbf{A}.$$

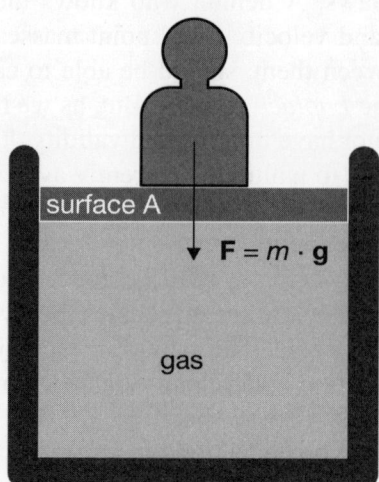

**FIGURE 2.1** Illustrating the definition of the pressure $P$ of a gas.

The SI unit of pressure is, accordingly, $1\,\mathrm{Pa} = 1\,\mathrm{N/m^2} = 1\,\mathrm{J/m^3}$ (Pascal). Pressure has the same units as energy density. Under standard conditions, the atmospheric air pressure is $P_{air} = 1\,\mathrm{atm} \approx 10^5\,\mathrm{Pa}$ ($= 1000$ hectopascals, as the meteorologists say).

The temperature is a new fundamental quantity, which cannot be derived from the mechanical quantities we have encountered up to now. Suitable measurement techniques for establishing a temperature scale follow from the gas laws (Section 2.1.3).

Gases, therefore, are characterized by the facts that they, unlike liquids and solids, fill the available volume homogeneously ($\rho = const$) and the force **F** they exert on the walls acts in the direction of the surface normal **A**. The pressure is, therefore, a scalar quantity (independent of the choice of spatial coordinates). In solids, on the other hand, **F** and **A** are generally not parallel, so that these two vectors are connected by a tensor equation.

## 2.1.1  The atomic hypothesis

In order to explain certain regularities in chemistry, around 1800 (following the philosophy of the ancient Greeks) the thesis was advanced that the matter surrounding us consists of *atoms*. According to this idea, matter does not fill space homogeneously, but has a *discrete structure*. Gases, liquids, and solids are, therefore, built up of the smallest particles, atoms and molecules. The mass $m$ of an atom is essentially specified in terms of the mass number $A$, so that

$$m \approx A \cdot 1.6 \times 10^{-27}\,\mathrm{kg}.$$

Here $A < 250$ is a natural number. (See Section 5.4.1.)

It follows from the density of solids and liquids that atoms can be visualized as small spheres with radii $r \approx 10^{-10}$ m. Since the density of a gas is some orders of magnitude lower than that of liquids and solids, the atoms in a gas move quite freely in space. For example, $1\,\mathrm{m^3}$ of air has a mass of about $1\,\mathrm{kg}$ and, therefore, a density $10^{-3}$ times that of water. Since water is highly incompressible, we can assume that the $H_2O$ molecules in water are densely packed. Thus, in the air the molecules are about 10 molecular diameters apart. Furthermore, the cohesiveness of liquids and solids shows that forces act between atoms which are much stronger than the force of gravity. The first law of thermodynamics (Section 2.3.2) shows that these are conservative forces (Section 1.3.1). The interatomic force between two atoms separated by a distance $r$ can, therefore, be described by a potential function $E_{pot}(r)$ (Figure 2.2).

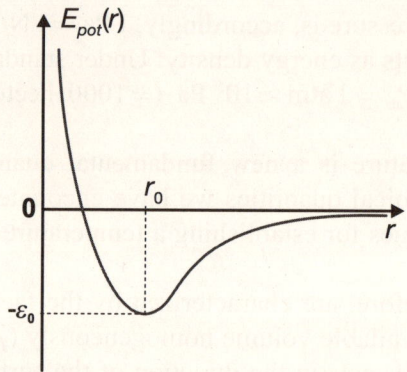

**FIGURE 2.2** Interatomic potential.

This intuitive picture of an atom in the sense of Newtonian mechanics is, nevertheless, only partly justified. For in many regards, atoms do not behave as small spheres (which have at least six degrees of freedom), but as point masses with only three degrees of freedom (Section 2.3.3). This peculiarity confirms that, in terms of thermodynamics, we can treat atoms as the smallest indivisible and structureless units of matter.

Hence, a gas consists of a very large, but finite, number of discrete particles; some gases are made up of atoms and others, of molecules. For example, under normal conditions ($P = 1.013 \times 10^5$ Pa $= 1$ atm, $T = 273$ K $= 0°$ C) 1 m$^3$ of air contains about $0.27 \times 10^{26}$ molecules. Because of the discrete structure of matter and the large number of particles of which macroscopic objects are composed, statistical average values are appropriate for describing the behavior of macroscopic objects. It appears that the thermal properties of matter, in particular, can be described under the assumption that the motions of atoms or molecules obey statistical laws.

## Problem 2.1

Discuss the interatomic force with the aid of Figure 2.2. In what range is this force attractive and in what range, repulsive?

> **Notes** Within the framework of thermodynamics, atoms can be treated as the indivisible smallest particles of matter. On one hand, they behave as mass points which can only undergo translational motion, and on the other, they resemble spheres with a radius on the order of $10^{-10}$ m. Hence, solids contain on the order of $10^{29}$ atoms/$m^3$.

## 2.1.2 The ideal gas

Proceeding from the atomic hypothesis, we shall describe a gas as an ensemble of very many identical particles which move in accordance with the laws of Newtonian mechanics in a vessel of volume $V$. If the gas is in thermal equilibrium, then the atoms of the gas move in completely disordered fashion. A thermal equilibrium state of the gas, therefore, corresponds to a statistical equilibrium state of the atomic ensemble, which is characterized in terms of *statistical averages* and *probability distributions* that are constant in time. Hence, the point is to determine the statistical averages corresponding to the macroscopic quantities $\rho$, $P$, and $T$.

In order to avoid complications and concentrate on the essentials, we shall consider an idealized limiting case of gases, namely the so-called *ideal gas,* rather than real gases. An ideal gas is an ensemble of particles which satisfies the following conditions:

---

**Ideal Gas**

---

- **Consists of very many particles.** This condition is always very well satisfied for macroscopic amounts of gas. For measuring macroscopic amounts of a gas the SI unit is 1 mol of the substance: 1 mole is the amount of matter in a system consisting only of atoms or molecules of a single type, a so-called unitary system, that contains the same number of individual particles as the number of atoms of the carbon isotope $^{12}$C (with mass number $A = 12$) in a sample with a mass of $0.12$ kg. That corresponds to about $N_A \approx 6.02 \times 10^{23}$ particles (the Avogadro number).
- **The particles act as point masses.** Thus, in particular, the spatial extent of the atoms is negligibly small. This condition is only satisfied well for highly rarefied gases in which the average distance $d$ between

neighboring atoms is much greater than the radii $R_A \approx 10^{-10}$ of the atoms.

- **There is no interatomic force.** This condition is approximately satisfied if the depth $\varepsilon_0$ of the interaction potential $E_{pot}(r)$ (Figure 2.2) is much smaller than the average kinetic energy $\langle E_{kin} \rangle$ of the atoms.
- **Collisions between atoms are always perfectly elastic.** For atomic gases whose particles have only three degrees of freedom, this condition is very well satisfied. This represents a special case of the generally accepted assumption that only conservative forces act between atoms.

Based on some simple considerations, the density $\rho$ and pressure $P$ of an ideal gas of this sort can be related to statistical averages of the ensemble. The density $\rho$ of the gas is obviously given by the product of the particle density $n$ (number of particles per m$^3$) and the mass $m$ of the atoms:

$$\rho = nm.$$

The pressure $P$ of the gas is equal to the momentum (since $\mathbf{F} = d\mathbf{p}/dt$ and $P = F/A$) transferred to the vessel walls by collisions with the wall per unit time and unit area. For example, because of these collisions the piston closing off the $z$ direction of the vessel in Figure 2.1 is not really at rest, but undergoes a slight vibration. Whenever an atom with velocity $\mathbf{v} = (v_x, v_y, v_z)$ strikes the piston, it bounces back from the surface with velocity $\mathbf{v} = (v_x, v_y, -v_z)$ and a momentum $\mathbf{p} = (0, 0, p_z)$, where $p_z = 2mv_z$, is transferred to the piston. The weight $F$ opposes these collisions. The number of collisions per unit time is $\frac{1}{2}nv_z A$, where $A$ is the area of the piston and the factor $\frac{1}{2}$ follows from the fact that only atoms with $v_z > 0$ strike the piston. If the effect of the collisions is balanced by the weight, then we have

$$\mathbf{F} = nmv_z^2 \mathbf{A}.$$

Note that in a statistical distribution not all the atoms have the same velocity $v_z$ and the statistical averages (denoted by angle brackets) are equal in thermal equilibrium, i.e., $\langle v_x^2 \rangle = \langle v_y^2 \rangle = \langle v_z^2 \rangle = \langle v^2 \rangle / 3$, so that we have

$$\frac{F}{A} = P = \frac{nm\langle v^2 \rangle}{3}.$$

Since the average kinetic of the atoms is $\langle E_{kin}\rangle = \frac{1}{2}m\langle v^2\rangle$, we then obtain

$$P = \frac{2}{3}n\langle E_{kin}\rangle. \qquad \text{(basic equation of the kinetic theory of gases)}$$

We still have to determine the relationship between temperature and the statistical averages of an ensemble of particles. Although the relationships of the density and pressure to the statistical averages follows from purely mechanical considerations, the statistical interpretation of temperature leads to fundamental difficulties. As we warned in the introduction to this chapter, Newtonian mechanics is a strictly deterministic theory. Statistical thermodynamics, on the other hand, deals with probability distributions and the statistical averages calculated from them. This dichotomy between determinism and chance leads to fundamental problems that cannot be discussed further here. Instead we shall rely on experimental knowledge. This shows that in natural phenomena chance plays an important role alongside deterministic effects and that the statistical approach is fundamentally justified. Given the situation outlined here, for the statistical interpretation of temperature discussed in the next section, we rely on the physics of ideal gases, where the laws of chance are also taken into account.

### Problem 2.2

Calculate the kinetic energy stored in $1\,\text{m}^3$ of air under normal conditions. What is the resulting average velocity for the air molecules? Is it greater or less than the speed of sound?

**Notes**   Gases consist of atoms (or molecules) which are largely able to move freely in space. The average distance $d$ between neighboring atoms is also significantly greater than the diameter $2r$ of the atoms. In the limiting case of an ideal gas, the pressure $P$ of the gas depends on the particle density $n$ and the average kinetic energy $\langle E_{kin}\rangle$ of the atoms, with

$$P = \frac{2}{3}n\langle E_{kin}\rangle.$$

The mass $m$ of an atom with mass number $A$ is roughly $m = A \cdot 1.6\times 10^{-27}\,\text{kg}$. The density $\rho$ of the gas is given by $\rho = nm$.

### 2.1.3 Temperature

The temperature of an object is measured with a thermometer. In the well-known mercury (Hg) and alcohol thermometers the thermal expansion of liquids is used to establish a temperature scale. Even when the same fixed reference points, such as the freezing and boiling points of water, are used, different liquids yield different temperature scales, as Figure 2.3 shows. The next question is, which temperature scale is best suited for the formulation of physical laws?

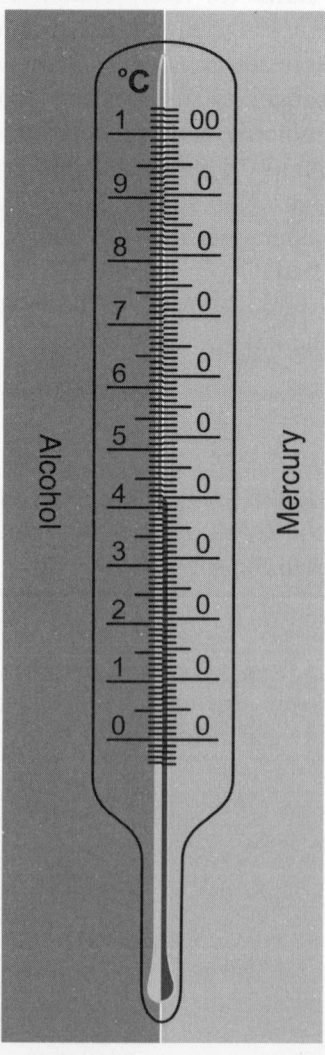

**FIGURE 2.3** Comparison of a mercury thermometer and an alcohol (spirit) thermometer.

The fundamental laws of thermodynamics take on an especially simple form if the temperature scale is established using a gas thermometer and the gases used for this purpose behave as ideal gases to the greatest extent possible. The best are the atomic noble gases such as helium (He) and neon (Ne). But, air also behaves in many respects almost as an ideal gas. Thus, we shall use it for a simple lecture demonstration.

## Experiment 2.1   The experiment of Boyle and Mariotte

A glass tube contains a ball that fits the cross section of the tube well enough that it can, on one hand, roll freely in the tube, but, on the other, is able to block the gas behind the ball in a segment of the tube with volume $V$ (Figure 2.4). The pressure $P$ in the tube can be varied using a pump or valve and can be read out from a manometer. We measure the length $l$ of the tube section bounded by the ball as a function of pressure $P$. Provided the temperature of the air in the tube does not vary during the measurement, this measurement shows that $l \propto 1/P$. Thus, for a fixed temperature the product $lP$ is constant.

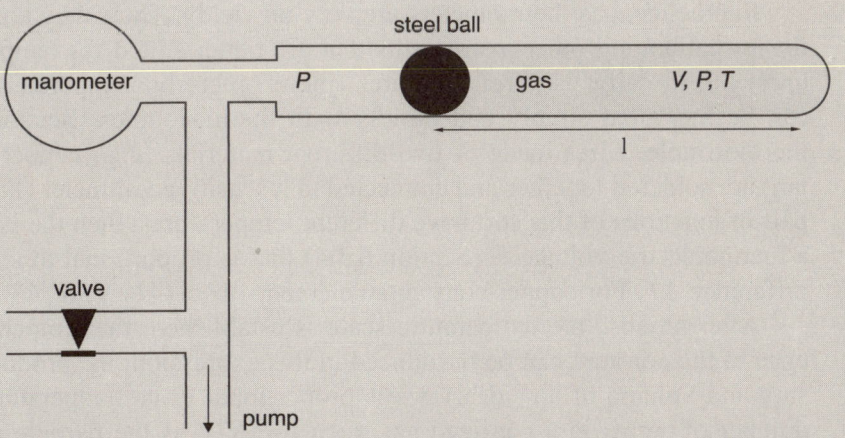

**FIGURE 2.4** The experiment of Boyle and Mariotte.

The result of this experiment is Boyle's (and Mariotte's) law:

*continued*

$$PV = const = f(T). \qquad \text{(Boyle-Mariotte law)}$$

If the temperature of the air is changed during the experiment, then the constant varies and, indeed, increases monotonically with the temperature $T$. Measurements on other, sufficiently ideal gases yield the same temperature dependence $f(T)$ for the constant in the Boyle-Mariotte law. This suggests using this temperature dependence to define an *absolute temperature scale*.

For the present, we shall define the *absolute temperature* as a quantity $T$ that is proportional to the product $PV$ of a gas whose behavior is sufficiently close to that of an ideal gas. In order to establish the constant of proportionality and, thereby, the SI unit $1\,\text{K}$ (Kelvin) for absolute temperature, the temperature $T(\text{H}_2\text{O})$ of the triple point of $\text{H}_2\text{O}$ is set at

$$T(\text{H}_2\text{O}) = 273.16\,\text{K}.$$

A temperature $t$ given in °C can be converted into absolute temperature (Kelvin) using the formula $T[\text{K}] = t[°\text{C}] + 273.16$.

In practice, gas thermometers are very unwieldy. Thus, they are used primarily for calibrating other instruments that are better suited for temperature measurements in daily use. Temperatures (more precisely, temperature differences) can be measured simply and rapidly with thermocouples (Section 6.3.4). In a thermocouple, wires made of two different materials (e.g., copper and constantan) are soldered together and connected to a sensitive voltmeter (Figure 2.5). If a pair of junctions of this sort have different temperatures, then the voltmeter reads a thermoelectric voltage $U$ (Section 6.3.4) that is proportional to the temperature difference $\Delta T$. For copper-constantan elements, $U \approx (41\,\mu\text{V/K})\Delta T$.

After an absolute temperature scale is established, the temperature dependence of the constant can be introduced in this expression: the product of the pressure and volume of an (ideal) gas is proportional to its temperature. If $N$ is the number of atoms in a confined gas, then $n = N/V$ is the particle density of the gas. Instead of $PV \propto T$, we can therefore write $P \propto nT$. The question remains of how the proportionality constant depends on the type of gas and, thus, for example, on the mass of its atoms. Measurements on different gases that behave as ideal gases to a sufficient degree, that is, gases with low enough particle densities $n$, show that the same proportionality constant can always be used.

This universal proportionality constant is the *Boltzmann constant k*. It has the value

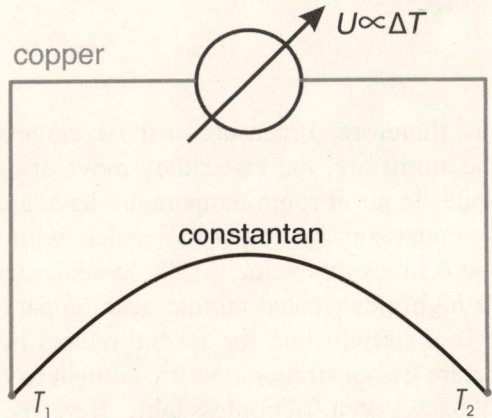

**FIGURE 2.5** A thermocouple.

$$k = 1.38 \times 10^{-23} \text{ J/K}. \qquad \text{(Boltzmann constant)}$$

From the Boyle-Mariotte law, therefore, we obtain

$$P = nkT. \qquad \text{(thermal equation of state of ideal gases)}$$

The thermal equation of state immediately implies

## Avogadro's Law

For equal pressure and equal temperature, all gases that behave more or less as ideal gases have the same particle density $n$.

Because $\rho = nm$, the densities $\rho$ of different gases vary as the mass $m$ of their atoms or molecules. The density of the noble gas He with its mass number $A = 4$ is one seventh that of nitrogen, whose molecules $N_2$ have a molecular mass number (molecular weight) $M = 28$. For this reason, balloons are filled with helium (Archimedes' principle).

A comparison of the thermal equation of state with the fundamental equation of the kinetic theory of gases (Section 2.1.2) yields the fundamental atomic significance of temperature:

$$\langle E_{kin} \rangle = \frac{3}{2} kT \,.$$

Temperature is, therefore, a measure of the average kinetic energy of the atoms. The lighter the atoms are, the faster they move at a given temperature. The $N_2$ and $O_2$ molecules in air at room temperature have a *thermal velocity* $v_{th}$ of about 500 m/s. By comparison, hydrogen molecules, with a molecular weight $M = 2$, move at almost 4 times this velocity. These velocities are easily measured today with the aid a highly evacuated atomic beam apparatus, in which the atoms are able to move in a straight line for several meters before colliding with another particle. In lecture demonstrations we are content to observe the motion of pucks with different masses on a frictionless table. Here, as well, light pucks move considerably faster than heavy ones.

### Problem 2.3

Calculate the particle densities $n$ and the average velocity of air (nitrogen) molecules and helium atoms under normal conditions. How do these values change when the gases are heated to the boiling point of water? At what temperatures will the atoms have the speed of a car moving at 100 km/h?

**Notes**    The average kinetic energy $\langle E_{kin} \rangle$ of the atoms of a gas is proportional to the absolute temperature $T$ of the gas:

$$\langle E_{kin} \rangle = \frac{3}{2} kT \,.$$

The proportionality constant $k$ in this formula, the Boltzmann constant $k = 1.38 \times 10^{-23}$ J/K, is a universal constant of nature.

The fundamental equation of the kinetic theory of gases thereby yields the thermal equation of state of an ideal gas:

$$P = nkT \,.$$

According to the thermal equation of state, at a specified temperature and pressure all (ideal) gases have the same particle density.

## 2.1.4  Probability distributions

The temperature $T$ and pressure $P$ of an ideal gas are correlated with number $n = N/V$ of atoms per volume and the average kinetic energy $\langle E_{\text{kin}} \rangle$ of the atoms. As long as the volume elements $V$ under consideration have enough particles, the number $N$ of atoms contained in $V$ fluctuates only slightly. (The amplitude of the fluctuations is roughly $\Delta N = \sqrt{N}$, Section 6.6.2.) Thus, the particle density $n = N/V$ has a well-defined value. On the other hand, the individual atoms of a gas in thermal equilibrium have very different velocities and, therefore, different kinetic energies. The velocity distribution is described using a *distribution function $f(v)$*. $Nf(v)dv$ gives the number of atoms in the volume $V$ whose velocities lie in the velocity interval $dv$. The distribution function $f(v)$ can be measured by studying the movement of the atoms in a vacuum. We shall make do with a measurement of the velocity distribution of ball bearings in a model experiment.

The model experiment shows, first of all, that the particle density falls off with height. The decrease is obviously a consequence of gravity's acting on the ball bearings. Only when the influence of gravity can be neglected is the particle density constant in the whole volume. Second, if the gas of ball bearings is sealed off above by a freely suspended piston, the experiment then shows how the weight of the piston opposes the collisions of the ball bearings and the gas pressure can then be explained.

Third, the gas of ball bearings can be used to measure the velocity distribution of the unordered flight of the ball bearings in all directions. To do this, we open a small hole on the side, through which the balls can leave the gas volume in a roughly horizontal direction. Depending on their velocities, the ball bearings fall into different compartments of a hopper located at different distances from the vessel wall. The distribution of the balls in the compartments illustrates the distribution function $f(v)$.

### Experiment 2.2    Maxwellian velocity distribution

The experimental setup is shown in Figure 2.6. A model gas consisting of ball bearings flying back and forth is produced by the vibration of a piston. (Fortunately, unlike the motion of the ball bearings, the thermal motion of atoms is silent. In addition, thermal motion does not have to be excited by a motor, since the atomic collisions are perfectly elastic.)

*continued*

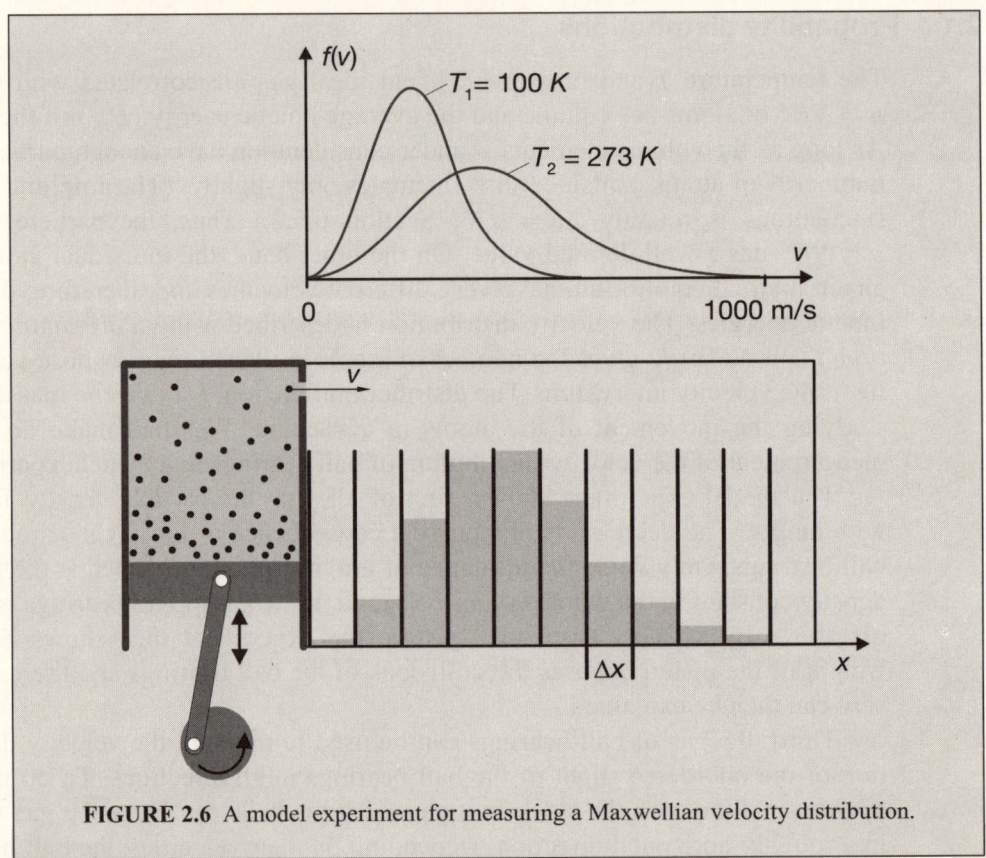

**FIGURE 2.6** A model experiment for measuring a Maxwellian velocity distribution.

It is harder to calculate the height and velocity distributions. Newtonian mechanics turns out to be inadequate here. In addition, a measure of probability must be established for finding a particle in given spatial and velocity intervals. For this, *phase space*, a product space of position and momentum space, is fundamental. Here we avoid detailed justifications for the probability distributions and settle for emphasizing a few features of the probability distributions that are important in making rough calculations.

The *Boltzmann barometric formula* describes the distribution of atoms in space. It can be derived using a few elementary arguments. The gas pressure $P$ falls off with increasing altitude in accordance with $\Delta P = \rho g \, \Delta h$, since the pressure decreases an amount $\Delta P$ for an increase $\Delta h$ in altitude because of the weight of the air column. For the density $\rho = nm$, the thermal equation of state for ideal gases (Section 2.1.3) gives $\rho = mP/kT$. Thus, for the gas pressure we obtain the differential equation $dP = -Pmg \, dh/DT$, which can be integrated for an assumed constant temperature:

$$P(h) = P_0 \exp\left(-\frac{mgh}{kT}\right). \qquad \text{(Boltzmann barometric formula)}$$

According to the barometric formula the gas pressure $P$ falls off exponentially with the potential energy $E_{\text{pot}} = mgh$ of the gas atoms from a pressure $P_0$ at $h = 0$. This exponential decrease in the probability with the energy is characteristic of thermal probability distributions. The exponential factor $\exp(-E/kT)$ is of fundamental importance here. It is also known as the *Boltzmann factor* after the physicist **Ludwig Boltzmann (1844–1906)**. A distinct drop in pressure with altitude is first observed when the potential energy of the atoms is on the order of the thermal energy $kT$.

---

### Problem 2.4

Calculate the height at which the air pressure has fallen by roughly half. What are the half-heights for hydrogen ($H_2$ molecules) and helium on hot summer days?

---

Like the height distribution, the *Maxwellian velocity distribution* of the atoms, $f(v)$, named after the physicist **James Clerk Maxwell (1831–1879),** is essentially determined by the Boltzmann factor. Here, of course, it is not the potential, but the kinetic energy $E_{\text{kin}} = \frac{1}{2}mv^2$ of the atoms which is involved. The probability $f(v)dv$ of finding an atom with velocity $v$ within a velocity interval $dv$ is not determined only by the Boltzmann factor, but also by a so-called *phase space factor*, which is proportional to $v^2$, must also be taken into account. To within a proportionality factor, we obtain

$$f(v)dv \propto v^2 \exp\left(-\frac{mv^2/2}{kT}\right)dv. \qquad \text{(Maxwellian velocity distribution function)}$$

If the probability distributions are known, then the statistical averages of the particle variables can be calculated. In particular, the Maxwellian distribution yields the average kinetic energy of the atoms, $\langle E_{\text{kin}} \rangle = \frac{3}{2}kT$.

---

### Problem 2.5

Estimate how many atoms in 1 mol of a gas have kinetic energies $E_{kin}$ that are ten times (or more) the average thermal energy of an atom. What altitude would hydrogen molecules with ten times the thermal velocity corresponding to $T = 300$ K at the earth's surface reach? (Neglect collisions with air molecules.)

---

Although we cannot see atoms or their motion directly, the motion of atoms can be observed indirectly under a microscope. This is because, in principle, all particles participate in the thermal motion. But for large particles the thermal motion is so slow that it cannot be perceived and atoms are not visible. Visible particles have a diameter $D$ at least on the order of the wavelength of visible light (Section 4.2.4), or $D \approx 1 \mu m$. The mass of particles of this sort is about $M \approx 10^{-15}$ kg, so their average thermal velocity at room temperature is on the order of a few mm/s. The velocity of a barely visible particle of this sort is, therefore, high enough for it to be observable under a microscope as *Brownian (molecular) motion.*

---

### Problem 2.6

Calculate the average thermal velocity of a small particle with a size of about 1 μm (e.g., pollen).

---

**Notes**    Thermal energy distributions are essentially determined by the Boltzmann factor $\exp(-E/kT)$. At an altitude $h$, where the potential energy $E_{pot} = mgh$ of the molecules in air equals their thermal energy $kT$, the air pressure is $P = P_0/e$ (where $e \approx 2.7$ is the Euler number, the base of the natural logarithms). In addition, the probability that an atom or molecule in a gas will have a high kinetic energy $E_{kin} \gg kT$ falls off exponentially with the Boltzmann factor.

## 2.2 STATES OF AGGREGATION

Real substances, such as hydrogen, oxygen, and nitrogen, are not always gaseous, but can also exist as liquids or solids. Likewise, most matter can exist in at least three different so-called *states of aggregation* (or *phases*). The best known are the three states of $H_2O$: ice, water, and water vapor. The temperature and pressure of a substance determine its state of aggregation. In general, the properties of a substance change continuously with pressure and temperature, but at certain values of the pressure and temperature discontinuous changes, known as *phase transitions*, take place. Because of these phase transitions, the different states of aggregation of a substance can be defined and distinguished from one another.

In this lecture the phases and phase transitions of substances which exist in three different states of aggregation will be discussed. Here it should be kept in mind that we shall deal exclusively with *unitary systems*, or systems consisting of only one species of atoms or molecules. Otherwise, the properties of the system will not only depend on the parameters of state of the system, i.e., its pressure, temperature, and volume, but also on the mixture proportions. For that reason, the description of multicomponent systems is more complicated.

In a unitary system with three states of aggregation, six different phase transitions can be distinguished. They are listed in Figure 2.7.

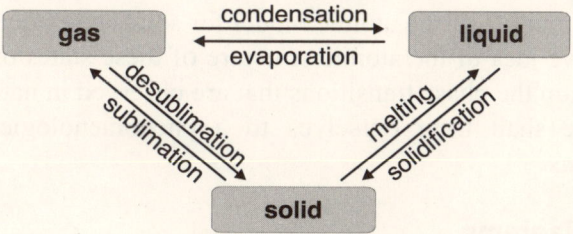

**FIGURE 2.7** Phase transitions.

## 2.2.1  Liquids and solids

Many properties of material objects can be explained under the assumption that all matter consists of atoms. Thus, the typical properties of gases can be interpreted in terms of atoms using the model of an ideal gas. There one assumes that there are no forces between the atoms. The structure of solid and liquid objects shows, however, that interatomic forces exist. As pointed out in Section 2.1.1, these are conservative forces with potential curves of the form shown in Figure 2.2 and a minimum $E_{\text{pot}}(r_0) = -\varepsilon_0$ at the equilibrium distance $r_0$. If the average kinetic energy $\frac{3}{2}kT$ of the atoms is much greater than $\varepsilon_0$, then the interatomic forces have only a marginal effect on the motion of the atoms and the ensemble of atoms behaves almost like an ideal gas. If, on the other hand,

$$kT \ll \varepsilon_0,$$

then the atoms adhere to one another and form liquids and solids.

In solids, the potential troughs in which the atoms lie are so deep that they can only oscillate back and forth, but do not have enough energy to escape them. The positions of the atoms relative to one another are, therefore, almost constant (more or less like the positions of the point masses in a solid object, Section 1.4.2). Hence, a solid is not easily deformed. The average distance between neighboring atoms in solids is roughly equal to the equilibrium distance $r_0$ for the interatomic potential. Thus, $r_0$ can be estimated from the particle density $n$ of a solid.

A similar situation occurs with liquids. Of course, in liquids the potential trough is not deep enough to fix the positions of the atoms. Here the average kinetic energy of the atoms is sufficient to rearrange them. A typical liquid is, therefore, easily deformable and quickly accommodates itself to the shape of a container, without changing its own volume significantly. Like solids, liquids are fairly incompressible, since in both these states of aggregation the atoms are densely packed.

This outline of the atomic model for solids and liquids does indeed provide a qualitative idea of the atomic structure of these states of aggregation, but it cannot explain the phase transitions that are observed in nature. In the following sections we shall limit ourselves to a phenomenological description of phase transitions.

## 2.2.2 Phase diagrams

Ideal gases obey the equation of state $P = nkT$. The three variables of state of a gas, the pressure, density, and temperature, not be set independently of one another, but are coupled through the equation of state. In the following, we shall consider not just gases, but also unitary systems that can exist in different states of aggregation. Usually, unlike gases, they do not occupy space uniformly, but can have different densities in different parts of space, depending on the state of aggregation. For this reason, in the following we shall use the total volume $V$ of the unitary system, rather than the particle density $n$, as the third variable of state. In particular, we shall refer to the volume $V_{\text{mol}}$ occupied by 1 mol of the substance. Hence, $V_{\text{mol}}$ contains $N_A$ atoms, where $N_A$ is the Avogadro number (Section 2.1.2),

$$N_A = 6.022 \times 10^{23}. \qquad \text{(Avogadro number)}$$

Since $n = N_A/V_{mol}$ for ideal gases, the equation of state for ideal gases has the form

$$PV_{mol} = RT. \qquad \text{(ideal gas equation of state)}$$

Here $R = N_A k = 8.31 \, \text{J} \cdot \text{K}^{-1} \cdot \text{mol}^{-1}$ is the so-called universal gas constant.

For real substances, the relationship among the three variables of state $P$, $V_{mol}$, and $T$ is a lot more complicated. For this reason, the functional relationship is displayed in a *phase diagram*. In daily life the pressure is often equal to the atmospheric pressure and, therefore, fixed at $P \approx 10^5$ Pa. Thus, it seems obvious to first consider the volume of a substance as a function of its temperature at constant pressure.

Figure 2.8 shows the typical variation in $V_{mol}(T)$, the molar volume at constant pressure $P$, for a unitary system. For an ideal gas it is a linearly increasing function. For most real substances the volume also increases with rising temperature. $H_2O$ is an exception. It reaches its minimum volume in the liquid state. $V_{mol}(H_2O)$ has an absolute minimum at a temperature of $4°C$.

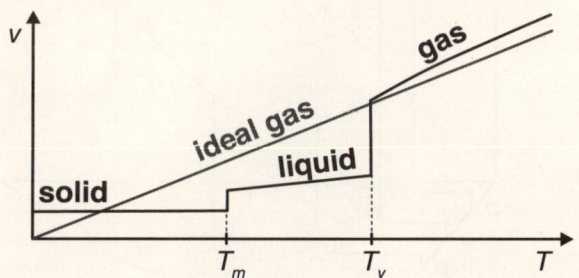

**FIGURE 2.8** A $V(T)$ diagram for a unitary system at $P = const$.

Since atoms in the solid and liquid states are densely packed and the volume of a single atom is about $10^{-29}$ m$^3$, the molar volume of solids and liquids is on the order of $V_{mol} \sim 10^{-5}$ m$^3$ = 10 cm$^3$. For gases, on the other hand, the molar volume is substantially larger. In general the molar volume varies continuously with temperature. Discontinuous changes are observed only at the melting and vaporization temperatures, $T_M$ and $T_V$; that is, during *phase transitions*. They are, therefore, also referred to as *phase shifts*.

If you compare the graphs for different pressures $P_1 < P_2$, you observe a displacement in the temperatures of the phase transitions, toward higher temperatures with increasing pressure when there is an increase in the volume with the phase transitions from solid to liquid or liquid to gas, and a displacement to lower temperatures when there is a decrease in volume. In any case, the temperature range $\Delta T_{\text{liq}} = T_M - T_V$ for the liquid phase decreases with falling pressure (Figure 2.9). Below the *triple point*, that is for pressures $P < P_T$, solids shift directly into the gaseous phase or *sublimate*. Furthermore, the magnitude of the volume change during a transition into the gaseous phase decreases with rising pressure. Above the *critical point*, or at pressures $P > P_K$, the liquid passes continuously into the gaseous phase. The two states of aggregation can then no longer be sharply distinguished from one another.

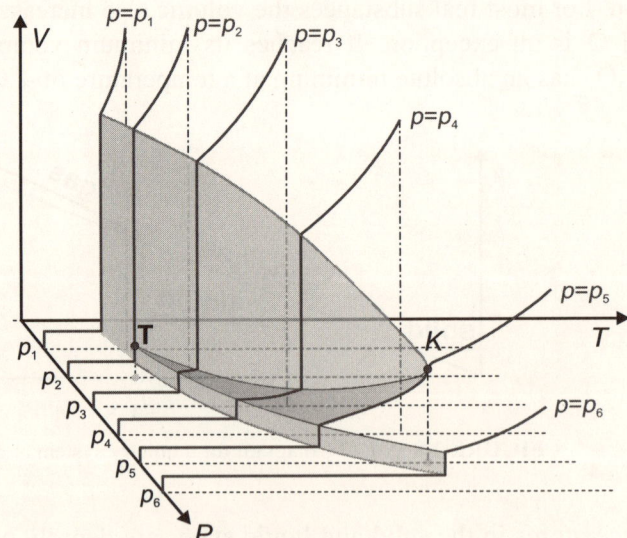

**FIGURE 2.9** $V(P,T)$ diagram for a unitary system.

The positions of the phase transitions in the $P$-$T$ plane are characteristic of unitary systems (Figure 2.10). The *sublimation curve* leading to the *triple point* $(T_T, P_T)$ marks the transition from the solid to the gaseous state. For pressures above the triple point, there is also a liquid phase, which is bounded by the *melting* and *vapor curves*. The vapor curve ends at the *critical point* $(T_K, P_K)$. A clear distinction between the liquid and gaseous phases is, therefore, only possible when the pressure lies within in the range $P_T < P < P_K$.

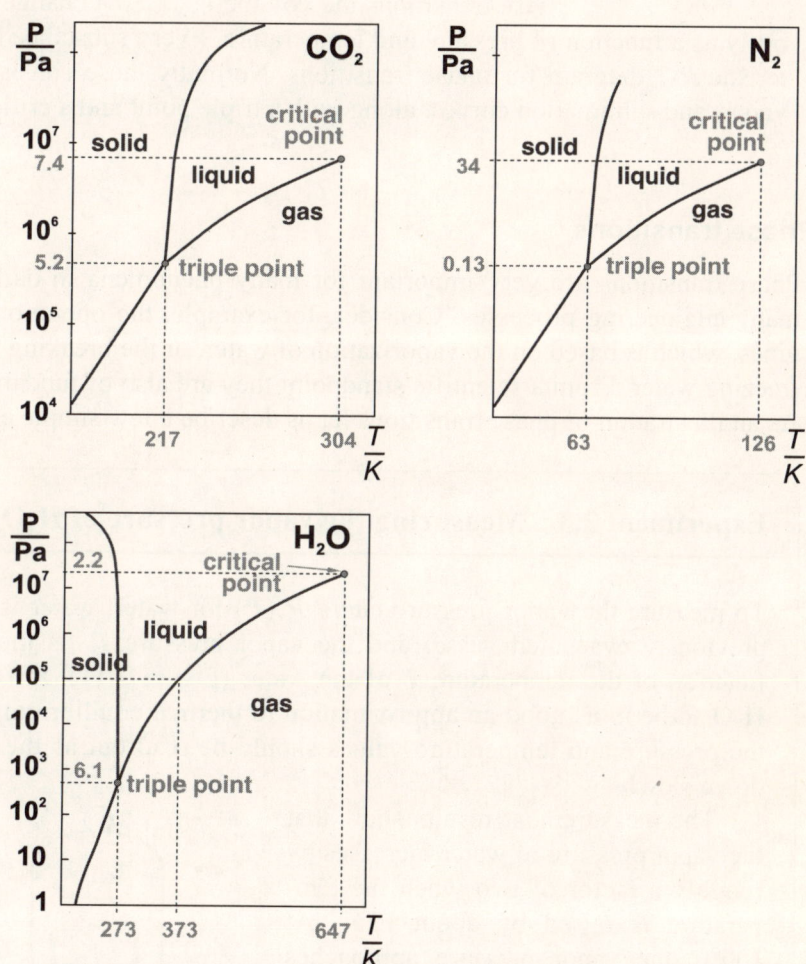

**FIGURE 2.10** *P-T* diagrams for the phase transitions of $CO_2$, $N_2$, and $H_2O$.

## Problem 2.7

Calculate the approximate volume change of $H_2O$ during evaporation of water at pressures $P = 10^5$ Pa and $P = 10^4$ Pa.

> **Notes**    Unitary systems typically occur in three different states of aggregation. They can be solid, liquid, or gaseous. During phase transitions the volume $V_{mol}(P,T)$ changes discontinuously as a function of pressure and temperature. Every substance has a characteristic $P$-$T$ diagram for phase transitions. Normally these consist of melting, vapor, and sublimation curves, along with a triple point and a critical point.

### 2.2.3  Phase transitions

Phase transitions are very important for many phenomena in daily life and for many engineering processes. Consider, for example, the operation of steam engines, which is based on the vaporization of water, or the breaking up of rocks by freezing water. From a scientific standpoint they are also of fundamental interest. As an illustration of phase transitions let us describe a few simple experiments.

---

## Experiment 2.3    Measuring the vapor pressure of H$_2$O

---

To measure the vapor pressure curve $P_V(T)$ for water, water is heated in a previously evacuated vessel and the vapor pressure $P_V$ is measured as a function of the temperature $T$ of the water (Figure 2.11). In order for the H$_2$O to be in as good an approximation to thermal equilibrium as possible, the pressure and temperature values should be read out as the water cools down slowly.

The measurement results show that the vapor pressure of water increases by roughly a factor of two when the temperature is raised by about 15 K. At 100°C the vapor pressure approaches the ambient air pressure under normal conditions and then begins to boil.

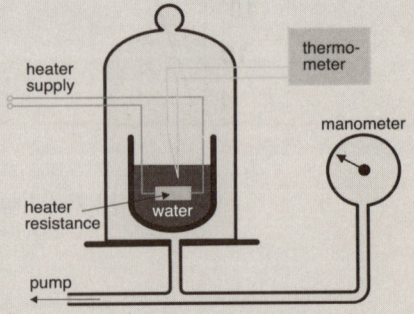

**FIGURE 2.11** Experimental setup for measuring the vapor pressure of water.

## Experiment 2.4    Regelation of ice

Since water has a roughly 10% higher density than ice, the melting temperature of ice decreases when the pressure is raised. Thus, ice melts if the pressure acting on it is raised sharply and refreezes when the load is released. This regelation (refreezing) of ice is what makes the movement of large glaciers possible and reduces the friction during skating. It also makes it possible for a thin wire (0.2 mm diameter) with a 10 kg weight on it to slip through a block of ice without cutting it.

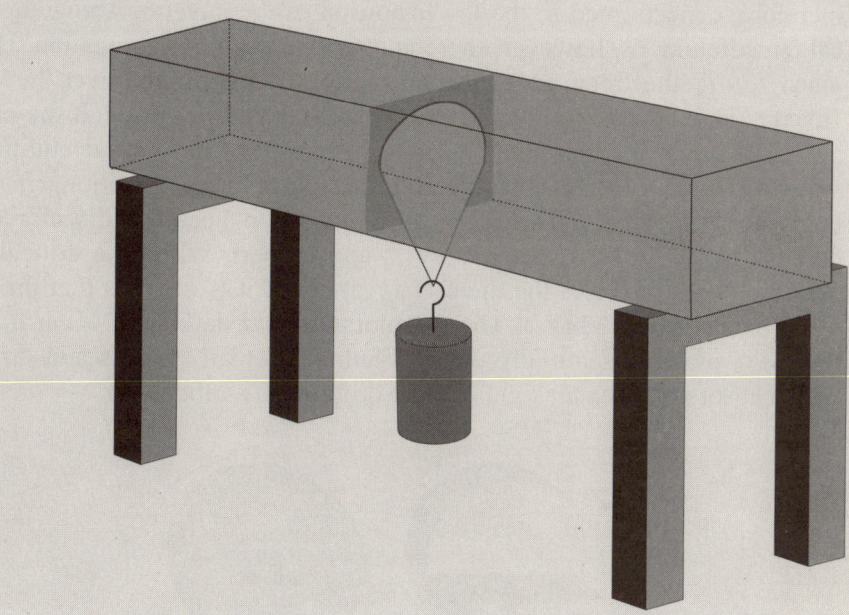

**FIGURE 2.12** Experimental setup for demonstrating the regelation of ice.

## Problem 2.8

Calculate the pressure the wire exerts on the ice if the block of ice is 0.1 m wide. By what factor is that pressure greater than 1 atm?

## Experiment 2.5  Critical point

In order to investigate the behavior of a substance at the critical point, we shall do an experiment with freon, whose critical point ($T_K = 354$ K, $P_K = 30 \times 10^5$ Pa) lies in a region that is easily accessible for experiments. As in the measurement of the vapor pressure curve for $H_2O$, we heat the freon in a closed vessel (a transparent glass cuvette). Below the critical temperature the liquid and gaseous phases of the freon are in thermal equilibrium. Thus, two phases can be seen, separated by a *meniscus* with the liquid below it and the gas above it (Figure 2.13). Most of the molecules are, therefore, concentrated in the lower portion of the cuvette. Above the critical temperature $T_K$, however, there is only one state of aggregation. Hence, when $T > T_K$ the freon molecules are uniformly distributed over the whole cuvette. Thus, dramatic changes take place on passing through the critical point. In order for the freon to be approximately in thermal equilibrium when it crosses the critical temperature, we observe the transition again as it cools down slowly. It is striking how the freon becomes discolored and darkens as the critical temperature is approached. When the critical temperature is attained and the meniscus reappears, it is obvious that the freon again exists in two phases. The discoloration and darkening occur because the freon molecules initially come together as small clusters and droplets, which absorb and scatter light like the droplets in a rain cloud.

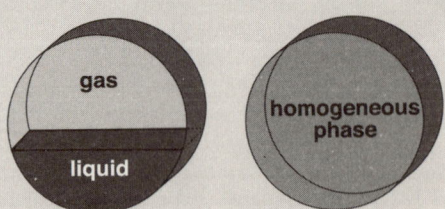

**FIGURE 2.13** A glass cuvette with freon below (a) and above (b) the critical temperature.

---

### Experiment 2.6    Three-phase equilibrium at the triple point

---

While only two states of aggregation can exist simultaneously along the phase transition curves, at the triple point, where the three curves meet, a substance can exist in all three phases at once. Since the triple point of a substance is a singularity in the $P$-$T$ plane, it is eminently suitable as a calibration point for establishing temperature scales. In particular, the triple point of $H_2O$ is used to establish the absolute temperature scale (Section 2.1.3).

Experimentally a relatively simple apparatus can be used to produce a three-phase equilibrium of nitrogen ($N_2$), whose triple point is at $T_T = 63\,K$ and $P_T = 13 \times 10^3\,Pa$. To do this, liquid nitrogen, which has a temperature $T = 82\,K$ at atmospheric pressure, is cooled by pumping to remove gaseous nitrogen from a vessel (Figure 2.11) which initially contains both gaseous and liquid nitrogen. Here the liquid nitrogen is contained in a good, thermally insulating Dewar flask (thermos bottle; Section 4.4.2). As the pressure is lowered, the liquid nitrogen evaporates adiabatically (Section 2.5.1). The liquid nitrogen is thereby cooled down. When the temperature $T_T = 63\,K$ is reached, the liquid nitrogen partially freezes, so that ice, liquid, and gas are present simultaneously in the vessel. Further cooling along the sublimation curve is possible only when no more liquid nitrogen is left.

---

**Notes**    At a given temperature $T$ and pressure $P$, a simple substance is usually either a solid, a liquid, or a gas. Only on the phase transition curves of a $P$-$T$ diagram is the substance not homogeneously distributed over the volume $V$. Then it usually exists in two states of aggregation. Only at the triple point, where $T = T_T$ and $P = P_T$, can all three phases coexist simultaneously in thermal equilibrium.

## 2.2.4  Real gases

The example of nitrogen shows that, especially at low temperatures, real gases do not actually behave as ideal gases. Even at the lowest temperatures an ideal gas undergoes no phase transitions, while all real gases condense. Helium has the lowest condensation temperature. At atmospheric pressure it is 4.2 K.

The real behavior of gases can be described amazing well if the ideal gas equation of state is supplemented by two terms which take into account the finite size of the atoms and the interatomic forces. Since the atoms have a finite volume $V_{at}$, they can only move freely within a volume $V_{mol} - b$ (reduced by the *covolume* $b = 4N_A V_{at}$). And since not only external forces, but also the attractive interatomic forces act on the gas atoms, the effective pressure holding the gas together is greater than the pressure $P$ on the vessel walls. The interatomic forces are more effective when the volume into which the atoms are being compressed is smaller. Thus, the correction term for the pressure falls off as $V^{-2}$. When these corrections are taken into account, the result is

$$\left(P + \frac{a}{V_{mol}^2}\right)(V_{mol} - b) = RT. \qquad \text{(van der Waals equation of state)}$$

The parameters $a$ and $b$ are chosen to fit the gas under discussion.

Figure 2.14 is a graphical representation of the isotherms ($T = const$) satisfying this equation of state. At sufficiently high temperatures the isotherms do not differ greatly from the hyperbolic isotherms of an ideal gas, but it is otherwise at

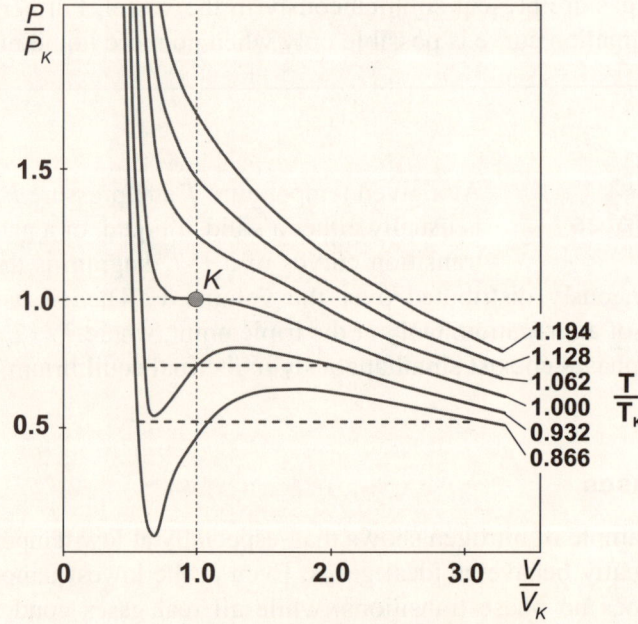

**FIGURE 2.14** A *P-V* diagram for the van der Waals equation.

low temperatures. There the isotherms have a minimum and a maximum. The minimum and maximum coincide for the isotherms $T = \frac{8}{27}R(a/b)$ at an inflection point with horizontal tangents. This temperature is the critical temperature $T_K$ of the gas.

A real unitary system follows the isotherms of the van der Waals equation only to a limited extent for isothermal processes. If $T < T_K$, then the process does not follow the calculated curve in the middle range of volumes, but, rather, it follows an equilibrium line which delimits two equal areas below and above the calculated isotherm. Only the volume of the unitary system changes along these equilibration lines. They describe the condensation of the gas at constant pressure and constant temperature. A condensation process of this sort for gaseous butane is demonstrated in the following experiment.

---

### Experiment 2.7   Liquefaction of butane

A glass cylinder is filled with butane and closed above with a moveable piston (Figure 2.15). Since liquid butane has a vapor pressure $P_V \approx 5 \times 10^5$ Pa at room temperature, at atmospheric pressure it is a gas. It can, however, be liquefied by raising the pressure. (The critical point of butane lies at $T_K = 425$ K and $P_K = 37 \times 10^5$ Pa.) In order to liquefy the gaseous butane, the piston just has to be pushed into the cylinder until the external force raises the pressure in the cylinder to $5\,\text{atm} \approx 5 \times 10^5$ Pa. Isothermal compression then follows the isotherm with $T/T_K \approx 0.7$ in Figure 2.14. If the volume is reduced further, the pressure in the cylinder remains constant and the gaseous butane will gradually be liquefied. The change of state now follows the equilibrium line. Once all of the butane is liquefied, the pressure rises steeply if the volume is reduced further.

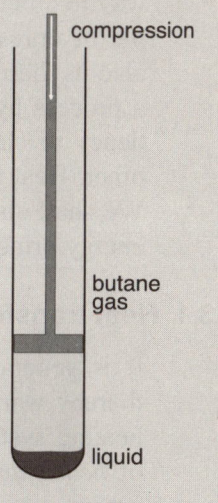

**FIGURE 2.15** Experimental apparatus for liquefying butane.

---

**Problem 2.9**

Calculate the force that has to be applied in order to liquefy butane. What is the largest cross section of the glass cylinder for which the piston can be operated by hand to liquefy butane?

---

| **Notes** | The phase transition, which occurs during condensation of real gases, obeys the van der Waals equation of state. As opposed to the ideal gas equation of state, that equation |

takes into account the facts that atoms have a finite size of about $10^{-10}$ m and the forces operating between the atoms are attractive at large distances ($r > r_0$).

## 2.3 ENERGY PRINCIPLE

In mechanics we must distinguish between dissipative and conservative forces in order to formulate an energy conservation law (Section 1.3.2). That energy principle holds in mechanics only if no dissipative forces are acting. Daily experience, however, also teaches us that when dissipative forces occur, especially in objects subject to friction, these objects are heated. This experience points the way to a better understanding of dissipative forces and leads to an energy principle of unrestricted validity. Not only the mechanical changes in a system, but also its thermal changes must be taken into account. *Temperature equilibration* is a process by which the temperatures of objects change. All objects have a tendency to change their temperature through the exchange of *heat* with one another. Heat transfer of this sort will be discussed next and studied quantitatively. We shall also introduce the concept of an *amount of heat*. Building on this, the energy principle can be formulated in a general way.

### 2.3.1 Heat transfer

It is generally known that a hot object is cooled in cooler surroundings and thereby warms up its surroundings. Thus, for example, the heating element of a heating system warms the surrounding space. For an experimental investigation of *heat transfer* of this sort between objects with different temperatures $T_1 \neq T_2$ we use a *calorimeter* (Figure 2.16). It is essentially a vessel that is well insulated thermally in order to greatly restrict heat exchange with the surroundings.

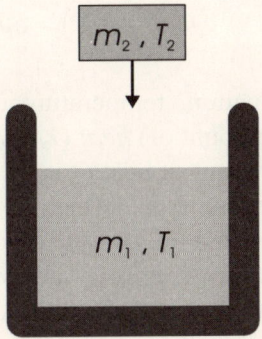

**FIGURE 2.16** Experimental setup for calorimetric measurements.

Let us first partially fill the calorimeter with water. The water temperature is $T_1$ and its mass is $m_1$. Now a second object, such as a metal block, with a known temperature $T_2$ and mass $m_2$, is placed in the water. *Temperature equilibration* then takes place between the water and metal. If $T_2 > T_1$ then the metal block is cooled and the water is warmed (Figure 2.17) until both objects have reached the same final temperature $T_f$. The conditions are chosen so that no phase transitions take place.

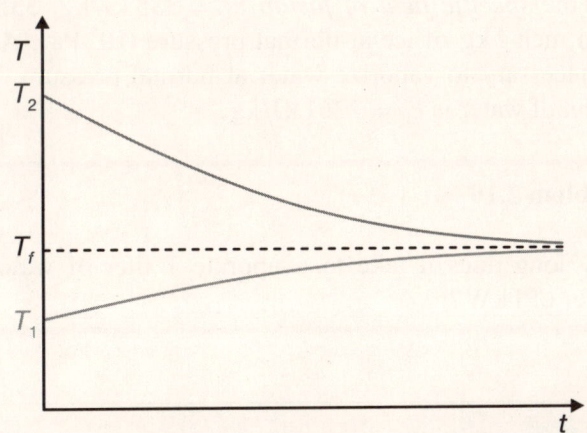

**FIGURE 2.17** The temperatures $T_1$ and $T_2$ as functions of time.

The measured temperature changes $T_2 - T_f$ and $T_f - T_1$ of the two objects are dependent, on one hand, on the mass ratio $m_1/m_2$ and, on the other, on the type of materials of which the objects consist. With the material constants $c_1$ and $c_2$ the following balance equation can be written for the temperature equilibration:

$$c_1 m_1 \left( T_f - T_1 \right) = c_2 m_2 \left( T_2 - T_f \right).$$

According to this equation, temperature equilibration can be interpreted as a process in which (an amount of) *heat* $Q_2 = c_2 m_2 \left( T_2 - T_1 \right)$ is given up by the hotter object and an equal amount of heat $Q_1 = Q_2$ is taken in by the colder object. Heat is a new physical quantity. It is defined by the calorimetric measurement procedure. In order to establish a unit for the amount of heat, we refer to the heat, 1 cal (calorie), required to warm $1 \, cm^3$ of water by 1 K (more precisely, from 14.5°C to 15.5°C). The first law of thermodynamics, which will be discussed in the next section, implies, however, that heat can be converted to work and work can be converted to heat in a fixed proportion, so that, in fact, 1 cal and 4.18 J are equivalent. Hence, these days both heat and work are measured in energy units [J]. Thus, the material constants $c_1$ and $c_2$, known as the *specific heat capacities* of the corresponding materials, have the units $[J \cdot kg^{-1} \cdot K^{-1}]$.

Up to now we have avoided cases where phase transitions take place during heat transfer between objects. Under this assumption, one object is cooled while the temperature of the second rises (Figure 2.17). On the other hand, if one throws enough ice cubes with a temperature of 0°C into lukewarm water, then the water just cools down to 0°C, while the ice cubes won't melt unless a temperature rise occurs. Heat obviously must be introduced in order to melt the ice. For ice, the *specific heat of fusion* $c_M = 335 \, kJ/kg$. 335 kJ is thus the heat required to melt 1 kg of ice at normal pressure ($10^5$ Pa). A much larger amount of heat is necessary to vaporize water at normal pressure. The *specific heat of vaporization* of water is $c_V = 2261 \, kJ/kg$.

---

**Problem 2.10**

How long does it take to evaporate 1 liter of water with a heat input power of 1 kW?

---

**Experiment 2.8    Melting a block of ice**

Heat is continuously applied to 1 kg of $H_2O$ that is initially in the form of ice. With isobaric ($P = const \approx 10^5$ Pa) warming the temperature increases in multiple steps (Figure 2.18). First the ice warms to 1°C and then melts at this temperature. Once the entire mass of ice is melted, the temperature of the water again rises until it vaporizes at 100°C. The temperature then stays constant again. Only when all the water has been evaporated is the water vapor heated above a temperature of 100°C when more heat is applied.

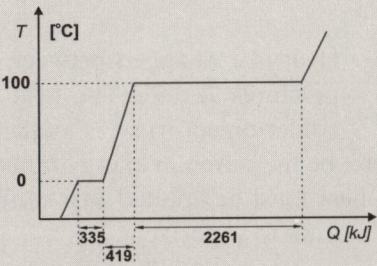

**FIGURE 2.18** The temperature of 1 kg of $H_2O$ as a function of the amount of heat $Q$ added to it.

Another simple experiment illustrates the significance of the heat of vaporization in daily life and for medical and engineering applications:

## Experiment 2.9   Evaporative cooling

A mercury thermometer is wrapped in cellulose and sprayed with a liquid which evaporates quickly in air, such as ethyl chloride (Figure 2.19). In a few seconds the temperature reading on the thermometer falls below the freezing point. In this experiment the ethyl chloride extracts the heat needed for evaporation from the thermometer. Hence the temperature of the mercury falls. (Since the vaporized ethyl chloride is continuously being volatilized, the multicomponent system under consideration here is not in thermal equilibrium.)

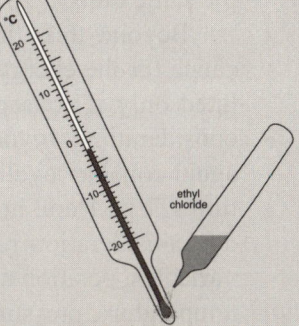

**FIGURE 2.19** Experimental setup for demonstrating evaporative cooling.

---

**Problem 2.11**

Calculate the average energy delivered to an $H_2O$ molecule (with mass number $A = 18$) when water is heated from 0°C to 100°C. Compare the result with the (change in) thermal energy $k\,\Delta T$.

---

**Notes**   Heat is exchanged between two objects with different temperatures $T_1 > T_2$. The heat moves from the warmer to the colder object. Heat is a quantity that is equivalent to energy and can, therefore, be measured in energy units [J].

In addition, heat must be applied to a solid in order to melt (or sublimate) it or to a liquid in order to vaporize it.

## 2.3.2  First law of thermodynamics

Many processes in daily life show that objects are heated not only when heat is delivered to them, that is, during temperature equilibration, but also when two objects rub rapidly against one another, as during the drilling of a hole. In frictional processes of this type, work $W$ is performed. At the same time, the heating represents an amount of heat $Q$. Both the work and the heat can be measured. Precise studies of processes of this sort show that in all processes involving both conservative and dissipative forces, work and heat are always converted in the same proportion. As noted in the preceding section, 4.18 kJ of work corresponds to 1 kcal of heat. Because of this equivalence, work and heat can be measured in the same units.

Beyond this, the equivalence of heat and work has a fundamental significance for the energy principle. In mechanics an energy principle could be formulated only if conservative forces alone are assumed to act in the system under consideration. In order to formulate a universally valid energy principle, we again consider cyclic processes, as we did before in the case of mechanics (Section 1.3.1). Here, however, we also take into account the fact that the system under consideration returns to its initial state both with respect to the mechanical variables, position and velocity, and with respect to the thermodynamic variables: temperature, pressure, and volume.

The *first law of thermodynamics* thus states that the sum $W_{in} + Q_{in}$ of the work and heat applied to the system in the course of a cyclic process is equal to the sum $W_{out} + Q_{out}$ of the work and heat lost during the cycle (Figure 2.20):

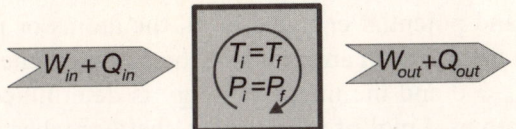

**FIGURE 2.20** Illustrating a thermodynamic cycle.

$$W_{\text{in}} + Q_{\text{in}} = W_{\text{out}} + Q_{\text{out}}. \qquad \text{(first law of thermodynamics)}$$

As in Section 1.3.2 we conclude from this that in any arbitrary process where the system goes from an initial state $A$ to a final state $Z$, the sum of all the applied (positive valued) and lost (negative valued) work $\delta W$ and heat $\delta Q$ is independent of the choice of path along which state $Z$ is reached from state $A$; that is,

$$\int_{A}^{Z} (\delta W + \delta Q) \text{ is independent of path.}$$

Because of this path independence, every state $Z$ of the system can be characterized by a variable of state, the energy:

$$E_Z = E_A + \int_{A}^{Z} (\delta W + \delta Q).$$

If we limit our discussion to unitary systems whose state is characterized solely by the thermodynamic state variables $T$, $P$, and $V$, then the energy can be identified as the *internal energy U* of the unitary system. In this case, therefore, we have

$$U_Z = U_A + \int_{A}^{Z} (\delta W + \delta Q).$$

For a closed thermodynamic system, this implies that its total energy, i.e., the sum of the kinetic, potential, and internal energies, is constant in time.

This energy principle provides the justification for the fundamental assumption of the atomic model of matter, namely, that only *conservative* forces act between atoms and that only *elastic collisions* occur in ideal gases. In terms of the atomic model, the internal energy of a unitary system is interpreted as the sum of the kinetic and potential energies of all the atoms or molecules. Since no forces act between the atoms in an ideal gas, the potential energy of the atoms can be set equal to $E_{\text{pot}} = 0$ and the internal energy is determined solely by the kinetic energy of the atoms. 1 mol of an ideal gas, therefore, has an internal energy of

$$U_{\text{mol}} = \frac{3}{2} N_A kT \quad \text{or} \quad U_{\text{mol}} = \frac{3}{2} RT. \quad \text{(caloric equation of state for an ideal gas)}$$

According to this caloric equation of state, the internal energy of an ideal gas depends only on the temperature of the gas, and not on its pressure or volume.

### Problem 2.12

Calculate the specific heat capacity of the (almost ideal) gases helium and neon (with mass numbers $A = 4$ and $A = 20$, respectively).

**Notes**

The total energy of a thermodynamic system is the sum of its kinetic, potential, and internal energies. The total energy of a closed thermodynamic system (i.e., one that exchanges neither work nor heat with its surroundings) is constant in time.

The internal energy $U$ of a unitary system is equal to the sum of the kinetic and potential energies of its atoms. Since no forces act between the atoms of an ideal gas, in an ideal gas $E_{\text{pot}} = 0$, so that $U_{\text{mol}} = \frac{3}{2} RT$.

### 2.3.3 Equipartition theorem

The caloric equation of state for ideal gases is consistent with measurements on the monatomic noble gases. For densities that are not too high and temperatures that are not too low, the interatomic force and the characteristic volume of the atoms should indeed have only a negligible influence on the behavior of the gases, so these gases will essentially behave as ideal gases. Measurements can be made, in particular, of the increase in internal energy as a gas is heated, and, therefore, of the heat capacity of the gas. If the volume is held constant as the gas is heated (so that no work is performed during heating), then the increase in internal energy is equal to the heat applied to the gas. According to the caloric equation of state, we expect that atomic gases should have a heat capacity per mol of

$$C_V = \frac{3}{2}R = 12.5 \, \text{J} \cdot \text{mol}^{-1} \cdot \text{K}^{-1}.$$

For the noble gases He, Ne, and Ar this value is in good agreement with the measured experimental values of the *molar heat capacity*. The molecular heat capacity of the molecular gases $H_2$, $N_2$, and $O_2$ has a different value,

$$C_V = \frac{5}{2}R = 20.8 \, \text{J} \cdot \text{mol}^{-1} \cdot \text{K}^{-1}.$$

Molecular gases have a higher molecular heat capacity than atomic gases, since molecules can perform rotational motions, in addition to the translational motions characteristic of atoms. Usually, i.e., under conditions that are not too extreme, the equipartition theorem of classical statistical mechanics holds:

### Equipartition Theorem

In statistical equilibrium the average kinetic energy of a single particle is equal to $kT/2$ times the number $f$ of its degrees of freedom:

$$\langle E_{\text{kin}} \rangle = \frac{f}{2}kT.$$

Atoms behave as point masses and thus have only the three translational degrees of freedom (for temperatures that are not too high). Diatomic molecules, on the other hand, behave as if they were dumbbells consisting of two point masses. Hence, they can also rotate about axes perpendicular to the axis of the dumbbell, so they have an additional two rotational degrees of freedom besides the three translational degrees of freedom, and $f = 5$.

A comment: the effective number of degrees of freedom of an atom or molecule can vary with temperature (Section 2.5.4). The fact that the internal structure of an atom or molecule does not affect the number of degrees of freedom can only be explained in terms of quantum mechanics (Section 5.1.3).

In general, the potential energy owing to the interatomic forces, and not just the kinetic energy of its atoms and molecules, will contribute to the internal energy of a substance. Thus,

$$U_{mol} = N_A \left( \langle E_{kin} \rangle + \langle E_{pot} \rangle \right).$$

Only the kinetic energy can be calculated using the equipartition theorem. The potential energy is determined in a different way. In real gases the potential energy of the atoms owing to the interatomic force varies in inverse proportion to the volume of the gas. The internal energy per mol of a real gas is, therefore, dependent on its volume $V_{mol}$:

$$U_{mol} = \frac{f}{2} RT - \frac{a}{V_{mol}}.$$

Here $a$ is the constant in the van der Waals equation (Section 2.2.4).

The effect of the interatomic force on the internal energy of a solid is greater. Here the atoms move constantly in potential troughs and, therefore, behave approximately as harmonic oscillators (Section 1.6.1). Since the kinetic and potential energies of harmonic oscillators are equal on the average, atomic solids such as copper or lead have an internal energy per mol equal to

$$U_{mol} = 3RT. \qquad \text{(Dulong-Petit law)}$$

The contribution of the potential energy to the internal energy explains, in particular, why large amounts of heat must be applied to a substance during phase transitions without a temperature increase taking place. Thus, during isothermal vaporization of a substance the atoms initially trapped in potential troughs must first acquire sufficient energy to escape the potential wells. But, according to the equipartition principle, the kinetic energy does not change. The heat of vaporization thus corresponds to the increase in potential energy.

**Problem 2.13**

From the specific heat of vaporization of $H_2O$ (Figure 2.18) estimate the depth of the potential trough of an $H_2O$ molecule (molecular weight $M = 18$) in (liquid) water.

**Notes**   At ordinary temperatures the average kinetic energy $\langle E_{kin} \rangle$ of the atoms or molecules of a substance is proportional to the temperature $T$ of the substance. Thus, during a phase transition, only the potential energy of the atoms or molecules changes, and not their kinetic energy. The kinetic energy of atoms corresponds to their translational motion (with $f = 3$ degrees of freedom). For diatomic molecules at ordinary temperatures, rotation about the axes perpendicular to the axis of the molecule must also be taken into account. These molecules, therefore, have $f = 5$ degrees of freedom. In general, the equipartition theorem holds:

$$\langle E_{\mathrm{kin}} \rangle = \frac{f}{2} kT.$$

## 2.3.4 Thermodynamic processes

The thermal equilibrium state of a unitary system with a specified substance can be described using two variables of state. For example, if $T$ and $V$ are known, the pressure $P$ can be obtained from a phase diagram or the thermal equation of state. Thermodynamic processes, which can be driven, for example, by heat or work, usually not only involve thermal equilibrium states, but also nonequilibrium states, such that pressure or temperature gradients can develop within the system. Only for processes that proceed sufficiently slowly can we assume that a given system is passing only through equilibrium states. Equilibrium processes of this sort will be discussed in the following.

Equilibrium processes of unitary systems in which one of the variables of state is constant are particularly easy to describe. Depending on whether $P = const$, $T = const$, or $V = const$, these processes are referred to as *isobaric*, *isothermal*, or *isochoric*. In addition, we can consider processes in which work but no heat, or heat but no work, are exchanged with the surroundings. In the first case, the process is referred to as *adiabatic*. In the latter case, an isochoric process takes place. Then work is done on or extracted from a unitary system only by compression or expansion, respectively. The work done in a volume change is given by

$$W = -\int_{A}^{Z} P\, dV.$$

During compression $dV < 0$, so that $W > 0$ and, by contrast, in an expansion $dV > 0$, which implies that $W < 0$.

As an example of a thermodynamic equilibrium process let us consider the *Joule-Thomson process*. It is often used in engineering for cooling gases to low temperatures and liquefying them (Section 2.5.3). In the experiment first performed by Joule and Thomson, a gas (e.g., $N_2$) in a glass tube is pressed through a porous wall (frit) (Figure 2.21). Let us assume that the gas in front of and behind the wall is in thermal equilibrium and can be described by the variables of state $P_1$ and $T_1$ or $P_2$ and $T_2$ and that the process is adiabatic with the glass tube serving as an ideal thermal insulator. Under these conditions, for example, if 1 mol of gas is pressed through the wall, work $W_1 = P_1V_1$ will be performed. On the other hand, on the other side of the wall the gas has to expand into the space, performing work $W_2 = P_2V_2$. For the work performed and the internal energies $U_1$ and $U_2$ of the gas in front of and behind the wall, the first law of thermodynamics implies that

$$P_1V_1 + U_1 = P_2V_2 + U_2.$$

Thus, in a Joule-Thomson process the variable of state $H = PV + U$ (*enthalpy*) is constant. For this reason a Joule-Thomson process is also referred to as an isenthalpic process.

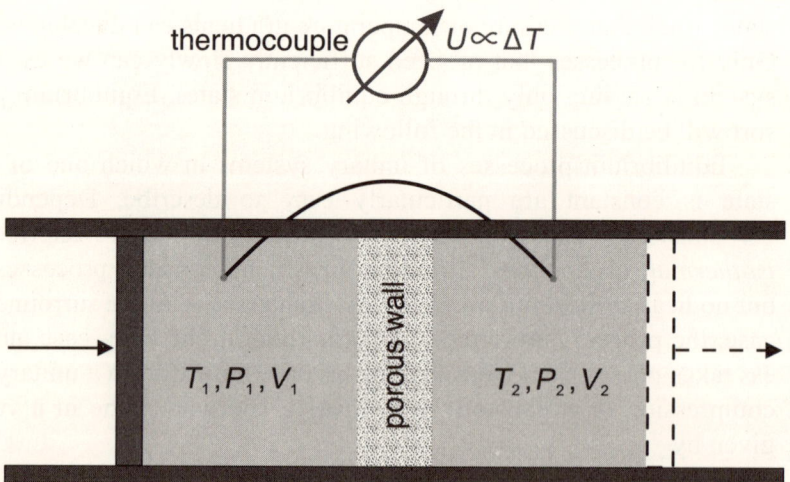

**FIGURE 2.21** Experimental setup for demonstrating the Joule-Thomson effect.

In this process it is obvious that $P_1 > P_2$. The question remains of how the temperature of the gas changes. The temperature change can be measured experimentally with a thermocouple. It can, however, also be calculated from the above equation if the thermal and caloric equations of state of the gas are known.

For ideal gases the equations of state are $PV_{\text{mol}} = RT$ and $U_{\text{mol}} = \frac{3}{2}RT$. The equality of the enthalpies implies that for ideal gases the temperature does not change in a Joule-Thomson process $(T_1 = T_2)$. For real gases, on the other hand, the temperature does change. Accounting for the characteristic volume of the atoms and the interatomic force as in the case of the van der Waals gases (Section 2.2.4), a first approximation (neglecting correction terms that are nonlinear in $a$ and $b$) for the enthalpy of the gas is

$$H(P,T) \approx \left(\frac{f}{2}+1\right)RT + \left(b-\frac{2a}{RT}\right)P.$$

Depending on the sign of the factor $(b - 2a/RT)$, a pressure drop at the porous wall is connected with either a rise or a fall in the temperature. This factor is zero at the *inversion temperature*

$$T_i = \frac{2a}{Rb}.$$

Below the inversion temperature the effect of the interatomic force on the Joule-Thomson process predominates over that of the characteristic atomic volume, so the temperature decreases with a pressure drop. In general, because $H = const$ the temperature change $dT$ and the pressure variation $dP$ are related by

$$\left(\frac{f}{2}+1\right)R\,dT \approx \left(\frac{2a}{RT}-b\right)dP.$$

---

### Experiment 2.10    Joule-Thomson effect

Nitrogen at room temperature flows from a pressure vessel through a glass tube with a porous wall (frit) (Figure 2.21). The temperature difference $\Delta T$ between the gas in front of and behind the wall is measured with a thermocouple. A temperature decrease of a few Kelvin is obtained. According to the above formula, the effect is larger at lower temperatures.

---

**Problem 2.14**

Verify the formula for the enthalpy of a van der Waals gas. The van der Waals constants for nitrogen are $a = 0.136 \, \text{J} \cdot \text{m}^3/\text{mol}^2$ and $b = 38.5 \times 10^{-6} \, \text{m}^3/\text{mol}$. Calculate the volume of an $N_2$ molecule and estimate the depth of the intermolecular potential operating between $N_2$ molecules. How big is the expected temperature change $\Delta T$ for a Joule-Thomson process at $T \approx 300 \, \text{K}$ if the pressure drop at the porous wall is $\Delta P = 1 \, \text{atm}$?

---

| **Notes** | In thermodynamic processes there are usually temperature and pressure gradients. Thermal equilibrium states occur only in the limiting case of an infinitely slow process. |
|---|---|

Strictly speaking, the equations of equilibrium thermodynamics are valid only in this limit.

## 2.4 ENTROPY PRINCIPLE

The atomic description of matter can lead to the idea that the thermal motion of atoms only obeys the deterministic laws of mechanics and, therefore, is exclusively affected by conservative forces. One consequence of this idea would be that all thermodynamic processes could, in principle, run backwards, i.e., that they would also be reversible. But this conclusion obviously conflicts with experience. For example, temperature equilibration always takes place between two bodies in contact which are isolated from their surroundings. The temperature difference of two such objects has never been observed to increase. No broken jug has ever reassembled itself, as in a film running backwards. Evidently, there is a difference between past and future in *equilibration processes*. Equilibration processes are irreversible and, therefore, manifest a time directionality. In this section we examine how the *irreversibility* of natural phenomena is embedded in the fundamental laws of physics. The deterministic fundamental laws of mechanics cannot explain irreversibility. The entropy principle alone provides an answer. It shows clearly that in the world of atoms chance rules along with determinism.

### 2.4.1 Reversible and irreversible processes

All mechanical processes in which no dissipative forces operate are reversible. In a film run backward, an undamped swinging pendulum moves exactly the same way as it does when the film is run forward. Only under the influence of dissipative forces do mechanical processes become irreversible. Then, the total mechanical energy in a closed mechanical system decreases with time. According to the first law of thermodynamics the mechanical energy is just converted into internal energy of the objects involved. Yet, when the internal energy is taken into account, the total energy of a closed system is still conserved. Nevertheless, all processes observed in nature are ultimately irreversible. We all get older, but never younger.

Purely thermodynamic processes, in which only the thermodynamic variables of state change, can also be reversible, at least, as in mechanics, when an idealized limiting case is considered. Ultimately, all thermodynamic processes are initiated by producing a thermodynamic disequilibrium. In order to perform work $W$ on a thermodynamic system, a pressure drop must be generated, and in order to apply an amount of heat $Q$, a temperature difference is required. In both cases, equilibration processes follow (Section 3.2), which return the system to a state of thermal equilibrium. These equilibration processes are irreversible; the pressure and temperature differences, of course, vanish on their own, but they never develop on their own. Thus, a thermodynamic process comes closer to the ideal of a reversible limiting case the more it avoids equilibration processes and, in particular, temperature and pressure drops. Only in the idealized limiting case, where only thermal equilibrium states operate, would the process be reversible.

It is also instructive to discuss the irreversibility of equilibration processes on the basis of atomistic models. Suppose, for example, that there is a temperature difference between two rooms filled with gas (Figure 2.22), so that the atoms of the warmer gas have a higher statistical average velocity than those of the colder gas. The average velocities of the atoms in the two gases become equal in the course of temperature equilibration.

From the standpoint of classical mechanics this process should be reversible. If the atoms were actually small spheres in the classical sense, then a sort of border patrol could be set up between the two rooms to ensure that the faster atoms could move in only one direction and the slower atoms only in the other. A *Maxwell demon* could perform the passport checks. It would only have to measure the velocity of the arriving atoms in good time and then decide whether the atoms should be allowed to pass or not.

This beautiful idea is not quite feasible, not so much because of the technical difficulty of measuring and making a decision fast enough, as because of the fundamental nature of the measurement process (Section 6.6.1). Here we shall make do with merely pointing out that the motions of the atoms obey the laws of chance, as well as the deterministic laws of classical mechanics. An ensemble of atoms thus tends, when left to itself, to approach a state characterized by the probability distributions of statistical physics (Section 2.1.4).

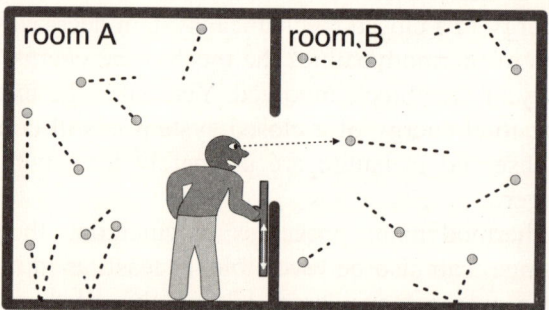

**FIGURE 2.22**  A Maxwell demon.

**Problem 2.15**

Run a film backward and try to find scenes which appear absurd and unreal. The processes taking place at those times are extremely irreversible.

**Notes**   In a world where the deterministic laws of classical mechanics were valid without restriction and only conservative forces operated, all processes would be reversible. In the real world a reversible process is an idealized limiting case. Thus, all observed processes are more or less irreversible.

## 2.4.2  Second law of thermodynamics

On a macroscopic level the irreversibility of natural processes follows from the second law of thermodynamics. According to the first law of thermodynamics (Section 2.3.2) work and heat are equivalent quantities. Work can be converted into heat and heat into work, as in a steam engine, for example. The second law, on the other hand, only deals with heat and makes it clear that heat and work are fundamentally different physical quantities. Although they are indeed measured in the same units, the measurement procedures for the two are radically different.

According to the first law it should be possible to convert heat into work without limit! If this were so, then there would be no energy supply problems in the world. For example, there might be a machine that extracts the ocean's heat and converts it into work to propel a ship (Figure 2.23) in accordance with the first law.

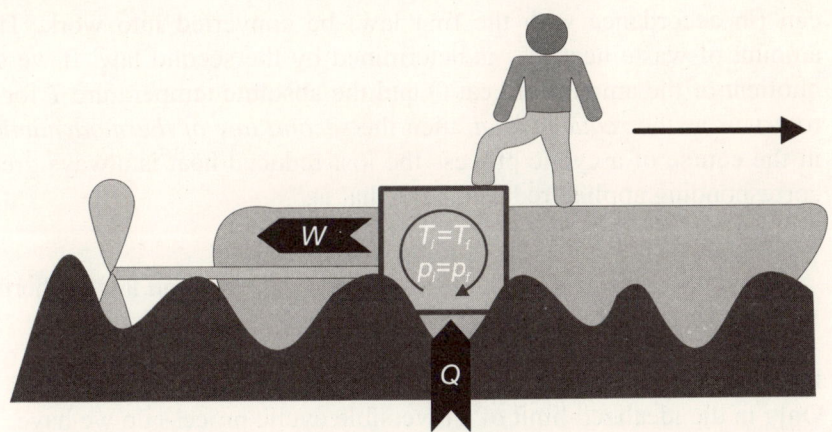

**FIGURE 2.23** An example of a perpetual motion machine of the second kind.

No one has ever succeeded in constructing a so-called *perpetual motion machine of the second kind* of this sort. On the contrary, the second law of thermodynamics has proven to be a universally valid law of nature. In order to put it into words, let us consider a cyclic process in which, as in a steam engine, heat is converted into work (Figure 2.24). Unlike with the first law, here the applied heat $Q_{in}$ is assumed to come from a heat reservoir at a (higher) temperature $T_>$.

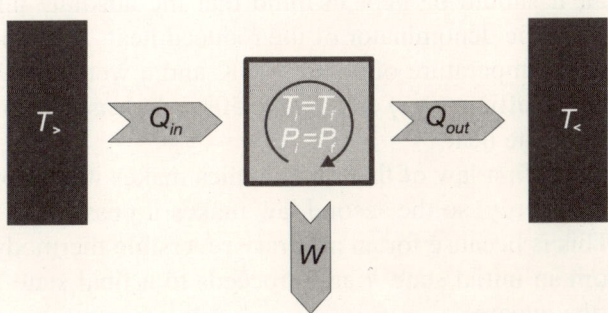

**FIGURE 2.24** Diagram of a thermodynamic cyclic process.

Experience teaches us that not all of the heat $Q_{in}$ can be converted into work, but part of it must be transferred as *waste heat* $Q_{out}$ to a heat reservoir at a lower temperature (usually the environment). Only the remainder

$$W = Q_{in} - Q_{out}$$

can (in accordance with the first law) be converted into work. The minimum amount of waste heat $Q_{out}$ is determined by the second law. If we designate the quotient of the amount of heat Q and the absolute temperature $T$ for a given heat reservoir as the *reduced heat*, then the *second law of thermodynamics* states that in the course of a cyclic process the lost reduced heat is always greater than the corresponding applied reduced heat, that is,

$$\left(\frac{Q}{T}\right)_{out} > \left(\frac{Q}{T}\right)_{in} \qquad \text{(second law of thermodynamics)}$$

For a heat engine, such as a steam engine, we thus have $(Q_{out}/T_<) > (Q_{in}/T_>)$. Only in the idealized limit of a reversible cyclic process do we have

$$\left(\frac{Q}{T}\right)_{out} = \left(\frac{Q}{T}\right)_{in}$$

The second law is especially consistent with the common perception that heat always flows from warmer to colder objects, as during temperature equilibration. Thus, in simple heat flow, $Q_{out} = Q_{in}$. The second law then implies that $T_{in} > T_{out}$. It follows further from the second law that in all heat engines a large portion of the heat produced in the burner is lost as waste heat (if it is not used for heating purposes, as it is in power plants with power-heating loops). In calculating the waste heat, it should be kept in mind that the absolute temperature of the heat reservoir is in the denominator of the reduced heat. A steam engine with a super-heated steam temperature of about 700 K and a wet steam temperature of 350 K has, at best, an efficiency $\eta = W/Q_{in} = 50\%$. At least half of the generated heat is given up as waste heat.

Just as the first law of thermodynamics makes it possible to define a variable of state, the energy, so the second law makes it possible to define a new variable of state. This is because for an arbitrary reversible thermodynamic process which begins from an initial state $A$ and proceeds to a final state $Z$, the second law implies that the integral

$$\int_A^Z \frac{\delta Q_{rev}}{T}$$

is independent of path. Here, as in the case of the first law, the amounts of applied and discharged heat $\delta Q$ are designated as positive and negative, respectively. Based on the path independence of the integral, a new variable of state, the *entropy*, can be defined:

$$S_Z = S_A + \int_A^Z \frac{\delta Q_{rev}}{T}. \qquad \text{(definition of entropy)}$$

Here the entropy $S_A$ of the initial state $A$ is again set arbitrarily.

It follows from the second law that the entropy does not change in a reversible adiabatic process, while, on the contrary, it increases in an irreversible adiabatic process. The entropy of a thermodynamic system can, however, be reduced if heat is withdrawn from the system (which is as reversible as possible). Here heat can also be transferred to warmer objects in accord with the second law provided sufficient work is expended. The heat pump is a machine that is suitable for this purpose (Section 2.4.3).

If the thermal and caloric equations of state for a unitary system are known, then it is possible to obtain the entropy of the system as an explicit function of the variables of state $T$ and $V$. For ideal gases (Section 2.1.2) the equations of state $PV = RT$ and $U = \frac{3}{2}RT$ and the relations

$$dU = \delta W_{rev} + \delta Q_{rev} \quad \text{and} \quad \delta W_{rev} = -P\,dV$$

yield the following expression for the integral $\int \delta Q_{rev}/T$:

$$\int_A^Z \frac{\delta Q_{rev}}{T} = \frac{3}{2}R\int_A^Z \frac{dT}{T} + R\int_A^Z \frac{dV}{V}.$$

Evaluating this integral yields the *entropy of an ideal gas*,

$$S(T,V) = S(T_A,V_A) + \frac{3}{2}R\ln\frac{T}{T_A} + R\ln\frac{V}{V_A}. \qquad \text{(entropy of an ideal gas)}$$

An atomistic interpretation of the entropy of an ideal gas makes the relationship between entropy and chance in atomic events clear. In terms of statistical mechanics, each state of an atomic ensemble can be assigned a probability $W$ that it will exist with a specified volume and prescribed temperature. The physicist Ludwig Boltzmann showed that the entropy $S$ and the probability $W$ are related in the following way:

$$S = k \ln W.$$

Finally, it should be pointed out that the close relationship between entropy and absolute temperature (implied by the second law) furnishes a justification for the establishment of the absolute temperature scale that is independent of the concept of an ideal gas. A justification of this sort is especially necessary for establishing the temperature scale near absolute zero, since in that temperature range no real substance behaves as an ideal gas (even approximately).

**Problem 2.16**

Calculate the change $\Delta S$ in the entropy of 1 mol of an ideal gas which is heated at constant pressure from 0°C to 100°C.

**Notes**

The second law of thermodynamics states that in all cyclic processes more reduced heat is given up than is applied; that is,

$$\left(\frac{Q}{T}\right)_{\text{out}} > \left(\frac{Q}{T}\right)_{\text{in}}.$$

### 2.4.3 Energy converters

The second law of thermodynamics is fundamental for all machines which convert heat into work, such as heat engines, or work into heat, such as refrigerators and heat pumps. We now illustrate the principle of thermal heat engines with the aid of some simple demonstration experiments.

## Experiment 2.11    Nitinol thermal heat engine

The design of this machine is similar to a reconstruction (Figure 2.25) of an historical attempt at a mechanical perpetual motion machine (Section 1.3.1). Outrigger weights are attached to a wheel that rotates about a horizontal axis (Figure 2.26).

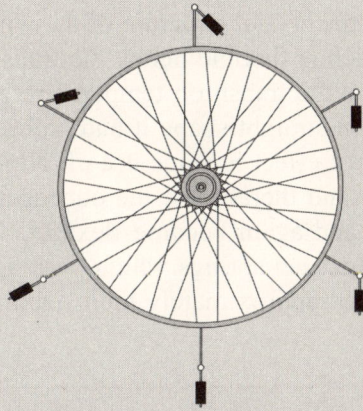

**FIGURE 2.25**  Example of a mechanical perpetual motion machine.

**FIGURE 2.26**  A nitinol thermal heat engine.

The mechanical perpetual motion machine is based on the naive idea that when they swing out during their downward motion the outrigger weights can produce a torque in the direction of the rotation of the wheel. Of course, the wheel cannot be driven in this way because the weights fall downward as they swing out and thereby lose potential energy. It follows directly from the energy principle that a machine of this sort cannot operate.

In the nitinol thermal heat engine, on the other hand, the outrigger weights gain energy as they are raised. Thus, this machine can actually do work. It really converts heat into work. A nickel-titanium-alloy (nitinol) wire loop is heated by immersion in the water bath, which is heated to about 70°C. This leads to a phase transition in which the crystal structure of the wire changes. Nitinol is a so-called memory metal, which is flexible below the transition temperature and can easily be deformed. Above the transition temperature it is substantially more rigid and tends to return to the straight shape it had before it was deformed. In the nitinol machine this memory effect is made use of. Nitinol strips heated in a water bath straighten out a bit and thereby lift the outrigger weights up slightly. The wires cool off in the air and again become flexible. The nitinol strips are again bent more strongly by the steel springs which push against them. When the outrigger weights have reached approximately their maximum height, they are then again raised.

### Experiment 2.12    Nitinol wire drive

Clever use of the phase transition of the memory metal nitinol is made in the nitinol wire drive shown in Figure 2.27. Here a nitinol wire serves as a V-belt for transmitting the motive power of a small brass wheel to a large plastic wheel. If the brass wheel is immersed in hot water, the nitinol wire, which is tightly wound around the wheel and heated above the transition temperature, tends to straighten out. This creates an unstable equilibrium, and a small shove is enough to set the wheels turning rapidly in one or the other direction.

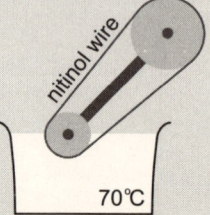

**FIGURE 2.27** Nitinol wire drive.

## Experiment 2.13    Dunking duck

Another phase transition is used to run the "dunking duck" shown in Figure 2.28. Here the head of the duck, which is covered with damp felt, is cooled by the evaporating water. This leads to a temperature difference between its stomach and its head. The highly temperature-dependent vapor pressure of the liquid in its stomach pushes the liquid into the duck's neck until it loses balance and dives into the water bowl with its beak in order to wet its head again. At the same time, pressure equilibration takes place between the head and stomach, so that the liquid flows back and the duck rights itself again.

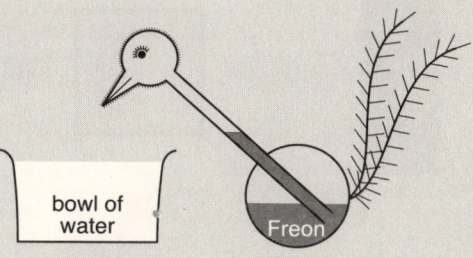

**FIGURE 2.28** Dunking duck.

As in these demonstration experiments, heat is converted into work in a heat engine. As shown in Figure 2.24, an amount of heat $Q_{in}$ is transferred from a thermal (heat) reservoir at a higher temperature $T_>$ to the heat engine and smaller amount of heat $Q_{out}$ is given up to a thermal reservoir at a lower temperature $T_<$. Only the difference is available as work $W$. The second law implies that the efficiency $\eta = W/Q_{in}$ cannot be greater than $(T_> - T_<)/T_>$, i.e.,

$$\eta < \frac{T_> - T_<}{T_>}.$$

Since the temperature difference between the thermal reservoirs in both of the demonstration experiments described above is small compared to the absolute temperature $T_>$, the efficiency of these processes is very low.

If, on the contrary, a machine is to transfer heat from a colder thermal reservoir to a hotter thermal reservoir, then work must be done on the machine (Figure 2.29). If the machine has the task of further cooling a colder thermal reservoir (refrigeration chamber) by removing heat from it, then it is a *refrigerator*. On the other hand, if the warmer thermal reservoir (heating chamber) is heated further, then have a *heat pump*. For the quantities indicated in Figure 2.29, the first two laws of thermodynamics yield the following inequalities:

$$Q_{\text{in}} + W = Q_{\text{out}} \quad \text{and} \quad \frac{Q_{\text{out}}}{T_>} > \frac{Q_{\text{in}}}{T_<}.$$

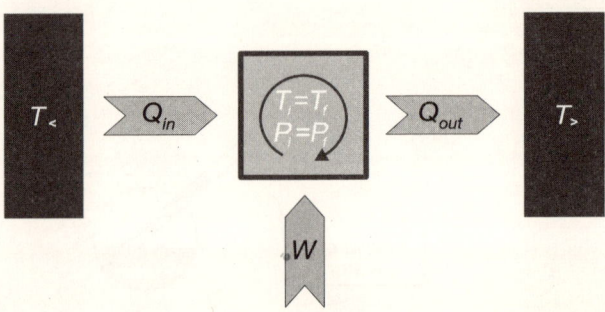

**FIGURE 2.29** Energy flow diagram for refrigerators and heat pumps.

The analog of the efficiency of a heat engine for refrigerators and heat pumps is an efficiency defined by $\varepsilon = Q/W$, where the numerator is the amount of heat extracted from the refrigeration chamber, $Q_{\text{in}}$, or delivered to the heating chamber, $Q_{\text{out}}$. The laws of thermodynamics give

$$\varepsilon = \frac{Q_{\text{in}}}{W} < \frac{T_<}{T_> - T_<}$$

for a refrigerator and

$$\varepsilon = \frac{Q_{\text{out}}}{W} < \frac{T_>}{T_> - T_<}$$

for a heat pump.

**Problem 2.17**

Find out at what temperatures a power plant in your neighborhood operates and what the plant's efficiency is. Compare the efficiency with the optimum value.

**Notes**   In all cyclic processes where heat from a heat reservoir at a higher temperature $T_>$ is converted into work, a portion of the heat input is lost as waste heat to a heat reservoir at a lower temperature $T_<$. The efficiency $\eta = W/Q_{in}$ of a heat engine of this type is always less than $(T_> - T_<)/T_>$.

### 2.4.4  Stirling process

Since thermodynamic processes of ideal gases in the reversible limit can be easily described theoretically and quantitatively, it is highly instructive to analyze and test experimentally those cyclic processes of ideal gases which are suitable for driving heat engines, refrigerators, and heat pumps. Another example that is also of practical interest, is the *Stirling process*.

In the reversible limit, the Stirling process consists of two isothermal and two isochoric subprocesses and is, therefore, represented in a $T(V)$ diagram as a rectangle (Figure 2.30). In the isothermal processes either $T = T_<$ or $T = T_>$ and in the isochoric processes $V = V_<$ or $V = V_>$.

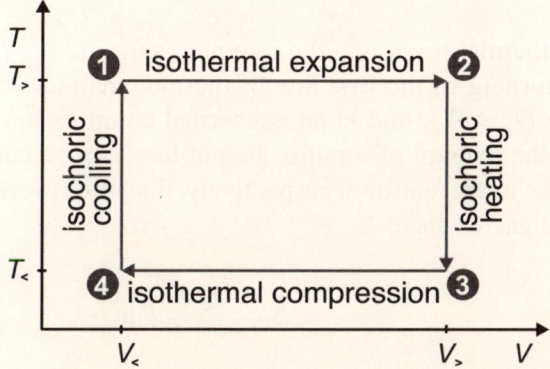

**FIGURE 2.30**  $T(V)$ diagram for the Stirling process.

In this cyclic process, heat is converted into work when the gas expands at $T_>$ and is compressed at $T_<$; otherwise, work is transformed into heat. This cyclic process can, therefore, be used both as a heat engine and as a refrigerator or heat pump. The amount of work that is performed or used per cycle becomes apparent when the cyclic process is represented in a $P(V)$ diagram (Figure 2.31). Then the work performed (sum of the lost ($W < 0$) and input ($W > 0$) work) per machine cycle is equal to the closed integral

$$W = -\oint P\,dV$$

and, therefore, to the area enclosed by the cycle in the $P(V)$ diagram. The work and heat exchange for ideal gases can be calculated for the isothermal and isochoric subprocesses. In doing this calculation we shall refer to 1 mol of the gas.

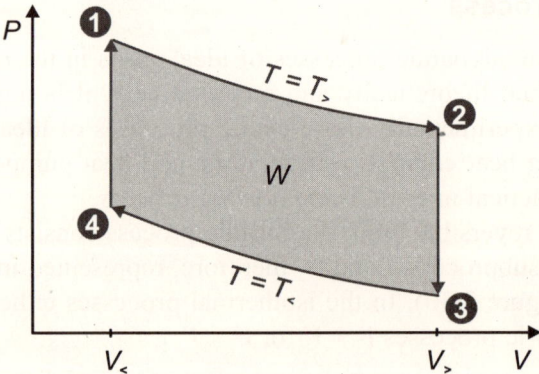

**FIGURE 2.31** $P(V)$ diagram for the Stirling process.

In isothermal processes the internal energy $U = \frac{3}{2}RT$ of an ideal gas is constant. According to the first law of thermodynamics, therefore, in an isothermal expansion $Q_{in} = W_{out}$ and in an isothermal compression $Q_{out} = W_{in}$. In the reversible limit, the amount of input or output heat can be calculated from the increase or decrease in the entropy, respectively. For the reversible isothermal processes of an ideal gas we have

$$\Delta S = \frac{Q}{T} = R\ln\frac{V_>}{V_<}.$$

Thus, for isothermal expansion at $T = T_>$,

$$Q_{\text{in}} = W_{\text{out}} = RT_> \ln \frac{V_>}{V_<}.$$

Correspondingly, for isothermal compression at $T = T_<$,

$$Q_{\text{out}} = W_{\text{in}} = RT_< \ln \frac{V_>}{V_<}.$$

In the two isochoric processes the internal energy of the ideal gas changes by the same amount $\Delta U$, since the internal energy of the ideal gas depends only on the temperature:

$$\Delta U = \frac{3}{2} R (T_> - T_<).$$

The heat brought in during isochoric heating will thus again be given up in the course of the cyclic process, specifically, during isochoric cooling. This heat is not exchanged with the surroundings in the Stirling process, but is temporarily stored internally.

Accordingly, over the entire cycle an amount of heat $Q_{\text{in}} = RT_> \ln (V_> / V_<)$ is supplied to the machine and an amount $Q_{\text{out}} = RT_< \ln (V_> / V_<)$ is extracted from the machine. For a heat engine based on the Stirling process, the above relations, according to the second law, give an efficiency

$$\eta = \frac{Q_{\text{in}} - Q_{\text{out}}}{Q_{\text{in}}} = \frac{T_> - T_<}{T_>}.$$

As expected, the reversible limiting case has the maximum efficiency.

One machine that runs approximately in a Stirling process of this type is the Stirling engine. It basically consists of a cylinder in which two pistons can move back and forth (Figure 2.32). The working (main) piston $A$ determines the magnitude of the volume and the displacement piston $V$ presses the gas either into the region of the heat reservoir at temperature $T_>$ or into the temperature region $T_<$.

In the isothermal processes the working piston moves and in the isochoric processes, the displacement piston. In the latter case the gas streams through the displacement piston. At the same time, the internal energy $\Delta U$ freed during cooling can be stored temporarily in the displacement piston, which also functions as a heat reservoir, in order to be supplied again during isochoric heating of the gas.

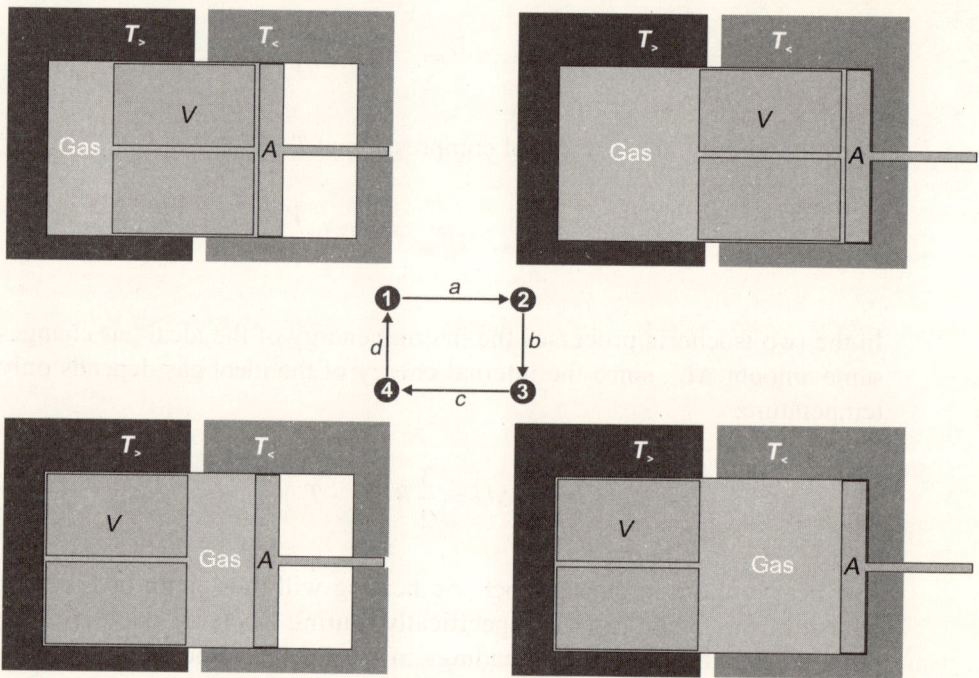

**FIGURE 2.32** Illustrating the operation of a Stirling engine.

During the isothermal processes, on the other hand, heat is extracted from or supplied to a heat reservoir and work of equal magnitude is performed by the working piston.

Even with optimal control of the pistons, pressure and temperature gradients do develop during operation of the Stirling engine. The cyclic process realized in the engine is, therefore, irreversible and the efficiency is less than the maximum. However, simple arrangements in which the pistons are actuated with a phase shift of $\pi/2$ by a cam disk show that even a plain Stirling engine driven by air will function and can propel a flywheel (Section 1.3.1). During operation the latter serves to store the mechanical energy which must be available during the compression processes. Here an electrically heated space at the head of the cylinder can serve as a heat reservoir and water cooling can be used at the lower end of the cylinder.

If the cam disk is driven externally by, for example, an electric motor, then the Stirling engine works as a refrigerator. The head of the cylinder is then colder than the water bath and can be cooled to temperatures below the freezing point.

If the direction of rotation of the drive is reversed, this apparatus works as a heat pump. Now the head of the cylinder will be heated to above the water temperature.

---

**Problem 2.18**

Calculate the work performed over a cycle by a Stirling engine operating with 1 mol of gas in the reversible limit.

---

| **Notes** | In a Stirling process, alternating isothermal and isochoric processes take place. Depending on the direction in which the cycle is run, heat will be converted into work or work into heat. In one case it operates as a heat engine and in the other, as a refrigerator or heat pump. |
| --- | --- |

## 2.5 LOW TEMPERATURES

The production of temperatures near and below the freezing point (0°C) is very important to many sectors of the economy and daily life, such as the refrigeration of food. For science and technology, there is even more interest in the production of temperatures near absolute zero, $T = 0$ K. On one hand, this offers the possibility of liquefying gases such as nitrogen, oxygen, hydrogen, or the noble gases, especially helium, once temperatures below the critical temperatures of these gases have been attained. These lie at temperatures $T < 100$ K. On the other hand, at low temperatures many materials have surprising properties, such as superconductivity and superfluidity, behavior which can only be understood in terms of quantum physics (Chapters 4, 5, and 6). An elementary example of such a quantum mechanical effect is the freezing out of degrees of freedom of the hydrogen molecule at temperatures near 100 K. We shall discuss this in the last section of this lecture. This gives a preliminary answer to the question of why atoms, which are usually represented as small planetary systems (Section 5.1.1) have only three degrees of freedom under normal conditions ($T \sim 300$ K). Beyond that, the freezing of degrees of freedom is closely related to the so-called *third law of thermodynamics*. The third law states that the absolute zero of the temperature scale can only be approached asymptotically and can never be reached exactly. Temperatures in the mK and μK ranges can be reached in experiments today.

### 2.5.1 Adiabatic changes of state

The basis for creating low temperatures is adiabatic changes of state, that is, processes in which no heat is exchanged with the surroundings. The internal energy of a thermodynamic system decreases during adiabatic changes of state, as long as work is performed by the system, as when a gas expands. In the limit of an ideal gas the internal energy is proportional to the gas temperature. Hence, the temperature decreases along with the internal energy. The temperature decrease in *reversible adiabatic expansion* of an ideal gas can easily be calculated. In this case the entropy of the gas does not change. In the course of an expansion from an initial volume $V_0$ to the volume $V$, we have (per mol)

$$S(V) - S(V_0) = \frac{3}{2} R \ln \frac{T}{T_0} + R \ln \frac{V}{V_0} = 0.$$

This implies that

$$\left(\frac{T}{T_0}\right)^{\frac{3}{2}} = \left(\frac{V}{V_0}\right)^{-1} \quad \text{or} \quad T^3 V^2 = const. \quad \text{(adiabatic change of state of an ideal gas)}$$

Hence, with increasing volume, the temperature of the gas falls.

For real gases, liquids, and solids the temperature is not proportional to the internal energy $U$. Rather, here the kinetic energy and potential energy of the atoms have to be taken into account separately. According to the equipartition theorem the temperature is proportional only to the kinetic part of the internal energy. Cooling will, therefore, be observed in processes in which the fraction of potential energy in $U$ increases at the expense of the kinetic energy fraction. A redistribution of this sort is important in the temperature drop in the Joule-Thomson process (Section 2.3.4). Very effective cooling takes place during phase transitions, especially during *adiabatic evaporation* of liquids or solids, since to do this the atoms or molecules of a substance must leave the potential troughs which bind them. Everyone knows about the cooling effect of evaporating water. In medicine, liquids that evaporate especially rapidly, such as alcohol or ethyl chloride, are used for cooling (Figure 2.19).

In Section 2.2.3 evaporation was used to cool liquid nitrogen to the temperature of the triple point of $N_2$. Evaporative cooling is even more dramatic when liquid $CO_2$ is allowed to flow out of a pressurized vessel.

## Experiment 2.14    Production of dry ice ($CO_2$ snow)

Since the temperature $T_T(CO_2) = 304$ K of the critical point of $CO_2$ is somewhat higher than ordinary room temperature (Figure 2.10), $CO_2$ can be stored as a liquid in pressurized tanks. Then the $CO_2$ has a vapor pressure of about $54 \times 10^5$ Pa. We now let this liquid $CO_2$ stream out of the tank through an outlet pipe and a throttle valve (Figure 2.33). Because of the sudden pressure drop much of the liquid $CO_2$ evaporates as it streams out. Some of it, however, cools down at the same time because of the rapid evaporation. A pressure equilibrium of the $CO_2$ is reached just in the neighborhood of the sublimation curve. This happens because at the temperature of the triple point $T_K = 216.6$ K the vapor pressure $P_K = 5.2 \times 10^5$ Pa is still substantially higher than the external air pressure. The cooling process thus sets in above the triple point, and solid $CO_2$ is produced. As in the case of freezing rain, ice crystals are formed, but these go directly into the gaseous state as they warm up. This material is referred to as $CO_2$ snow or dry ice.

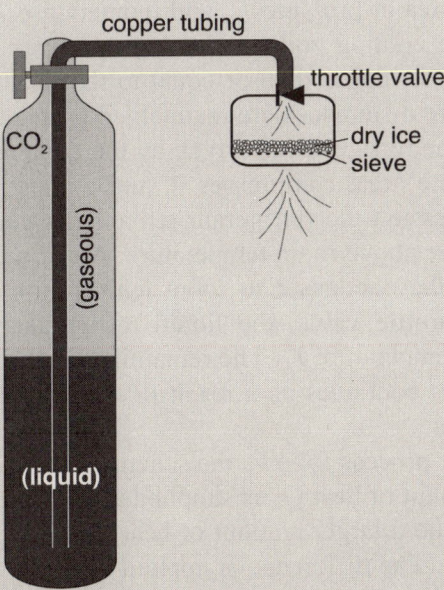

**FIGURE 2.33** Experimental setup for producing dry ice.

---

### Problem 2.19

Calculate the increase in entropy of 1 mol of ideal gas in a Joule-Thomson process in which the gas pressure falls from $P_1$ to $P_2$. (Note that the Joule-Thomson process is an *irreversible* adiabatic process.)

---

| **Notes** | In adiabatic processes there is no heat exchange with the thermodynamic system under discussion and its surroundings. In a reversible adiabatic process the entropy of the thermodynamic system does not change. |
|---|---|

## 2.5.2 Compression refrigerators

The refrigerators and freezer chests used to store food are in most cases driven by a compression refrigerating machine. In these machines a refrigerant (e.g., ammonia $NH_3$) passes through a cycle in which the cooling processes essentially relies on adiabatic evaporation of the refrigerant. The major components of a compression refrigerator of this type are shown in Figure 2.34. It consists of a low-pressure area at pressure $P_<$ and temperature $T_<$, which determines the temperature in the cooling zone, and a high-pressure area with pressure $P_>$ whose temperature $T_>$ is higher than or equal to that of the surroundings. The pressure and temperature differences are maintained by the compressor and throttle valve. The cycle of the refrigerant is driven by the compressor which draws vapor from the low-pressure area, compresses it, and pumps it into the high-pressure area. During compression the refrigerant remains gaseous and is adiabatically heated to a temperature above room temperature. Air or water cooling is used to cool the refrigerant in the condenser to room temperature and condense it. As it flows through the throttle valve, the liquid refrigerant is partially vaporized and is cooled to the temperature $T_<$. The remaining liquid collects in the evaporator. The cooling process continues as a result of the evaporation forced by the pumping out of the vapor.

This cyclic process follows the scheme illustrated in Figure 2.29 for a refrigerator. An amount of heat $Q_{in}$ is supplied at temperature $T_<$ to the refrigerant from the cold area and a larger amount of heat $Q_{out}$ is given off to the surroundings at temperature $T_>$. The difference is applied as work on the refrigerant in the compressor.

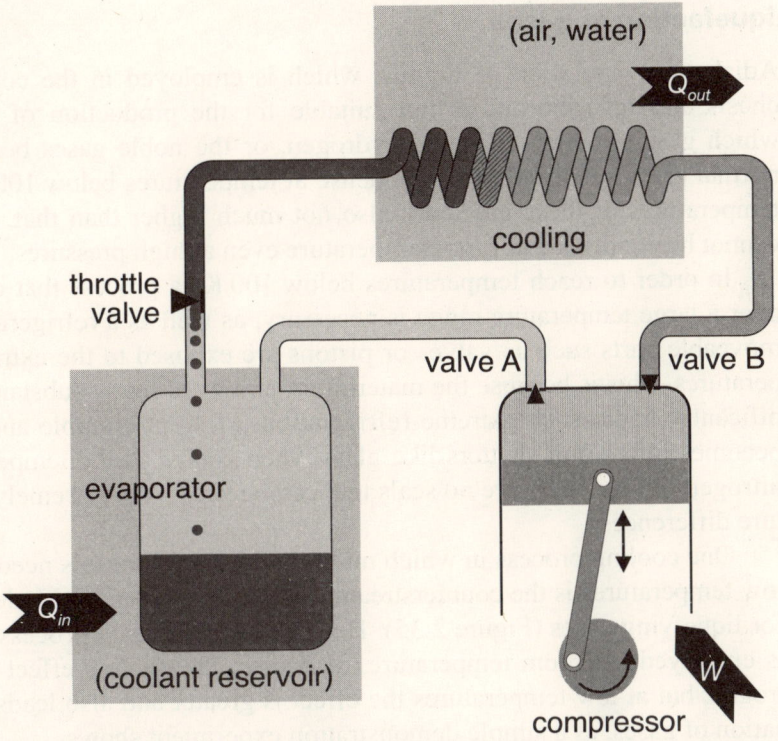

**FIGURE 2.34** Illustrating the operation of the compression refrigerating machine in an ordinary refrigerator.

## Problem 2.20

Calculate the maximum possible efficiency of refrigerators which have a cold zone temperature approaching the freezing point of $H_2O$ and surroundings at a temperature of 20°C.

| Notes | Refrigerators based on the adiabatic evaporation of a suitable refrigerant can be used for deep freezing of food. The temperature of the cold area ($T > 200$ K) is, however, still well above absolute zero. |

### 2.5.3 Liquefaction of gases

Adiabatic evaporation of liquids, which is employed in the cooling of freezer chests and refrigerators, is not suitable for the production of temperatures at which gases such as nitrogen, hydrogen, or the noble gases become liquid. At normal pressures these gases condense at temperatures below 100 K. The critical temperatures of these gases are also not much higher than that. Therefore, they cannot be condensed at room temperature even at high pressures.

In order to reach temperatures below 100 K, a process that cools efficiently over a large temperature range is necessary, as well as a refrigerator in which no moveable parts such as valves or pistons are exposed to the extremely low temperatures. This is because the material properties of many substances change significantly under such extreme refrigeration. Flowers crumble and plastic tubing becomes brittle and shatters like glass when cooled to the temperature of liquid nitrogen, 77 K. There are no seals that can be used over extremely large temperature differences.

One cooling process in which no moveable parts or seals need to be cooled to low temperatures is the counterstream process developed by Linde. It can be used for liquefying gases (Figure 2.35). Here a Joule-Thomson process (Section 2.3.4) is employed. At room temperature, of course, the cooling effect is small for nitrogen, but at low temperatures the effect is greater and also leads to the condensation of gases, as a simple demonstration experiment shows.

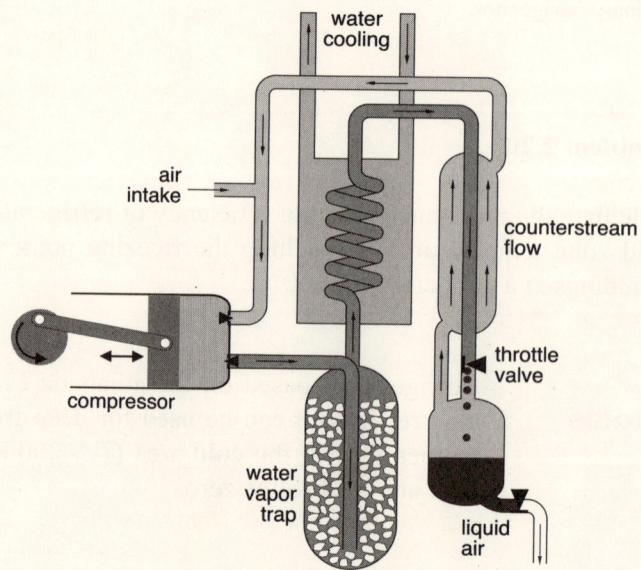

**FIGURE 2.35** Sketch of a gas liquefaction installation with a counterstream flow system.

## Experiment 2.15   Liquefaction of nitrogen

Nitrogen flows out of a pressurized tank through a roughly 2 m long copper tube, which is coiled up into a spiral, and on into the open air. The gas is cooled by sudden expansion, as in the Joule-Thomson experiment (Section 2.3.4), by a few °C. A more impressive result is obtained if the same experiment is done with nitrogen that has been precooled (Figure 2.36). For this, we cool the copper tubing with already liquefied nitrogen. The gas is allowed to flow slowly through the tubing, so it is well cooled but still gaseous. With rapid flow and correspondingly greater expansion at the end of the tubing, the Joule-Thomson effect operates, and part of the outflowing nitrogen will be liquid.

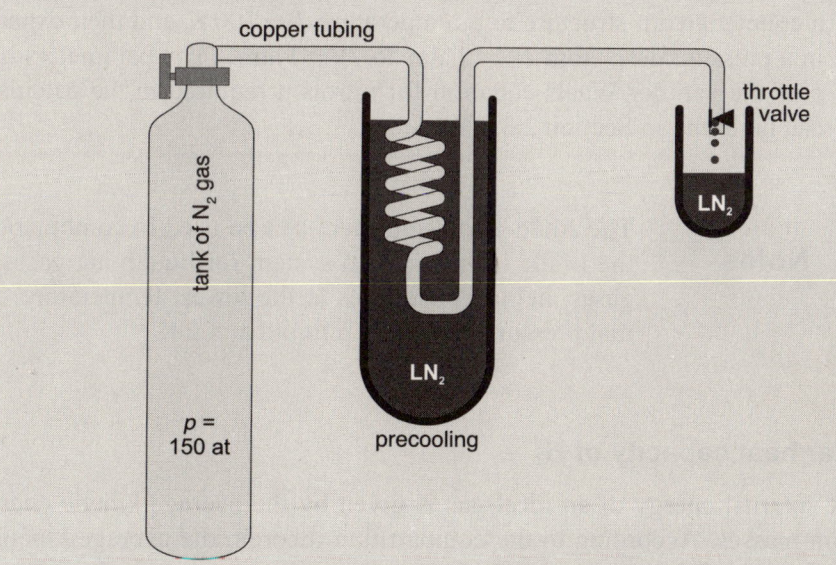

**FIGURE 2.36** Experimental setup for liquefaction of nitrogen after precooling.

In order to employ the Joule-Thomson effect on a commercial scale for liquefaction of gases, therefore, good precooling must be arranged. This precooling is provided by the Linde counterstream system (Figure 2.35). In the Linde process the gas runs through a cycle similar to that for the refrigerant in a compression refrigerator, but remains gaseous throughout the entire process. The counterstreaming flow occupies the place of the throttle valve. On streaming out of the jet, the gas is cooled by the Joule-Thomson effect and then, as it streams back through the counterstream structure, it precools the gas flowing backward to the jet. In this way, when the refrigerator is set into operation, an up-to-down temperature gradient develops in the counterstream structure, and liquid nitrogen is produced when a temperature that is only slightly above the condensation temperature of the gas is attained near the jet.

Nitrogen and oxygen can be cooled and condensed in a single step with the counterstream procedure. Gases such as hydrogen and helium condense at still lower temperatures, so they have to be precooled beforehand, since the inversion temperature (Section 2.2.4) of these gases is lower than the usual ambient temperature (300 K).

---

**Problem 2.21**

Calculate the temperature change $\Delta T$ of nitrogen when it is precooled in a counterstream structure to a temperature $T = 100$ K and then expanded in a pressure drop from $P_> = 2$ atm to $P_< = 1$ atm. (The parameters $a$ and $b$ of the van der Waals equation for nitrogen required in the calculation can be found in Section 2.3.4.)

---

| **Notes** | The Joule-Thomson effect can be used in combination with the Linde counterstream system for liquefying gases. Of all gases, helium condenses at the lowest temperature. At normal pressure, helium is a liquid at 4.2 K. |
|---|---|

## 2.5.4 Molar heat capacity of H₂

The internal energy of an ideal gas is given by the average kinetic energy of its point masses. According to the equipartition theorem the average kinetic energy per degree of freedom is $kT/2$. Point masses have only the three translational degrees of freedom, so that the internal energy $U_{mol}$ of an ideal gas is given by $U_{mol} = \frac{3}{2}RT$. As everyone knows, this formula also holds for atomic gases, although atoms are not point masses, but are small spheres with diameters on the order of $10^{-10}$ m. As rigid bodies they should also undergo rotational motions and, therefore, have six degrees of freedom according to the laws of Newtonian mechanics. As small planetary systems, consisting of an atomic nucleus and electrons (Section 5.1.1), they should have many more degrees of freedom.

From the standpoint of classical mechanics the restriction of the motions of atoms to translations in space is odd and puzzling. This is a clear indication of the differences between the macroscopic and atomic worlds. In Chapters 5 and 6 we shall show that Planck's quantum of action $h$, which has no role in classical mechanics, is key to the understanding of atomic physics. It has the dimensions of angular momentum (Section 1.4), so it is of immediate significance for rotational motions. Studies of the molar heat capacity of molecular hydrogen at low temperatures make this relationship clear.

At normal pressure, hydrogen remains in a gaseous state until a temperature of 27 K is reached. The internal energy $U_{mol}$ of hydrogen is thus essentially determined by the kinetic energy of its molecules. Based on the equipartition theorem (Section 2.3.3) we expect $U_{mol} = \frac{5}{2}RT$, since diatomic molecules should have three translational and two rotational degrees of freedom. Hence, the molar heat capacity $C_V$ (measured at constant volume) should be $C_V = \frac{5}{2}R$, a value that is confirmed by measurements at room temperature. At low temperatures, however, substantially lower values are measured. At temperatures below 100 K, $H_2$ molecules behave as point masses with only three degrees of freedom (Figure 2.37). The rotational motions seem to be frozen (or quenched).

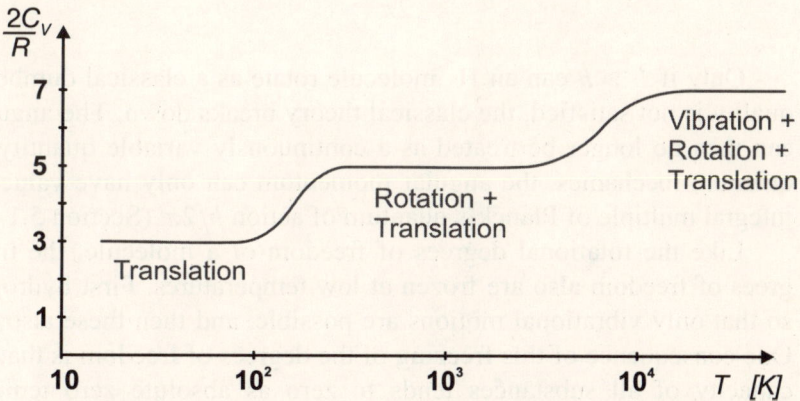

**FIGURE 2.37** Temperature dependence of the molar heat capacity of $H_2$.

This *freezing of degrees of freedom* is a quantum mechanical effect. We cannot, of course, explain it here in detail, but can make its relationship to Planck's quantum of action more obvious with a simple estimate. To do this, we calculate the average angular momentum $L(H_2)$ a hydrogen molecule should have according to classical mechanics based on its thermal motion. According to Section 1.5.2, we have

$$\langle E_{\text{rot}} \rangle = \frac{\langle L^2 \rangle}{2J}.$$

On the other hand, at 100 K, $\langle E_{\text{rot}} \rangle = kT$ has a value of about $1.4 \times 10^{-21}$ J. The moment of inertia $J$ of the dumbbell shaped $H_2$ molecule is $J = \frac{1}{2} m_p d^2$. With the proton mass $m_p = N_A^{-1} \cdot 10^{-3}$ kg $= 1.6 \times 10^{-27}$ kg and an interatomic distance between the two protons of $d = 0.7 \times 10^{-10}$ m, we obtain $2J = 0.8 \times 10^{-47}$ kg·m². This yields an average angular momentum for the rotational motion of the hydrogen molecule of

$$L(H_2) = \sqrt{2J \langle E_{\text{rot}} \rangle} \approx 1.05 \times 10^{-34} \text{ J·s}.$$

The average angular momentum at this temperature is, therefore, of the same order of magnitude as Planck's constant

$$\hbar = \frac{h}{2\pi} = 1.05 \times 10^{-34} \text{ J·s}.$$

Only if $L \gg h$ can an $H_2$ molecule rotate as a classical dumbbell. If this inequality is not satisfied, the classical theory breaks down. The angular momentum can then no longer be treated as a continuously variable quantity. According to quantum mechanics, the angular momentum can only have values which are an integral multiple of Planck's quantum of action $h/2\pi$ (Section 5.1.3).

Like the rotational degrees of freedom of a molecule, the translational degrees of freedom also are frozen at low temperatures. First hydrogen condenses, so that only vibrational motions are possible, and then these also are frozen out. One consequence of this freezing of the degrees of freedom is that the molar heat capacity of all substances tends to zero as absolute zero temperature is approached. Thus, as the temperature decreases it becomes ever harder to cool a substance efficiently. This difficulty is expressed in the *third law of thermodynamics* (Nernst's theorem) which states that although it is possible to come arbitrarily close to $T = 0$ K, absolute zero can never be reached.

---

### Problem 2.22

Estimate the magnitude of the angular momentum of nitrogen and oxygen molecules at $T = 300$ K. Why is the rotational motion of the molecules not frozen out at this temperature?

---

**Notes** Only the angular momentum of macroscopic objects can vary quasicontinuously, since it is many orders of magnitude greater than Planck's quantum of action. The angular momenta of atoms and molecules, on the other hand, are not always much greater than $h$. Hence, the quantization of angular momentum (and the fact that it can only vary in steps of $h/2\pi$) shows up physically in the rotational motion of atoms and molecules.

# Chapter 3

# Equilibration Processes and Waves

## Summary

- Wave motion in linear media
- Equilibration processes
- Sound waves in gases
- Electricity and magnetism
- The electromagnetic field
- Electromagnetic waves

Extended objects usually do not behave either as rigid bodies (Section 1.4.2) or as ideal gases in thermal equilibrium (Section 2.1.2). Both are idealizations. In real objects, mechanical deformations and temperature and density gradients can be caused by various processes, depending on the experimental conditions. On one hand there are processes involving macroscopically observable motions. They obey the dynamic laws of mechanics. Wave motions, in particular, are of fundamental significance in science and engineering. On the other hand, there are processes that rely on the thermal motion of atoms and molecules, which are based on the laws of chance. These processes are fundamentally different from wave motions. A typical example is temperature equilibration, in which the temperatures of two objects gradually approach one another.

Wave motions and equilibration processes do not only occur in extended media. Waves can also propagate in a vacuum. Today communication links can be set up to distant satellites using radio waves. The propagation of electromagnetic waves is based on the laws of electrodynamics. In the second part of this chapter we shall give an introduction to the foundations of electrodynamics in order to explain the propagation of electromagnetic waves in a vacuum. Equilibration processes which take place through a vacuum, such as temperature equilibration by thermal radiation, can, on the other hand, only be explained in terms of quantum mechanics (Section 4.4.4).

## 3.1 WAVE MOTION IN LINEAR MEDIA

Everyone is aware of the appearance of water waves. In general, they are irregular, with different distances between neighboring wave peaks or troughs. In a water bath, however, it is possible to excite periodic water waves. Depending on whether the periodically up-and-down oscillating exciter is a point or a line, waves with circular or straight wave fronts will be produced (Figure 3.1). The familiar water waves propagate along the water's surface, so the oscillations in height are functions of two position coordinates and time. Other waves, such as sound waves, usually propagate in (three-dimensional) space. Thus, the pressure oscillations associated with the propagation of sound waves are functions of three spatial coordinates and time. The mathematical description of waves propagating along a one-dimensional medium, such as waves on a string, is simpler. In order to become acquainted with the fundamental properties of waves, we must first consider waves which propagate along linear media in one dimension.

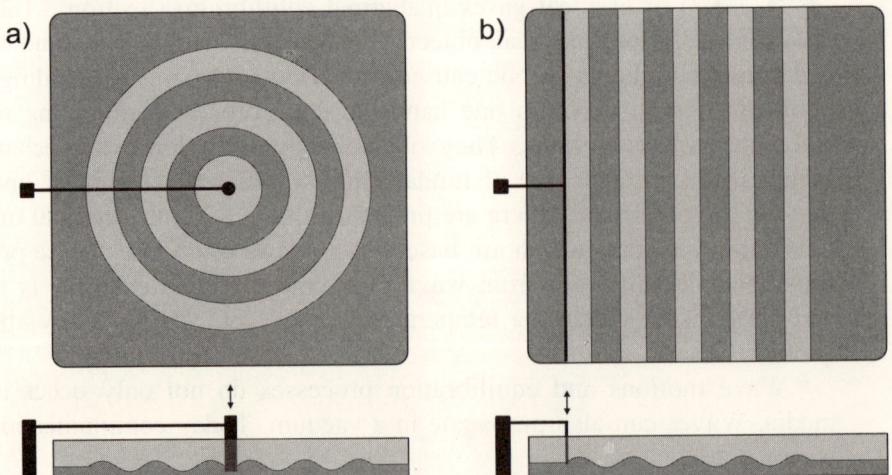

**FIGURE 3.1** Water waves generated by point (a) and line (b) exciters.

### 3.1.1  Linear chains

We begin with a purely mechanical system that is governed by the laws of point mechanics, the linear chain. It consists of a finite number $N$ of balls of equal mass arranged along a line and joined by springs (Figure 3.2). If the ball at position $x_n$ is displaced transversely, e.g., in the $z$ direction, then it will be pulled back to its rest position by the stretched springs and execute an oscillation $z(x_n, t)$. However, the neighboring balls are also set into motion by the springs as they are brought into tension, so that ultimately all the balls in the chain move.

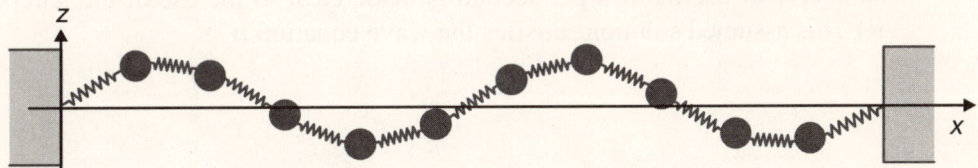

**FIGURE 3.2**  An instantaneous picture of an oscillating linear chain.

As for a system of coupled oscillators (Section 1.6.4), the equation of motion follows from the second axiom of Newton, $\mathbf{F} = m\mathbf{a}$. For a sphere at position $x_n$ the $z$ component of the acceleration vector is $a_z = \partial^2 z(x_n, t)/\partial t^2$. The restorative force is from the tension in the springs. In the rest position the forces of the two springs attached to the sphere balance out, so that the sphere remains at rest. But, in addition, when the sphere lies in a line with its two neighboring spheres, i.e., the curve $z(x, t)$ formed by the chain at the point $x_n$ is not curved, so the second derivative $\partial^2 z(x, t)/\partial x^2$ vanishes at the position of the sphere, the forces *acting* on the sphere also cancel out. A force $F_z$ that accelerates the sphere exists only if $\partial^2 z(x, t)/\partial x^2 \neq 0$. It works in the positive direction when $\partial^2 z(x, t)/\partial x^2 > 0$ and in the negative direction when $\partial^2 z(x, t)/\partial x^2 < 0$. For a displacement that is not too large, we can assume as a first approximation that $F_z$ and, therefore, the acceleration are proportional to the curvature of the curve formed by the chain. This approximation yields the following equation of motion for the chain

$$\frac{\partial^2 z(x,t)}{\partial t^2} - c^2 \frac{\partial^2 z(x,t)}{\partial x^2} = 0. \qquad \text{(wave equation)}$$

This partial differential equation is fundamental for the theoretical description of the propagation of waves along linear media and is, therefore, called the *wave equation*. The (positive) constant of proportionality $c^2$ obviously has the dimension $(\text{m/s})^2$. Thus, $c$ is a velocity. Its physical interpretation follows from the solutions of the wave equation. First we shall seek solutions that are periodic in the position and time. To do this we make the initial assumption that

$$z(x,t) = A\exp(i(kx - \omega t)).$$    (harmonic wave)

Here $A = |A|\exp(i\alpha)$ is the amplitude of the wave, with magnitude $|A|$ and phase $\alpha$, while $k$ and $\omega$ determine the periodicity of the wave in space and time, i.e., the wavelength $\lambda$ and frequency $\nu$. $k = 2\pi/\lambda$ is the *wave number* and $\omega = 2\pi\nu$ is the *(angular) frequency*. (Whether the frequency is the angular frequency $\omega$ or the number $\nu$ of oscillations per second is made clear in the use of the letters $\omega$ and $\nu$.) This assumed solution satisfies the wave equation if

$$\frac{\omega}{k} = c.$$

Thus, $c$ is the velocity at which a state of the oscillation with constant phase $\varphi = kx - \omega t$ propagates in the $x$ direction. Then $kx - \omega t = const$ implies that $\partial x/\partial t = \omega/k = c$. Thus, $c$ is called the *phase velocity* of the wave. In general, a value $\omega > 0$ is chosen for the angular frequency $\omega$. For $k$, on the other hand, negative values are also allowed. A wave with $k < 0$ propagates in the direction of the negative $x$-axis.

---

### Problem 3.1

Show that the functions $\sin(kx - \omega t)$, $\cos(kx - \omega t)$, and $\exp(i(kx - \omega t))$ with $\omega/k = c$, as well as all functions $z(x - ct)$, are solutions of the wave equation.

---

The phase velocity $c$ of the wave motion of a linear chain depends, in general, on the wave number $k = 2\pi/\lambda$ (Section 3.1.4). Here we are concerned only with waves with $\lambda \gg a$, where $a$ is the distance between neighboring spheres. Then $c = const$ to a good approximation. The phase velocity $c$ is given in terms of the natural frequency $\omega_0$ at which a sphere between two fixed neighboring spheres would oscillate and the separation $a$ by

$$c = \omega_0 a.$$

Like the motion of an undamped pendulum (Section 1.6.1), the wave motion of a system of point masses between which only conservative forces act is reversible.

The reversibility of the motion can be recognized from the fact that the equation of motion only contains second derivatives with respect to time. The square of the time interval $\partial t$ is independent of sign. Thus, the motion seen in a film running backward obeys the same differential equation as the motion seen in the same film running forward.

**Notes**

A linear chain of point masses is a mechanical model system for one dimensional media with a continuous distribution of mass. The equation of motion for these media is the wave equation:

$$\frac{\partial^2 z(x,t)}{\partial t^2} - c^2 \frac{\partial^2 z(x,t)}{\partial x^2} = 0. \qquad \text{(wave equation)}$$

The travelling waves

$$z(x,t) = A \exp(i(kx - \omega t))$$

are periodic solutions of the wave equation.

## 3.1.2 Resonators

Travelling waves are, strictly speaking, solutions of the wave equation for infinitely extended media. In finite media there are usually *boundary conditions* which must be taken into account. If a chain is fixed at both ends, like a violin string, then no travelling waves develop on the chain during periodic excitation, but standing waves do:

$$z(x,t) = A \sin(kx) \exp(-i\omega t).$$

The spheres then oscillate with different amplitudes. Antinodes appear at positions $x$ for which $\sin(kx) = 1$ and nodes, at positions for which $\sin(kx) = 0$. Nodes of the standing waves must, in particular, appear at both ends of the chain. A chain of length $L$, therefore, has a countable series of characteristic oscillations (eigenmodes). A standing wave is possible only when $L = n\lambda/2$ is an integer or half-integer multiple of the wavelength. The fundamental mode occurs for $n = 1$. Modes with $n > 1$ are referred to as harmonics.

A linear chain consisting of $N$ spheres has $N$ eigenmodes, given that each sphere has only one degree of freedom (thus, for example, only a transverse oscillation in the $z$ direction is possible). The eigenmode with the highest frequency has the shortest possible wavelength $\lambda = 2a$. Then neighboring spheres oscillate in counterphase. With a continuous distribution of mass, as in a violin string, $\lambda$ can be arbitrarily small and, therefore, $n$ can be arbitrarily large.

The natural mode of a string with $\lambda_n = 2L/n$ can be excited if a point mass on the string is periodically moved up and down at the corresponding frequency $\omega_n = c/\lambda_n$. If the excitation frequency is varied, it turns out that, as with the forced oscillations of harmonic oscillators (Section 1.6.3), resonant changes in the amplitude and phase of the forced oscillations of the string will occur at all the characteristic frequencies $\omega_n$. In this case the string acts as a linear *resonator*.

The air column in a flute can be excited to resonant oscillations (Figure 3.3) in a fashion analogous to a violin string. The characteristic modes of the air column can have nodes or antinodes at the ends of the column. Only at a closed end a node develops and at an open end, an antinode.

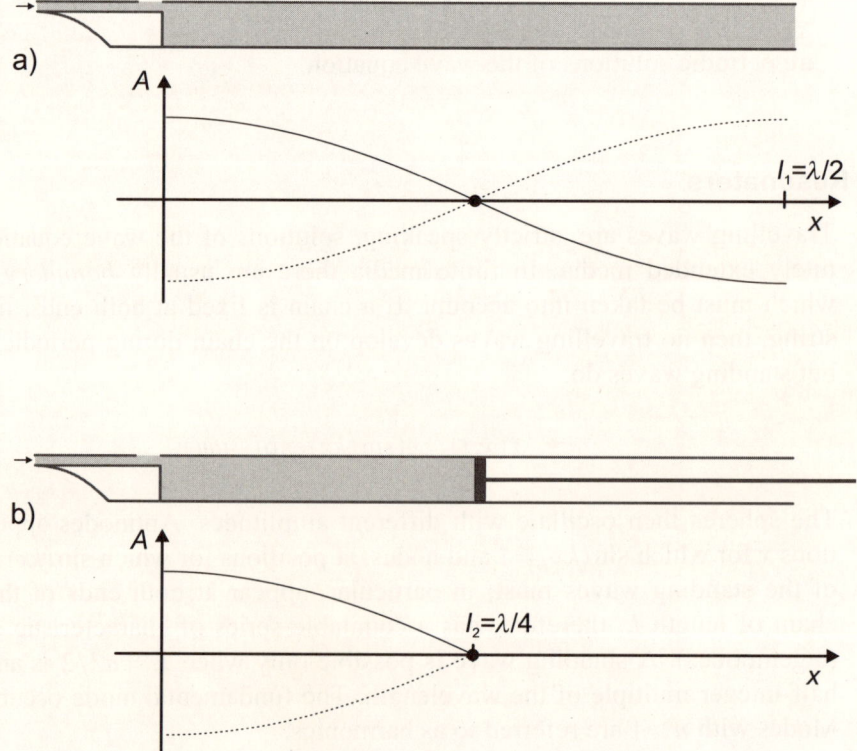

**FIGURE 3.3** Fundamental oscillation mode of an open (a) and closed (b) flute.

| | |
|---|---|
| **Notes** | The standing waves on a fixed string of length $L$ have wavelengths $\lambda = 2L/n$. Thus, a resonator of this type has a discrete series of characteristic frequencies |

$$v_n = n\frac{c}{2L}, \text{ with } n = 1,2,3,\ldots.$$

### 3.1.3 Superposition

Standing waves can be understood as the overlaying of two travelling waves propagating in opposite directions. Since $\sin(kx) = (\exp(ikx) - \exp(-ikx))/2i$, we have

$$A\sin(kx)\exp(-\omega t) = \frac{A}{2i}(\exp(kx - \omega t) - \exp(-kx - \omega t)).$$

According to this formula, an incident wave at a fixed end of the linear resonator will be reflected with a phase factor $\exp(i\pi) = -1$, so that a node develops there. At an open end, on the other hand, an incident wave is reflected without a phase change, so that there, as with the function $\cos(kx)$ at $x = 0$, an antinode develops.

The representation of standing waves as the overlaying of forward and backward propagating waves is a special case of a general procedure for finding and analyzing solutions of the wave equation. Since the wave equation is a linear differential equation, not only are the travelling waves $\exp(i(kx - \omega t))$ solutions of the wave equation, but so are all superpositions of travelling waves with arbitrary amplitudes $A_n$; that is,

$$z(x,t) = \sum_{n=1}^{\infty} A_n \exp(i(k_n x - \omega_n t)).$$

This *superposition principle*, which applies to the solutions of the wave equation, permits the construction of a great variety of solutions. For an infinitely long string with a continuous mass distribution, all functions of the form $z(x,t) = z(x - ct)$ are solutions of the wave equation.

As an illustration let us consider the superposition of two travelling waves with equal amplitudes, whose wave vectors $k_1$ and $k_2$ differ by only a small amount:

$$z(x,t) = A[\exp(i(k_1 x - \omega_1 t)) + \exp(i(k_2 x - \omega_2 t))].$$

On setting $k_{1,2} = k \pm \Delta k$ and $\omega_{1,2} = \omega \pm \Delta \omega$, with $\Delta k \ll k$, we can write $z(x,t)$ in the form

$$z(x,t) = A \exp(i(kx - \omega t)) \cdot 2 \cos(\Delta k\, x - \Delta \omega t).$$

---

### Problem 3.2

Show that these two formulas for $z(x,t)$ describe the same function.

---

Figure 3.4 is an instantaneous picture of this function. The amplitude of the superposed waves is modulated periodically, so that a sequence of groups of waves develops. This type of modulation of a wave is known as a *beating* (Section 1.6.4). Up to now we have assumed that the phase velocity $c$ is a constant for the wave motion, so we have $\Delta \omega / \Delta k = \omega / k = c$. The groups of waves, therefore, move at the same velocity as the phases of the superposed waves. Different propagation velocities occur, however, in dispersive media (Section 3.1.4).

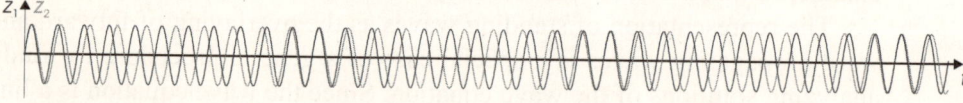

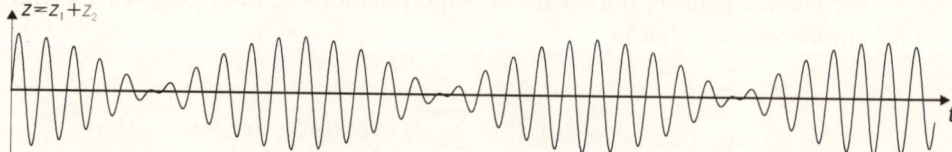

**FIGURE 3.4** An instantaneous picture of the beating produced by the superposition of two travelling waves.

**Notes**   The *superposition principle* applies to the solutions of the wave equation: if $z_1(x,t)$ and $z_2(x,t)$ are solutions of the wave equation, then every linear combination $z(x,t) = a_1 z_1(x,t) + a_2 z_2(x,t)$ of these two solutions is also a solution of the wave equation.

### 3.1.4 Dispersion

Thus far, we have assumed that the phase velocity $c$ at which waves propagate in a medium is independent of their frequency and wavelength. For many media, this assumption is not satisfied. Even for a linear chain $c = const$ is only an approximation valid for $\lambda \gg 2a$. In fact, the discrete structure of the linear chain causes $c$ to depend on $k$, so that $\omega$ is not proportional to $k$. The usually nonlinear relation $\omega(k)$ between $\omega$ and $k$ is called a *dispersion relation*. For a linear chain it has the form (in Figure 3.5)

$$\omega(k) = 2\omega_0 \sin(ka/2).$$

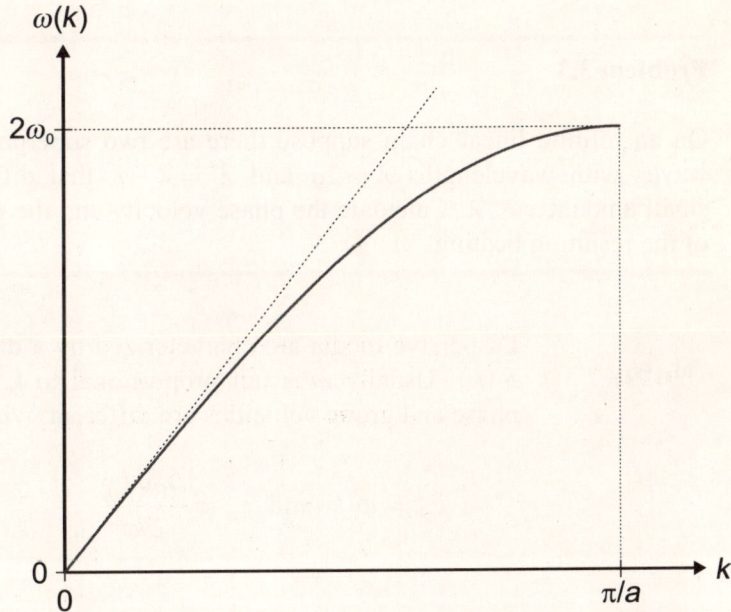

**FIGURE 3.5** The dispersion relation for a linear chain.

As we pointed out in Section 3.1.1, therefore, the phase velocity $c = \omega_0 a$ is a constant only in the limits of $\lambda \gg 2a$ or $ka \ll \pi$. Dispersion usually has to be taken into account for the propagation of waves in many other media besides the linear chain. The dispersion of light waves in glass and water (Section 4.1.2) is well known. It causes the spectral decomposition of sunlight into the colors of a rainbow. Here we are dealing with waves in three-dimensional media. In linear media, dispersion is best recognizable through the propagation of groups of waves. In dispersive media the groups of waves produced by the superposition of waves with a very small difference in their wave vectors will have a different velocity than that of the phases of the individual waves. Then it is necessary to distinguish between the *group velocity* $c_{gr}$ and the *phase velocity* $c_{ph}$. For the example discussed in Section 3.1.3, when the dispersion is included, we obtain $c_{gr} = \Delta\omega/\Delta k$. In the limit of small difference frequencies we obtain the general formula

$$c_{gr} = \frac{\partial \omega(k)}{\partial k}.$$

If $\omega$ and $k$ are not proportional to one another, then the group velocity differs from the phase velocity $c_{ph} = \omega/k$.

---

**Problem 3.3**

On an infinite linear chain suppose there are two superposed travelling waves with wavelengths $\lambda = 2a$ and $\lambda' = \lambda - \varepsilon$ that differ by a very small amount $\varepsilon \ll \lambda$. Calculate the phase velocity and the group velocity of the resulting beating.

---

**Notes**

Dispersive media are characterized by a dispersion relation $\omega(k)$. Usually $\omega$ is not proportional to $k$. In this case the phase and group velocities are different, with

$$c_{ph} = \omega/k \quad \text{and} \quad c_{gr} = \frac{\partial \omega(k)}{\partial k}.$$

## 3.2 EQUILIBRATION PROCESSES

Unlike the wave motions described in Section 3.1, equilibration processes are irreversible. Wave motions and equilibration processes thus obey differential equations with correspondingly different structures. The wave equation relates the spatial curvature of the wave function to the second derivative with respect to time. The resulting proportionality constant $c^2$ is the square of a velocity. Equilibration processes, by contrast, are described by a partial differential equation that relates the spatial curvature of a function of state to the first derivative with respect to time. The resulting proportionality constant $D$ (the diffusion constant) has the dimensions $[\text{m}^2 \cdot \text{s}^{-1}]$. Reversible wave motion can be illustrated by a simple mechanical model such as the linear chain. In order to illustrate an irreversible equilibration process, we shall draw on the simple thermodynamic model of an ideal gas (Section 2.1) in which the laws of chance as well as the laws of mechanics come into effect. This will show that the diffusion constant depends on the thermal motion of the atoms, specifically on the average thermal speed $v_{th}$ and the so-called *mean free path l* of the atoms.

Equilibration processes are ultimately the cause of friction in moving objects. The relationship between friction and equilibration phenomena shows up especially clearly in the internal friction of flowing gases. Thus, in the first section we discuss the foundations of fluid mechanics.

## 3.2.1 Flows

Streams, ocean currents, and winds are examples of flowing liquids and gases. Flows of this sort arise from differences in height or pressure and temperature gradients. With decreasing height, however, the water in a stream does not flow increasingly faster, as the energy theorem of mechanics demands. Rather, the streaming water is slowed down by internal friction and by friction with the stream bead. Heat is produced in this way.

Because of the friction, the water flows at different speeds depending on whether it is near the shore or in the middle of the stream. A flow is thus described by a *flow field* $\mathbf{v}(\mathbf{r}, t)$, which specifies the velocity $\mathbf{v}$ with which the water moves at time $t$ at every point $\mathbf{r}$ in the stream.

As a simple example of flow, we consider the flow of water through a straight pipe with a constant circular cross section (Figure 3.6). Under stationary conditions (time independent motion), the water flows fastest in the center of the pipe (at $r = 0$). Toward the edge, however, at $r = R$ the flow velocity $v(r)$ falls to zero. The field of the velocity vectors directed along the pipe has a corresponding parabolic dependence on $r$:

$$v(r) \propto (R - r)^2.$$

The constant of proportionality depends on the pressure drop and the internal friction of the water. Since both the cross-sectional area and the maximum velocity of the flow increase as $R^2$, the flux $\Phi$ (the amount of fluid passing through unit area per unit time) of the water is proportional to $R^4$; thus,

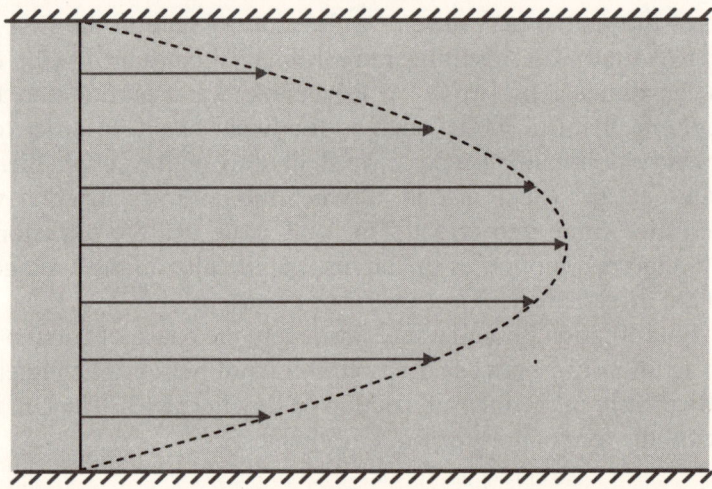

**FIGURE 3.6** Flow through a straight pipe with a circular cross section.

$$\Phi \propto R^4. \qquad \text{(Hagen-Poiseuille law)}$$

The flow field is somewhat more complicated if the cross section varies along the pipe (Figure 3.7). At a narrowing of the pipe the flow must speed up and at a widening it must slow down, since the total flow $F$ is constant along the pipe. Thus, for incompressible media $v_{max} \propto R^{-2}$.

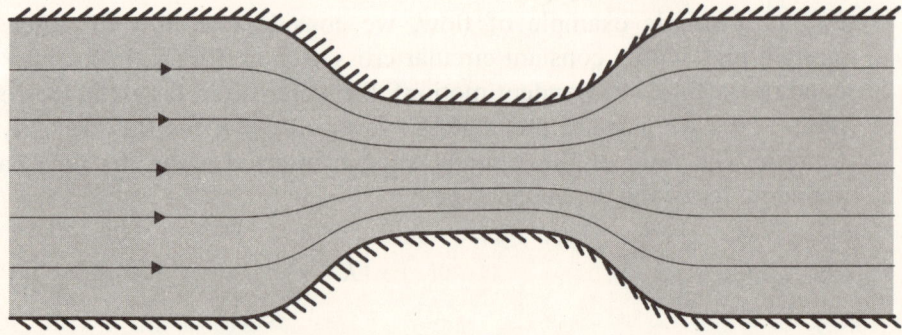

**FIGURE 3.7** Flow (of an ideal fluid) through a pipe with a narrow throat.

In this case the flow field is best represented by streamlines. The direction of the streamlines indicates the direction of the velocity and the distance between streamlines is inversely proportional to the magnitude of the velocity.

## Experiment 3.1    Static pressure of flowing liquids

It is interesting to measure the pressure of a flowing liquid at the wall of a pipe. Measurements of this sort can be made easily with a series of fluid manometers mounted on the pipe (Figure 3.8). The height $h$ of the column of liquid in a manometer indicates the pressure $P = \rho g h$ (more precisely, the overpressure relative to the air pressure $P_0 \approx 10^5\,\text{Pa}$) of the flowing liquid. Here $\rho$ is the density of the liquid and $g$ is the acceleration of gravity. For a pipe of constant cross section the pressure of the liquid falls off steadily in the direction of the flow. Here the pressure drop is obviously the pressure needed to maintain the flow against the impeding internal friction of the liquid. For a pipe with a variable cross section the pressure at a narrowing of the pipe is strikingly lower than at a section of the pipe with a wide cross section. This pressure reduction follows from the energy theorem of mechanics.

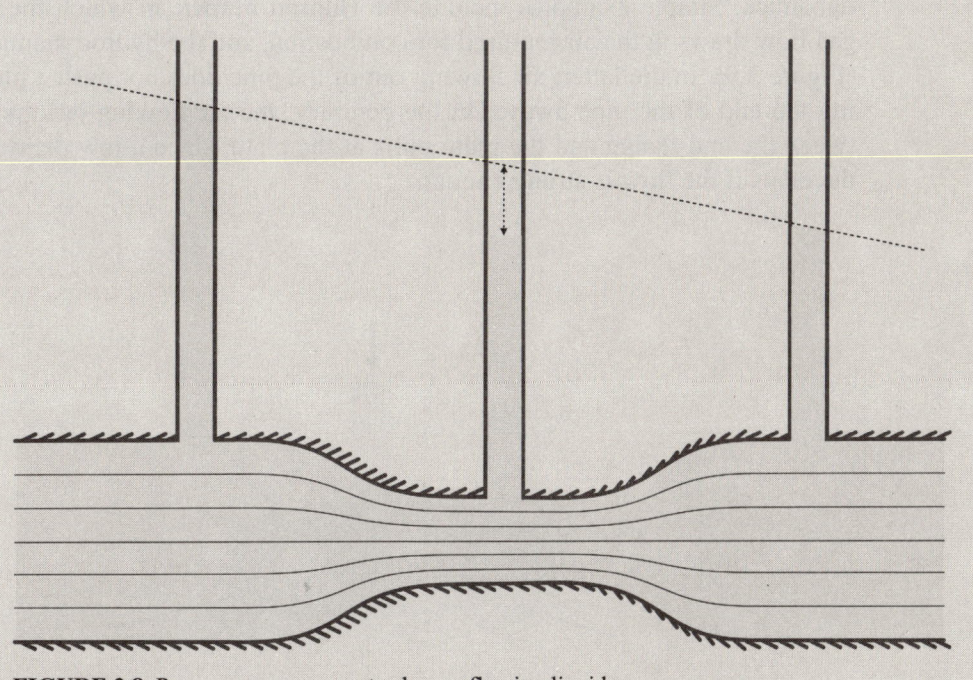

**FIGURE 3.8** Pressure measurements along a flowing liquid.

The kinetic energy density $u_{kin} = E_{kin}/V$ of a liquid flowing at velocity $v$ is

$$u_{kin} = \frac{1}{2}\rho v^2$$

and the potential energy density $u_{pot} = E_{pot}/V$ of the liquid is equal to the pressure $P$. Neglecting internal friction, that is, for a so-called *ideal fluid*, we then have

$$\frac{1}{2}\rho v^2 + P = const. \qquad \text{(Bernoulli equation)}$$

The Bernoulli equation implies that with an increasing flow velocity the static pressure drops. At a narrowing of the pipe the flow is accelerated by a pressure drop, while at a widening, it is slowed down by a pressure rise.

The Bernoulli relation between pressure and flow velocity explains many natural phenomena and is used in many branches of technology, especially aerodynamics. Simple examples include the Bunsen burner, in which the emerging gas flow draws in the air required for combustion, and the hydrodynamic paradox (Figure 3.9). In the latter, air flowing out of the pipe does not push a plate covering the end of the pipe away. On the contrary, the air flowing out the sides between the end flange and the plate pulls at the plate, since a low pressure region develops if the flow is strong enough.

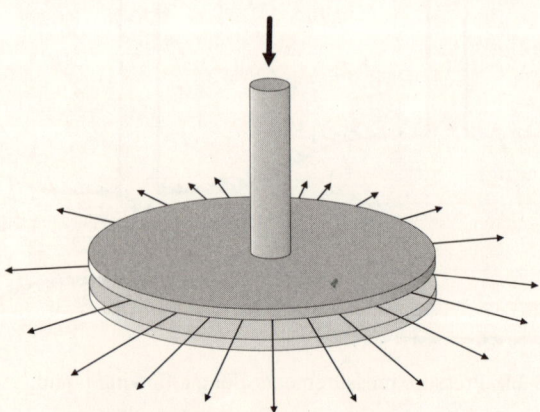

**FIGURE 3.9** A hydrodynamic paradox.

The Bernoulli equation is valid only under the assumption that the internal friction which always appears in flows can be neglected. In gases that almost behave as an ideal gas, the thermal motion of the atoms also leads to internal friction. For example, a falling sphere is slowed down by friction in the air. Here the frictional force $F_F$ is proportional to the velocity $v_S$ and to the radius $R$ of the sphere, i.e.,

$$F_F = 6\pi\eta v_S R. \qquad \text{(Stokes' law)}$$

The proportionality constant which shows up here is the *coefficient of viscosity* $\eta$. It is caused by the thermal motion of the atoms or molecules in a gas (Section 3.2.2).

---

**Problem 3.4**

Calculate the velocity with which a drop of water of radius $r = 1\,\mu m$ falls in air under normal conditions. The viscosity of air is $\eta = 17.4 \times 10^{-6}\,\text{Pa} \cdot \text{s}$.

---

| Notes | Flowing liquids and gases are described by a flow field $\mathbf{v}(\mathbf{r}, t)$ which associates a velocity $\mathbf{v}$ of the moving material with every point $\mathbf{r}$ in space and time $t$. If internal friction is neglected, the Bernoulli equation is satisfied: |
|---|---|

$$\frac{1}{2}\rho v^2 + P = const.$$

## 3.2.2 Random motion of atoms

The thermal motion of atoms is governed by the laws of chance. In the case of gases, we can, indeed, assume that the atoms fly in straight lines for short distances in accordance with the principle of inertia (Newton's first law; Section 1.2.2). But in each collision they change their direction irregularly. Therefore, only statistical statements can be made about their thermal motion.

The *mean free path (length) l* is the decisive parameter for all equilibration processes in gases. It depends, on one hand, on the particle density $n$ of the gas and, on the other, on the size of the atoms. Previously, we have assumed that in an ideal gas the atoms behave as point masses. Now we shall take their size into account and treat them like small billiard balls with a radius $R$ on the order of $R \sim 10^{-10}$ m. Two spheres collide if the distance $b$ between their lines of flight is less than twice the radius of the spheres (Figure 3.10). Thus, the spheres are said to have a (circular) *collision cross section* $\sigma = 4\pi R^2$. For atoms this is on the order of $\sigma \approx 10^{-19}$ m$^2$. Since each sphere has a volume $V_0 = n^{-1}$ within which it can move freely, on the average the spheres can move a linear distance

$$l = \frac{1}{\sigma n}$$

before they again are deflected in a collision with another sphere. Assuming that the changes in direction during collisions are purely random, so that all collision angles occur with equal probability, the atoms follow an irregular zigzag motion (Figure 3.11).

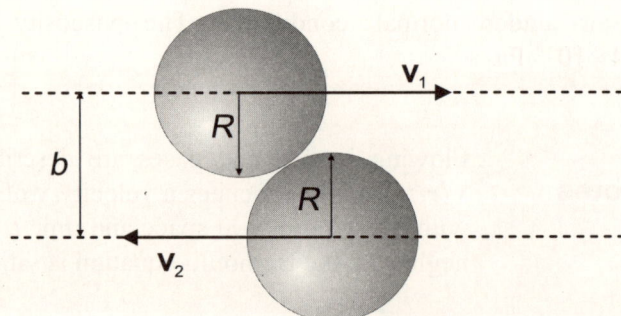

**FIGURE 3.10** For the definition of the collision cross section σ of spherical particles.

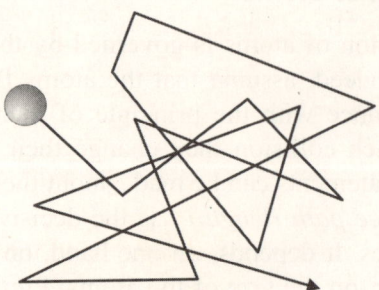

**FIGURE 3.11** Random walk of an atom.

The time evolution of all equilibration processes in gases is essentially determined by the product of the mean thermal velocity $v_{th}$ and the mean free path $l$. The higher the thermal velocity and the longer the mean free path are, the faster the equilibrium will be attained. As for the process of diffusion (Section 3.2.4), we define the *diffusion coefficient D* for gases as

$$D = \frac{1}{3}lv_{th}.$$

It has the dimensions $[m/s^2]$. Its order of magnitude can easily be estimated. Under normal conditions $n = P/kT \approx 2.5 \times 10^{25}$ m$^{-3}$, $l = (\sigma n)^{-1} \approx 0.2 \times 10^{-6}$ m, and $v_{th} = \sqrt{3kT/m} \approx 500$ m/s. We then obtain $D \approx 0.2 \times 10^{-4}$ m$^2$/s, in agreement with the experimental value. (In this estimate the greatest uncertainty is in the value of the cross section $\sigma$. This is because atoms and molecules are not billiard balls. Here we have taken $\sigma = 3 \times 10^{-19}$ m$^2$.)

The viscosity $\eta$ of gases can be calculated from the diffusion coefficient. The frictional force acting on a falling sphere is not just determined by the kinematic parameters of the motion of the atoms, but also depends on the mass $m$ of the atoms and the momentum transferred in collisions with the sphere per unit time. The theory of internal friction thus yields the following for the viscosity of gases:

$$\eta = \rho D. \qquad \text{(coefficient of viscosity for gases)}$$

Here $\rho = nm$ is the density of the gas. Since the mean free path and, thereby, $D$ are inversely proportional to the particle density $n$, the coefficient of viscosity $\eta$ is independent of the particle density of the gas. Of course, this result cannot be absolutely correct. Otherwise, a sphere in an evacuated glass tube at pressures of a few Pa would experience the same frictional force as at normal pressure. This theory of internal friction can be applied to the case of spheres only if $l \ll R$, i.e., if $l$ is small compared to the radius $R$ of the sphere.

---

**Problem 3.5**

Estimate the mean free path $l$ of atoms in a high vacuum with a residual gas pressure of $P \approx 10^{-4}$ Pa.

---

**Notes**

The product of the mean free path $l$ and the thermal velocity $v_{th}$ of the atoms is decisive for all equilibration processes in gases.

### 3.2.3  Temperature equilibration

In order to investigate the laws governing the evolution of equilibration processes in time and space, let us consider temperature equilibration between two objects with different temperatures (Section 2.1.3). The two objects might, for now, be two (large) heat reservoirs at different temperatures $T_>$ and $T_<$ linked by a heat-conducting rod (Figure 3.12). Heat then flows with the temperature drop from $T_>$ to $T_<$.

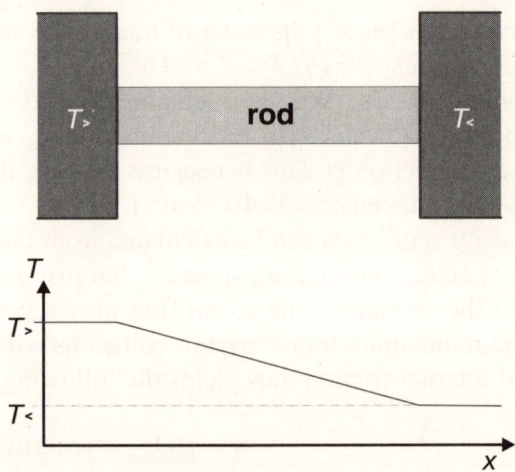

**FIGURE 3.12** Illustrating the conduction of heat.

First, we shall consider the time independent case. Let the heat reservoirs be so large that their temperatures are essentially unchanged by the heat flow. A temperature drop develops along the rod that is constant in time. Hence, in a homogeneous rod with a uniform cross section everywhere, the temperature falls with constant gradient. Now $dQ$ is the amount of heat which flows through a rod with a cross-sectional area $A$ during a time $dt$. The heat flux $i = (dQ/dt)/A$ is then proportional to the temperature difference $\Delta T = T_> - T_<$ and is inversely proportional to the length $\Delta x$ of the rod:

$$i = \lambda \frac{\Delta T}{\Delta x}.$$

The proportionality constant $\lambda$ is a constant of the material and is known as the *(specific) thermal conductivity*.

The evolution of the temperature is of more interest when temperature equilibration takes place between objects with finite heat capacity. In this case the heat flux and temperature evolution are not stationary. The temperature of the rod is then a function $T(x,t)$ of position $x$ and time $t$, and the heat flux $i(x,t)$ at position $x$ is determined by the temperature gradient $\partial T/\partial x$ at that point at time $t$, which is therefore also a function of $x$ and $t$:

$$i(x,t) = -\lambda \frac{\partial T(x,t)}{\partial x}.$$

A second relationship between the two functions $T(x,t)$ and $i(x,t)$ arises if we study the influence of the heat flux on the temperature of the rod. For this, we consider a segment of the rod of length $dx$ at position $x$. Its temperature varies if the incoming and outgoing heat fluxes are different, that is, when $i(x+dx,t) - i(x,t) = (\partial i/\partial x)dx \neq 0$. Then, in time $dt$ an amount of heat $dQ = (\partial i/\partial x)dx\,dt\,A$ flows into the segment $dx$ of the rod. According to Section 2.3.1 this heat input leads to a rise in the temperature of the rod segment by $dT = dQ/cm$. Here $c$ is the specific heat and $m = \rho A\,dx$ is the mass of a rod segment with density $\rho$. A gradient in the heat flux leads, therefore, to a time variation in the temperature:

$$\frac{\partial T(x,t)}{\partial t} = \frac{1}{c\rho} \cdot \frac{\partial i(x,t)}{\partial x}.$$

Since, on the other hand, the temperature gradient determines the heat flux, we obtain the following differential equation for $T(x,t)$:

$$\frac{\partial T(x,t)}{\partial t} = \frac{\lambda}{c\rho} \cdot \frac{\partial^2 T(x,t)}{\partial x^2}. \qquad \text{(differential equation for temperature equilibration)}$$

Like all equilibration processes, temperature equilibration is also caused by the random motion of atoms and is, consequently, irreversible. This means that the equilibration process is described by a differential equation which contains the first derivative with respect to time, as opposed to the wave equation. As with the wave equation, here the time variation of the function is related to the (spatial) curvature of the temperature function.

Temperature equilibration, like temperature itself, is closely related to the thermal motion of atoms. In gases, therefore, temperature equilibration is, like internal friction, dependent on the mean free path and the thermal velocity of the atoms. Apart from a numerical factor $f$ of order 1, the *thermal diffusivity* $\lambda/(c\rho)$ in the equation for temperature equilibration is equal to the diffusion coefficient $D$, i.e.,

$$\lambda = fc\rho D.$$

---

### Problem 3.6

Show that during temperature equilibration the entropy of the closed system under consideration increases, so that the process is irreversible.

---

**Notes**

During temperature equilibration the temperature $T(x,t)$ is a function of position and time. It satisfies the partial differential equation

$$\frac{\partial T(x,t)}{\partial t} = \frac{\lambda}{c\rho} \cdot \frac{\partial^2 T(x,t)}{\partial x^2}.$$

The first derivative with respect to the time contained in this equation is a characteristic of irreversible processes.

## 3.2.4 Diffusion in gases

If two different gases or liquids are brought together, for example, by removing a partition, then the two substances will mix thoroughly because of the thermal motion of their atoms and molecules. Mixing of this sort even takes place when the two substances have the same temperature and pressure. It is referred to as *diffusion*.

---

### Experiment 3.2  Diffusion of gases

---

As an illustration let us study the diffusion of gases through a porous wall (Figure 3.13). A vessel made of porous ceramic and filled with air is sealed with a liquid manometer (a U-tube filled with water). If a glass container is turned upside down over the ceramic vessel and filled with helium (helium is lighter than air, so it is trapped in the upside-down glass container), then the pressure in the ceramic vessel initially increases for a few seconds and then falls off to the ambient air pressure.

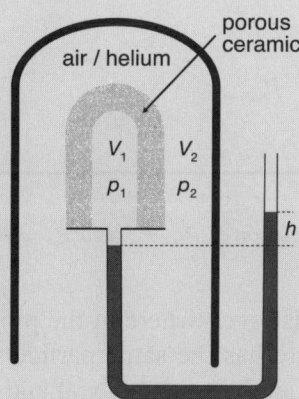

**FIGURE 3.13** Experimental setup for demonstrating the diffusion of gases.

In order to explain this process, let us consider diffusion in a vessel with two chambers, which are separated from one another by a porous wall. One of the chambers, with volume $V_1$ is filled with helium and the other, with volume $V_2$ is filled to the same pressure with air. Both gases diffuse through the porous wall. Thus, with time, an equilibrium state develops in which both chambers are filled with helium and air in the ratio $V_1 : V_2$. Since the mass number $A = 4$ of helium is substantially lower than the molecular weight $M \approx 28$ of air ($N_2$), the thermal velocity of the He atoms is about a factor of 2.7 higher than that of the nitrogen molecules. The partial pressure $P_{He}$ of helium, therefore, rises faster in the chamber that was originally filled with air than the partial pressure $P_{air}$ of the air decreases in that chamber (Figure 3.14). At early times the sum of the two partial pressures rises in that chamber, while at early times it decreases in the other chamber.

The equilibration of a concentration difference by diffusion is, like temperature equilibration, an irreversible process. Under corresponding experimental conditions, therefore, it satisfies the same partial differential equation. For simplicity, let us consider a linear configuration, as we did for temperature equilibration; that is, a pipe in which gases $A$ and $B$ are mixed. The same pressure and the

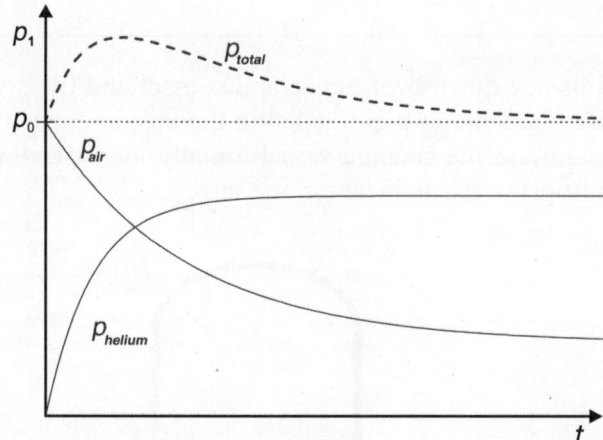

**FIGURE 3.14** Time variation of $P_{\text{He}}$, $P_{\text{air}}$, and $P_{\text{tot}} = P_{\text{He}} + P_{\text{air}}$ in a vessel initially filled with air.

same temperature exist everywhere in the pipe. Then, because $P = nkT$ (Section 2.1.3), the gas mixture has the same particle density $n = n_A + n_B$ everywhere in the pipe. But the partial densities $n_A(x,t)$ and $n_B(x,t)$ will vary with time if there is an initial density gradient along the pipe (which lies in the $x$-direction). Under these conditions the partial pressures obey a differential equation that has the same structure as the differential equation for temperature equilibration:

$$\frac{\partial n_A(x,t)}{\partial t} = D \frac{\partial^2 n_A(x,t)}{\partial x^2}. \qquad \text{(diffusion equation; Fick's law)}$$

This is a linear differential equation and can be solved by assuming a solution of the form $n_A(x,t) = \exp(ikx - t/\tau)$. This is a solution if $k^2\tau = D^{-1}$. However, all superpositions of these functions are solutions of the differential equation, as is, for example, the undulatory distribution function $n_A(x,t) = n_0 + n_1 \cos(kx)\exp(-t/\tau)$ with $n_1 < n_0$. The amplitude of an undulatory distribution function of this sort dies out with time constant $\tau$. In gases $D = v_{\text{th}}l/3$, so that

$$\tau = \frac{\lambda^2}{4\pi^2 v_{\text{th}} l}.$$

The decay time $\tau$ also increases with the square of the wavelength $\lambda = 2\pi/k$ of the undulatory distribution.

The solutions of the diffusion equation describing the dispersion of impurity atoms injected into a carrier gas at a time $t = 0$ at a position $x = 0$ are also of physical interest. In this case the distribution function $n_G(x,t)$ is a Gaussian bell curve (Figure 3.15),

$$n_G(x,t) = n_0(t)\exp\left(-\frac{x^2}{2L^2(t)}\right). \qquad \text{(Gaussian distribution)}$$

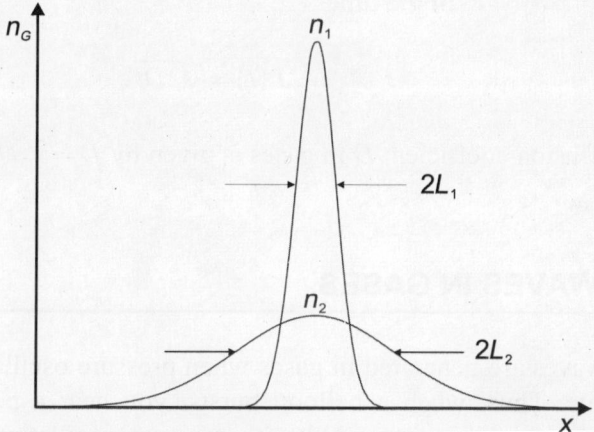

**FIGURE 3.15** Gaussian distribution functions owing to diffusion.

The two parameters $n_0(t)$ and $L(t)$ in this distribution function vary as the square root of time. When the amplitude $n_0$ of the distribution decreases, the width $L$ increases:

$$n_0(t) = \frac{N}{2\sqrt{\pi Dt}}$$
$$L(t) = \sqrt{2Dt}.$$

Here $N$ is the number of injected atoms. The increase in the width $L$ with the square root of the time is a reflection of the random motion (Figure 3.11) of the atoms. The wider the distribution is, the lower the rate $dL/dt = D/L$ at which the atoms advance outward will be.

**Problem 3.7**

Show that the Gaussian distribution is a solution of the diffusion equation.

| | |
|---|---|
| **Notes** | The particle concentration of an impurity gas injected at $x = 0$ is dispersed in a carrier gas by means of diffusion. The width $L$ of the distribution increases as the square root of the time: |

$$L(t) = \sqrt{2Dt}.$$

The diffusion coefficient $D$ in gases is given by $D = v_{th}l/3$ (Section 3.2.2).

## 3.3 SOUND WAVES IN GASES

Sound waves are generated in gases when pressure oscillations are produced at a single site. Thus, when a balloon bursts, you hear a pop. The propagation of sound waves is also associated with pressure oscillations $P(x,t)$ in space and time. Since $P = nkT$ the particle density $n$ varies along with the pressure. The pressure gradients that develop in the gas induce oscillating flows of particles in the gas which then produce density oscillations. The gas atoms, however, do not move only because of these particle flows. At the same time, motion owing to thermal diffusion takes place. Wave propagation and diffusion compete here. We shall consider both processes in the following.

### 3.3.1 The propagation of sound

We first consider sound waves with long wavelengths, i.e., those for which the wavelength $\lambda$ is much greater than the mean free path $l$ of the atoms. Under these conditions diffusion processes can be neglected. Over distances that are not too large, during sound propagation there is very little conversion of the macroscopic stream flows of the gas into thermal motion. Thus, the propagation of sound waves is nearly an adiabatic process. In the adiabatic limit the dynamic equations of hydrodynamics and the equation of state for an ideal gas yield a wave equation for the oscillations in the pressure $P(x,t)$:

$$\frac{\partial^2 P}{\partial t^2} - c_S^2 \left( \frac{\partial^2 P}{\partial x^2} + \frac{\partial^2 P}{\partial y^2} + \frac{\partial^2 P}{\partial z^2} \right) = 0. \qquad \text{(wave equation)}$$

This equation has the same structure as the wave equation for a linear chain (Section 3.1.1). As opposed to the linear chain, three spatial dimensions have to be taken into account in the propagation of sound. Thus, the time variation in the pressure is determined by the curvature of the function $P(x,t)$ in all three spatial directions. The phase velocity $c_S$ for sound propagation in an ideal gas is given by

$$c_S = \sqrt{\frac{5P}{3\rho}}.$$

Since $P = nkT$ and $p = nm$, the velocity of sound is on the order of the thermal velocity of the atoms. In air under normal conditions, $c_S = 330 \text{ m/s}$.

Plane waves are simple periodic solutions of the wave equation. For the pressure oscillations $P_1(\mathbf{r},t) = P(\mathbf{r},t) - P_0$ superimposed on the normal ambient background pressure $P_0$, we assume that

$$P_1(\mathbf{r},t) = P_1 \exp(i(\mathbf{k} \cdot \mathbf{r} - \omega t)).$$

In addition, usually $P_1 \ll P_0$. These waves have plane wave fronts. They propagate transversely to the wave fronts in the direction of the *wave vector* $\mathbf{k}$. As in the case of a linear chain, the assumed solution satisfies the wave equation if $\omega/k = c_S$. Here $k = |\mathbf{k}|$.

Most sound waves emerge from a central sound source. The wave fronts are then spherical, rather than plane. For calculating these spherical waves, it is advantageous to write the pressure as a function of polar coordinates, $P = P(r,\theta,\varphi)$ (Section 1.1.1), and convert the wave equation to polar coordinates:

$$\frac{\partial^2 P}{\partial t^2} - c_S^2 \left( \frac{1}{r} \frac{\partial^2 (rP)}{\partial r^2} + \frac{1}{r^2 \sin\theta} \frac{\partial}{\partial \theta} \left( \sin\theta \frac{\partial P}{\partial \theta} \right) + \frac{1}{r^2 \sin^2\theta} \frac{\partial^2 P}{\partial \varphi^2} \right) = 0.$$

Spherically symmetric solutions, in which the solutions $P(r,t)$ depend only on $r$ and not on the angular coordinates $\theta$ and $\varphi$, are especially simple. They satisfy the wave equation

$$\frac{\partial^2 P}{\partial t^2} - c_s^2 \frac{1}{r} \frac{\partial^2 (rP)}{\partial r^2} = 0.$$

Solutions of this equation are given by spherical waves emerging from a source at the center:

$$P(r,t) = P_0 + \frac{\Delta P}{kr} \exp(i(kr - \omega t)).$$

The amplitude $P_1 = \Delta P / kr$ of this sound wave falls off with distance $r$ from the source. The intensity $I(r)$ of the sound waves (power per unit area; Section 3.3.3) is proportional to the square of the amplitude. Thus, $I(r) \propto r^{-2}$.

The drop in amplitude of these spherical waves with distance $r$ from the source is consistent with the energy theorem. Thus, the intensity integrated over a spherical surface, $P = I(r) \cdot 4\pi r^2$, is the energy transported per unit time through that surface. It is independent of $r$ and equals the power radiated by the sound source. The power $P$ falls off with increasing distance only if damping processes are included.

**Problem 3.8**

Show that a spherically symmetric spherical wave $P(r,t)$ whose amplitude is proportional to $1/r$ is a solution of the wave equation.

The random thermal motions of the atoms, which have been neglected thus far, cause damping of a wave. The decay time $\tau = 3(\lambda/2\pi)^2 / (v_{th} l)$ of an oscillatory pressure distribution was calculated in Section 3.2.4. In corresponding fashion, the thermal motion of the atoms also causes attenuation of sound waves. When damping is taken into account, the power $P(r) = I_0 \exp(-r/R)$ falls off exponentially with distance. The range $R = \tau c_S$ is thus on the order of

$$R \sim \frac{(\lambda/2\pi)^2}{l}.$$

Thus, the range of sound waves increases as the square of the wavelength and is inversely proportional to the mean free path $l$ of the gas atoms.

### Experiment 3.3     A bell in a vacuum

Under normal conditions the mean free path in air is $l \approx 0.2\,\mu m$ (Section 3.2.2). It is many orders of magnitude smaller than the wavelength of ordinary sound waves. Thus, even a high pitched bell with a relatively short wavelength can be heard over long distances. For a wave with $\lambda/2\pi = 1\,cm$, $R \sim 500\,m$. At a pressure $P \sim 10\,Pa$, on the other hand, $R \sim 5\,cm$. A bell placed in an evacuated bell jar can, therefore, no longer be heard outside the jar (Figure 3.16). But the ringing of the bell can gradually be heard more loudly as air is slowly let into the evacuated jar.

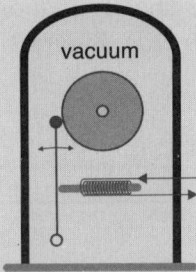

vacuum

**FIGURE 3.16** A bell in an evacuated bell jar.

---

**Notes**

The plane waves

$$P_1(\mathbf{r},t) = P_1 \exp(i(\mathbf{k}\cdot\mathbf{r} - \omega t))$$

are solutions of the wave equation. These waves span all of space and extend from $t \to -\infty$ to $t \to \infty$. They propagate in the direction of the wave vector $\mathbf{k}$ at the phase velocity $c_S = \omega/k$. The plane wave fronts are perpendicular to $\mathbf{k}$.

### 3.3.2  Chladni sound figures

Two- and three-dimensional resonators exist by analogy with the one-dimensional resonators in linear media (discussed in Section 3.1.2) where standing waves can develop along the medium. The characteristic frequencies of these resonators depend on the boundary conditions and on the shape of the resonators.

It is not as easy to calculate the characteristic frequencies as it was for a string fastened at both ends. An impression of the characteristic oscillations of a round disk is provided by *Chladni sound figures*. If sand is spread over a disk that is excited resonantly, the node lines of the characteristic oscillations become visible.

---

### Experiment 3.4    Resonances of a metal plate

A metal disk is excited to its characteristic oscillations using a loudspeaker (Figure 3.17). Sand spread over the plate will be pushed entirely out of the regions where the plate oscillates strongly and will accumulate at the node lines of the characteristic oscillations. Some examples of the resulting Chladni figures are shown in Figure 3.18.

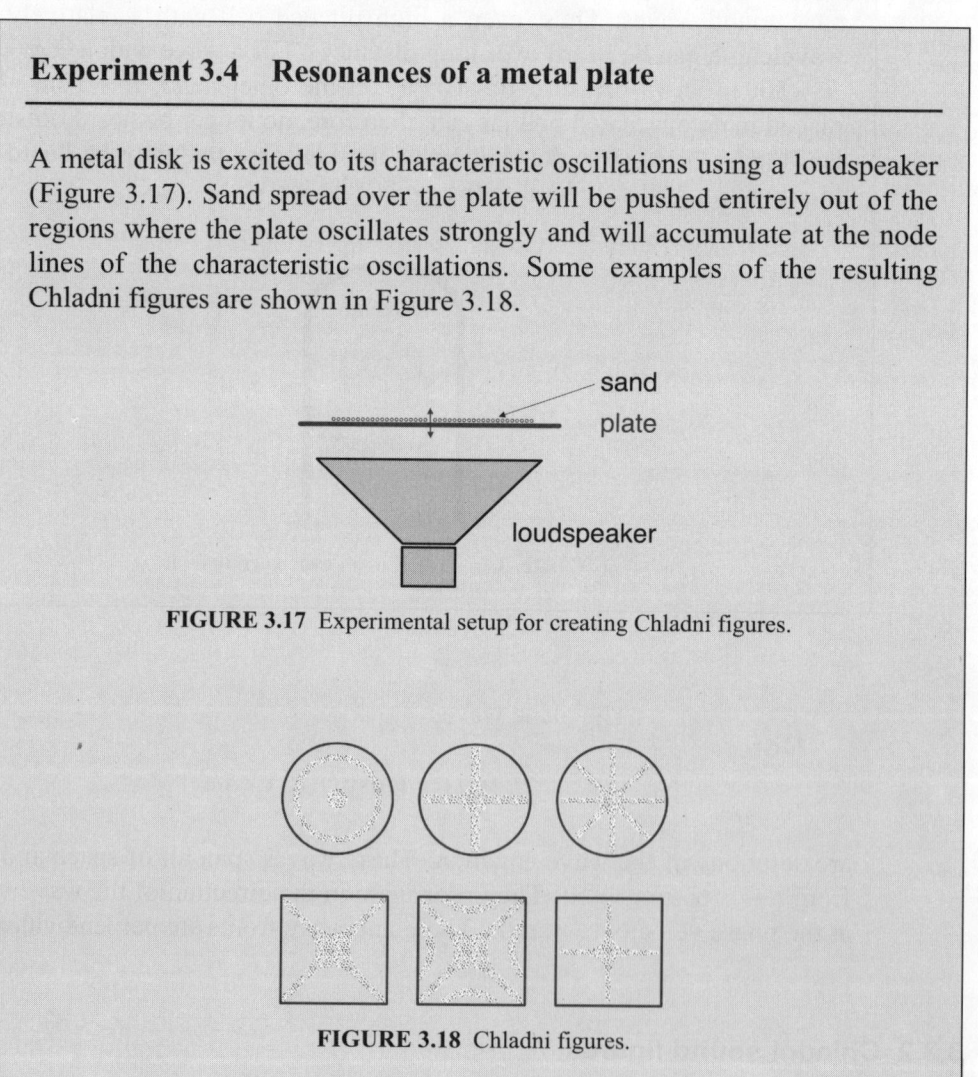

**FIGURE 3.17** Experimental setup for creating Chladni figures.

**FIGURE 3.18** Chladni figures.

---

Hollow cavities used as three-dimensional resonators can also be excited to their characteristic oscillations. The spatial eigenmodes of these resonators have nodal surfaces. Some of the characteristic oscillations of a spherical resonator with radius $R$ are spherical waves with the following wave functions:

$$P_n(r,t) = \frac{P_n}{kr}\cos(k_n r)\exp(-i\omega_n t).$$

These satisfy the boundary condition $P_n(R,t) = 0$ if $k_n R$ is an integral multiple of $\pi$. However, there are other characteristic oscillations that depend on the polar angles $\theta$ and $\varphi$, as well as the radial coordinate $r$.

---

**Problem 3.9**

A hollow sphere of radius $R = 0.1\,\mathrm{m}$ is filled with air and excited to resonant oscillations. Calculate the frequencies of the spherically symmetric eigenmodes.

---

| **Notes** | Two- and three-dimensional resonators have, as do one-dimensional resonators, have a countable set of eigenmodes. Many characteristic oscillations can correspond to a |

given characteristic frequency. They differ from one another in the shape of their nodal surfaces.

## 3.3.3 Energy and information transfer

For energy or news to be transferred from one place to another, oil must be carried in huge tankers over the world's oceans or letters sent (at least formerly). Energy transport and information transfer are associated with the transport of matter in these cases. However, energy and information can also be transferred by waves, thereby avoiding the transport of matter. From time immemorial sound waves have served animals and mankind for communication and the energy transport associated with pressure waves in explosions can be an impressive experience. Likewise, energy and information can be transferred by water, light, and radio waves. In sound and water waves, matter is made to move back and forth locally, while the propagation of light and radio waves is not coupled to matter so they can propagate in a vacuum. However, light and radio waves obey a wave equation of the same type as that for sound waves and are likewise associated with energy transport. Using the propagation of sound in gases as an example, we now illustrate energy and information transfer by waves.

Wave propagation is a process that is continuous in space and time. Thus, rather than the energy of individual objects, here we seek to determine the *energy density u* in an extended medium. Since the application of Newtonian mechanics to a continuous, extended medium is not easy, we limit the discussion of energy transport to a few general considerations. Thus, let us consider a standing plane sound wave with amplitude $2P_1$ in a resonator with a gas pressure $P$,

$$2P_1(x,t) = 2P_1 \sin(kx)\exp(-i\omega t).$$

It can be regarded as the superposition of two travelling waves with amplitudes $P_1$ moving in opposite directions. In a standing wave, states with maximum potential energy and maximum potential energy of the medium alternate in the course of its local back and forth motion. The potential energy is primarily concentrated in the antinodes and the kinetic energy, in the nodes. When the potential energy is at a maximum, $u_{kin} = 0$. At these times, the gas density in the wave peaks has risen by a fraction $P_1/P$ and in the wave troughs it has decreased by the same fraction. Then the increase in the potential energy of the gas relative to the equilibrium state of the gas is given by (to within a numerical factor of order unity) by the product of the pressure change $P_1$ and the fraction $P_1/P$ of the displaced gas. The maximum potential energy density in the standing sound wave is thus of order $u_{pot} \sim P_1^2/P$. Thus, the energy density $u$ of a standing wave is proportional to the square of the amplitude $P_1$, as is the energy of a harmonic oscillator (Section 1.6.1).

There is no energy transport in a standing wave. On the other hand, in a travelling wave, energy is transported at the phase velocity in the direction of the wave vector. The energy density of a travelling wave can be found by decomposing a standing wave into two travelling waves moving in opposite directions. The two travelling waves each have half the energy density of the standing wave. The energy density of a travelling wave of amplitude $P_1$ is, therefore, given by

$$u \sim \frac{1}{2} \cdot \frac{P_1^2}{P}.$$

It moves at the phase velocity $c_S$ of the wave. The energy transport that takes place during wave propagation can thus be characterized in terms of the *intensity* (power per unit area or energy flux) $I = c_S u$ of the wave.

Because of the energy transport associated with wave propagation, signals and information can be transmitted over large distances by waves. The transmitter and receiver are tuned to the same frequency. In order to transfer information, the amplitude, frequency, or phase of the oscillations from the transmitter are modulated. In this way, wave groups are formed as during the superposition of waves with adjacent frequencies (Section 3.1.3). The modulated wave emitted by the transmitter excites oscillations in the receiver. After demodulation of these oscillations, the information from the transmitter is available to the receiver. Sound waves have been used by animals since the dawn of history for this form of information transfer. And, with the discovery of electromagnetic waves (Section 3.6.1), worldwide telecommunications became possible in accordance with the same principle.

---

### Problem 3.10

A violin produces sound waves with a power $P = 1\,\mathrm{mW}$. Calculate the amplitude $P_1$ of the sound waves at a distance $r = 10\,\mathrm{m}$.

---

| **Notes** | The energy density $u$ of a wave is proportional to the square $|A|^2$ of its amplitude $A$: |
|---|---|

$$u \propto |A|^2.$$

## 3.3.4 Doppler effect

Let us again consider a transmitter and a receiver which are communicating via sound waves. Only when the sender and receiver are at rest relative to the gas in which the sound waves propagate does the receiver hear the pitch of the transmitter. If, on the contrary, the receiver or transmitter (source) are moving with velocities $\mathbf{v}_E$ or $\mathbf{v}_S$, respectively, relative to the gas in which the sound waves are propagating, then the transmitter and receiver frequencies will generally be different from one another.

If the transmitter is moving, the waves emerging from the transmitter no longer form concentric spherical wave fronts like those emitted from a transmitter at rest. Rather, the centers of the spherical wave fronts are shifted relative to one another in accordance with the velocity of the transmitter (Figure 3.19). Thus, the distance between neighboring wave fronts in the direction of the transmitter's motion is shortened by a factor of $(c_S - v_S)/c_S$ and is lengthened in the opposite direction. Hence, the receiver hears a higher frequency if the source is moving toward it and a lower frequency if the source is moving away from it; that is,

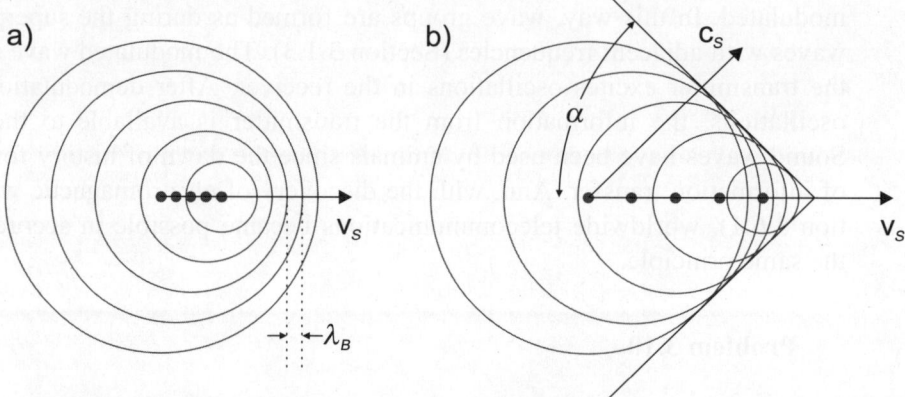

**FIGURE 3.19** Wave fronts of sound from a moving source: (a) $v_S < c_S$, (b) $v_S > c_S$.

$$v_E = \frac{c_S}{c_S - v_S} v_S. \qquad \text{(Doppler shift for a moving source)}$$

If $v_S > c_S$ the result is a *sonic boom*. The wave fronts form a *Mach cone*, at which successive wave troughs and peaks overlap constructively (Figure 3.19). The aperture angle $\alpha$ of the Mach cone is given by

$$\sin \alpha = \frac{c_S}{v_S}.$$

A somewhat different relationship between the frequencies of the source and receiver is obtained if the receiver is moving at a velocity $v_E$ toward a stationary source. In this case, the wave fronts do indeed form concentric rings around the stationary transmitter, but the wave fronts reach the receiver over shortened time intervals. Thus, the frequency perceived by the receiver is higher by a factor of $(c_S + v_E)/c_S$; that is,

$$v_E = \frac{c_S + v_E}{c_S} v_S. \qquad \text{(Doppler shift for a moving receiver)}$$

The frequency perceived by the receiver is lowered if the receiver moves away from the transmitter. In this case, $v_S < 0$. If $v_S = -c_S$, then $v_E = 0$.

## Problem 3.11

A car travelling at a velocity $v = 100$ km/h passes you. By what percentage does the pitch of the engine noise change? Compare your result with experimental values.

**Notes** For velocities of a transmitter (source) and receiver that are not too large ($v_{S,E} \ll c_S$), the relative velocity of the transmitter and receiver is essentially $v_{rel} = |\mathbf{v}_S - \mathbf{v}_E|$, but this is not true for the movement of the transmitter or receiver relative to the medium in which the sound waves are carried. In that case, the transmitter and receiver frequencies differ by approximately

$$\Delta\nu \approx \frac{v_{rel}}{c_S} \nu_S.$$

## 3.4 ELECTRICITY AND MAGNETISM

Daily experience teaches us that in order to have an effect on an object, we have to touch it or we need an intermediary medium in order to affect it indirectly. For example, we need air in order to blow out a candle or to understand each other when we speak. However, there are long-range effects, such as gravitation, which operate between stars or through empty interstellar space. On earth, people have known for thousands of years about the apparently mysterious long-range effects of magnetism, the force acting between magnetic materials, such as iron, cobalt, and nickel, and the effects of electrical forces, which act between rubbed insulators, such as cat fur, amber, and styrofoam. Everyone can do some simple experiments at home in order to become acquainted with these forces: picking up spilled nails with a magnet, for example, or moving shreds of paper with rubbed styrofoam.

More extensive experiments show that these forces also operate in a vacuum. For example, in the experiment on free fall (Section 1.1.2), we moved objects in a vacuum. This modest understanding led to the development of the physics of electricity and magnetism in the 19th century, which, like the invention of steam engines and internal combustion engines, unleashed a technological revolution. We shall discuss the physical foundations of electricity and magnetism in this lecture.

### 3.4.1 Charge and current

Unlike gravitation, which is always an attractive force, attraction and repulsion are observed in experiments when insulators are rubbed. The strength of the electrical force is also independent of the masses of objects, but depends on a property known as *electrical charge*. The electrical charge of an insulator can obviously be changed by rubbing. In order to explain how both attraction and repulsion are possible, we assume the existence of positive and negative charges. Experiments show that like charges repel and opposite charges attract.

In general, we make a distinction between *insulators* and *electrical conductors*. Electrical charges stay in one place on insulators, while charges can move freely in electrical conductors. If two electrical conductors are connected by a conducting wire, charge transfer will generally take place. An *electric current* of charge carriers flows from one object to the other until an equilibrium state is again achieved. This current can be compared to the stream of water that flows when a sluice gate is opened. Before the gate is opened, the heights of the water inside and outside the gate are different. After it is opened, the flow of water comes to rest only when the heights of the water on both sides of the gate are the same. Correspondingly, when the two conductors are separate their charge carriers on the two are initially at different *potentials*. After they are joined by a conductor, positive charge carriers flow from the metal with the higher potential to the metal with the lower potential and negative charge carriers flow in the opposite direction. The charges come to rest only when the potential difference has been balanced. A potential difference between two conductors is also referred to as an *electrical voltage*. When we speak of the voltage of a conductor, it is understood that the voltage is relative to the potential of the earth, or the *ground potential*.

---

### Experiment 3.5    Charge transport

In order to clarify the effect of voltages on charge carriers, let us consider an isolated conducting sphere suspended at the midpoint between two metal plates to which a voltage of 10 kV is applied (Figure 3.20). The two plates form a *capacitor*. When a voltage is applied, the capacitor is charged, that is, the surfaces of the two plates which face one another are loaded with positive or negative charges, respectively. If the sphere touches one of the plates, e.g., the positive pole, then the sphere will also be positively charged. It is then at the same potential as the plate. Since like charges repel, the sphere will be accelerated back toward the negative pole and attracted to it. There it becomes negatively charged and is again attracted to the positive pole. This process continues onward. The charge transfer that takes place with the pendulum motion of the sphere can (on the average in time) be measured as an electrical current. If the plates of the capacitor are disconnected from the voltage source after they are charged, the electrical current leads to equilibration of the potential between the plates and the pendulum finally comes to rest.

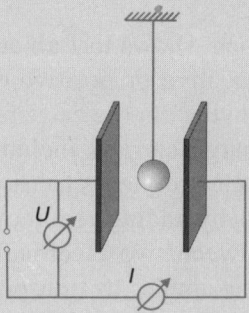

**FIGURE 3.20** A conducting sphere between the plates of a capacitor.

In this experiment the electrical force acting between charges accelerates the sphere. If the plates are oriented horizontally, the sphere can be set to oscillating by a suitable choice of the experimental parameters. Then the electrical force exactly compensates the force of gravity acting on the sphere (Figure 3.21). In 1909 Robert Millikan started a series of experiments in which he used a microscope to observe the motion of small droplets of oil in a capacitor of this sort. He showed that the charges on the droplets cannot take arbitrary values, but are always integral multiples of a smallest charge $e$. This smallest unit of charge is the *elementary charge $e$* and has the value

$$e = 1.602 \times 10^{-19} \text{ C}.$$

The SI unit of charge, $1\,\text{C} = 1\,\text{A} \cdot \text{s}$ (Coulomb), is thus equal to $0.6 \times 10^{19}$ elementary charges. The unit of electrical current is also established along with the unit of charge. The electrical current is the amount of charge flowing through a conductor per unit time. The unit of electrical current is therefore $1\,\text{C/s} = 1\,\text{A}$ (Ampere).

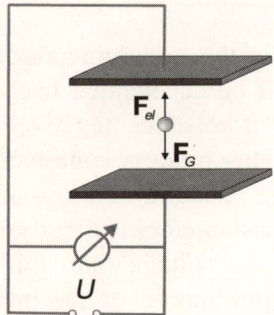

**FIGURE 3.21** Illustrating the principle of Millikan's experiment.

Millikan's experiment shows that all atomic particles are either neutral or can carry one (or more) negative or positive elementary charge. In particular, electrons, which are the moving charge carriers in metallic conductors, have a charge equal to $-e$. Other charge carriers include ions, which are responsible for the transport of charge in electrolytes (electrically conducting liquids). Ions are created from neutral atoms by adding or removing electrons (Section 5.3).

The voltage $U$ between two electrical conductors is also equivalent to the work $W$ required, for example, to transport one elementary charge $e$ from one conductor to the other. The forces between electrical charges are conservative (Section 3.4.2). Thus, the integral of the work with respect to distance is independent of the path, but is proportional to the magnitude of the charge that is transported. Accordingly, the voltage between two conductors is defined as $U = W/e$. The SI unit of voltage is $1\,V$ (Volt), with $1\,V = 1\,J/(A \cdot s)$.

In electrical engineering the units of energy [J] and power [W] are often reduced to the electrical units of current [A] and voltage [V]. Based on the definitions of the units [V] and [A], we have $1\,J = 1\,V \cdot A \cdot s$ and $1\,W = 1\,V \cdot A$.

In atomic and solid state physics, on the other hand, the elementary charge is the decisive unit of charge. Thus, in these branches of physics there is some advantage in using the unit [eV] (read as e-V or *electron volts*) as the unit of energy, with

$$1\,eV = 1.602 \times 10^{-19}\,J.$$

$1\,eV$ is the kinetic energy of a charge carrier with charge $e$ after it has fallen through a potential difference $U = 1\,V$.

---

### Problem 3.12

A 100 W incandescent bulb is connected to the commercial electrical grid. What is the electrical current flowing through the bulb?

---

| **Notes** | Positive and negative electrical charges exist. The SI unit of charge is [A·s]. The smallest unit of charge occurring in nature is the elementary charge |
|---|---|

$$e = 1.6 \times 10^{-19} \text{ A·s}.$$

All other charges are integral multiples of the elementary charge.

Electrical currents arise when charges move. The SI unit of electrical current is the Ampere [A].

## 3.4.2 Coulomb force and Lorentz force

At first, the forces which act between magnetic materials and those which make it possible for rubbed glass rods to attract scraps of paper seem to have little to do with one another. The first experiments that showed that electricity and magnetism are connected were performed in 1820 by Hans Christian Oersted. He showed that magnetized needles would become oriented in the vicinity of electric currents. The magnetized needles oriented themselves perpendicular to the electrical conductor in such a way that the directions of the needles combine to form concentric circles around the conductor (Figure 3.22). If the direction of the current is reversed, the magnetized needles also reverse direction. The direction of the current and the orientations of the north poles of the needles correspond to a righthanded screw.

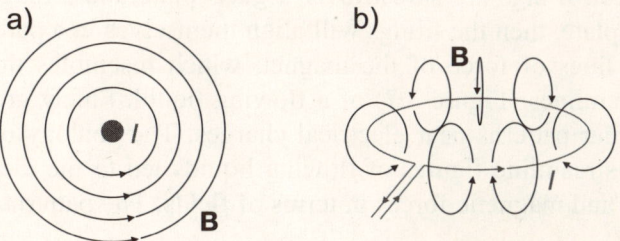

**FIGURE 3.22** Magnetic field patterns associated with electrical currents.

The magnetism associated with electric currents also has the characteristic that two parallel current-carrying conductors will attract or repel each other, depending on whether the two currents are flowing in the same or opposite directions (Figure 3.23). The force between two electrical currents is used today to define and realize the SI unit [A] of electrical current.

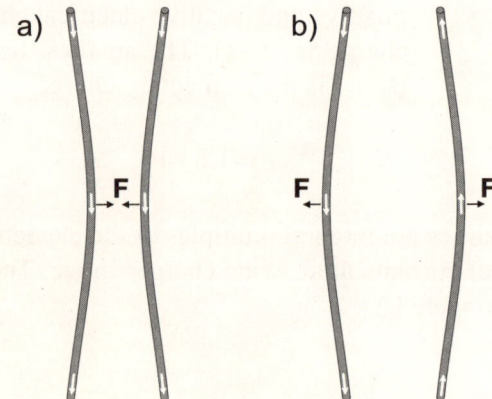

**FIGURE 3.23**  Attraction and repulsion of current-carrying conductors.

Around 1830, when Michael Faraday began to study the forces operating between magnets, electrical currents, and electrical charges, he used *lines of force* to visualize the operation of these long-range forces in space. The idea of electric and magnetic fields later developed from Faraday's lines of force. The lines of force or *field lines* of magnets, currents, and charges can be made visible in some simple demonstration experiments.

## Experiment 3.6    Field lines

If iron filings are spread over a glass plate and a magnet is brought under the plate, then the filings will align themselves in a pattern corresponding to the lines of force of the magnet, which resembles the pattern formed by streamlines (Figure 3.7) of a flowing liquid. Finely ground grain will form similar patterns near electrical charges. The analogy of these patterns with the streamline figures of flowing liquids led to the idea of describing electric and magnetic forces in terms of fields. The patterns formed by electrical and magnetic fields in a few simple configurations of electrical currents and charges are shown in Figures 3.22 and 3.24, respectively.

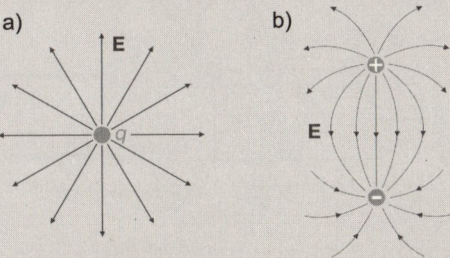

**FIGURE 3.24**  Electric field patterns associated with electrical charges.

As in the streamline patterns of flowing liquids, the direction of the lines of force indicates the direction of the field vectors, and the density of the lines of force is proportional to the magnitude of the field vector. The field $\mathbf{E}(\mathbf{r})$ near charges and the field $\mathbf{B}(\mathbf{r})$ near electrical currents are derived from the forces exerted on a test charge, such as an elementary charge $e$, located at position $\mathbf{r}$ by the charges and currents. These forces can be experimentally demonstrated using the experimental apparatus shown in Figure 3.25.

## Experiment 3.7   Deflection of electron beams

Electrons produced by thermal emission (Section 4.3.3) at a cathode are allowed to move into an evacuated glass flask and are accelerated by a voltage $U_B$ between the cathode and a stop with a hole in it. This produces an electron beam beyond the stop. The electrons pass through an electric field $\mathbf{E}$ between the plates of a capacitor or the magnetic field $\mathbf{B}$ created by a set of two current-carrying coils. The electrons will be deflected by the fields. The deflection of the electron beam can be observed if the electrons are allowed to strike a fluorescent screen.

*continued*

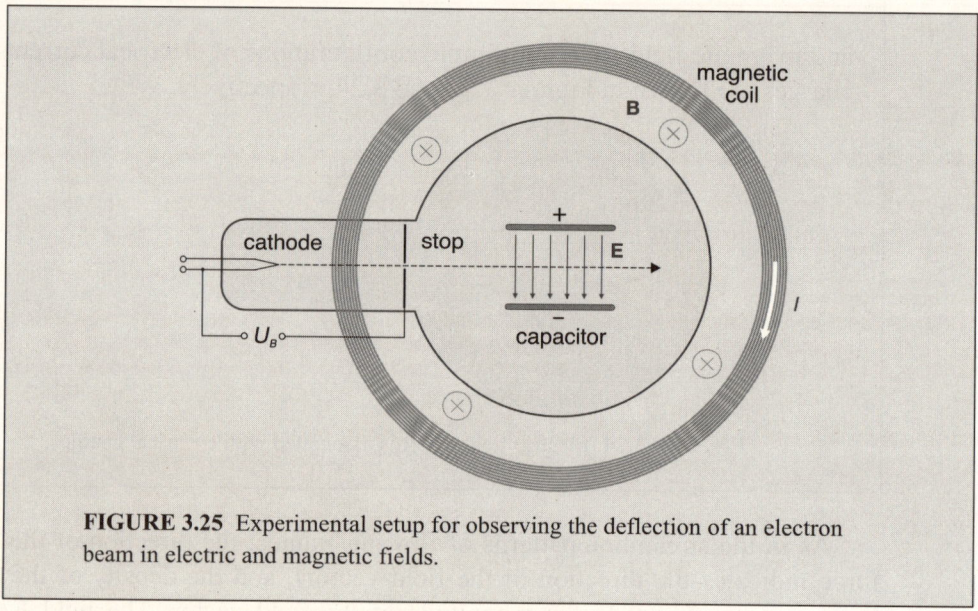

**FIGURE 3.25** Experimental setup for observing the deflection of an electron beam in electric and magnetic fields.

This experiment shows that the electrons are deflected in an electric field perpendicular to the direction of the beam toward the positive plate of the capacitor. This is because a force $\mathbf{F}(\mathbf{r})$ acts parallel to the field and the charge on the electrons is negative. Since the force acting on a charge $q$ in this arrangement is proportional to the charge $q$, one defines the field strength $\mathbf{E}(\mathbf{r})$ at position $\mathbf{r}$ by setting $\mathbf{E}(\mathbf{r}) = \mathbf{F}(\mathbf{r})/-e$. The SI unit of electric field strength is thus $1\,\mathrm{V/m}$ and the *Coulomb force* acting on an electron is, therefore,

$$\mathbf{F}_C = -e\mathbf{E}. \qquad \text{(Coulomb force)}$$

Furthermore, this experiment also shows that in a magnetic field $\mathbf{B}(\mathbf{r})$ perpendicular to the beam, the electrons will be deflected perpendicular to the beam and to the magnetic field. Here the deflecting force is not only proportional to the charge on the particle, but also to its velocity $\mathbf{v}$. For the *Lorentz force* acting on an electron in a magnetic field we therefore set

$$\mathbf{F}_L = -e(\mathbf{v} \times \mathbf{B}). \qquad \text{(Lorentz force)}$$

With this formula, the field $\mathbf{B}(\mathbf{r})$, the so-called *magnetic induction*, is defined. The SI unit [T] (Tesla) for the magnetic induction is thus

$$1\,\mathrm{T} = 1\,\mathrm{V \cdot s/m^2}.$$

## Problem 3.13

Electrons accelerated by a voltage $U_B = 10\,\text{kV}$ move through the field configuration of Figure 3.25 in a straight line if the ratio $E/B$ of the electric to the magnetic field is $E/B \approx 6 \times 10^7$ m/s, for example, if $B = 1\,\text{mT}$ and $E = 6 \times 10^4$ V/m. Calculate the specific charge $e/m$ (charge to mass ratio) of the electrons from these values. With the known value of the elementary charge $e = 1.6 \times 10^{-19}$ C, you can then calculate the mass $m$ of the electrons and the speed at which the electrons fly through the field configuration. How large is this speed compared to $c$, the speed of light?

Based on the definition of the field vectors, the fields in the neighborhood of charges and currents can now be determined quantitatively. The radially directed electric field surrounding a point charge q is given by the *Coulomb field* $\mathbf{E}(\mathbf{r})$:

$$\mathbf{E}(\mathbf{r}) = \frac{1}{4\pi\varepsilon_0} \cdot \frac{q}{r^2} \cdot \frac{\mathbf{r}}{r}.$$

The magnitude of the magnetic field $\mathbf{B}(\mathbf{r})$ surrounding a current $I$ flowing in a straight line is

$$B(r) = \frac{\mu_0}{2\pi} \cdot \frac{I}{r}.$$

The proportionality constants $\varepsilon_0$ and $\mu_0$ which show up here are known as the (dielectric) *permittivity* and (magnetic) *permeability*, respectively, of free space. The vacuum permeability

$$\mu_0 = 4\pi \times 10^{-7} \text{ V·s/(A·m)}$$

is set arbitrarily to fit the definition of the SI unit [A] for electric current. The vacuum permittivity is obtained from measurements of the field strength in the neighborhood of charges:

$$\varepsilon_0 = 8.854 \times 10^{-12} \text{ A·s/(V·m)}.$$

Both constants are related to the velocity of light, by

$$c^2 = \frac{1}{\varepsilon_0 \mu_0}.$$

This relationship is a first indication of the electromagnetic nature of light (Section 3.6).

The electric and magnetic fields of point charges and straight conductors, respectively, can each be characterized by two fundamental properties that can be generalized to other time-independent electric and magnetic fields:

---

**The Lines of Force of the Fields Produced by Time-Independent Charge and Current Distributions**

---

■ All the lines of force of electric fields begin and end exclusively on charges.
■ There are no closed electric field lines.
■ Only closed magnetic field lines exist.
■ Every closed magnetic field line encloses an electric current that produces the magnetic field.

---

In terms of these properties, electric and magnetic fields resemble the velocity fields of incompressible fluid flows.

---

**Notes**

Electrical charges produce a radially directed electric field. The lines of force begin and end on charges. An electric field exerts a Coulomb force on a test charge $-e$ equal to

$$\mathbf{F}_C = -e\mathbf{E}.$$

Electric currents produce a tangentially directed magnetic field. This field surrounds the current with field lines that close on themselves. A magnetic field acts on a test charge $-e$ moving at velocity $\mathbf{v}$ with a Lorentz force

$$\mathbf{F}_L = -e(\mathbf{v} \times \mathbf{B}).$$

### 3.4.3 Electromagnetic induction

Oersted's observation in 1820 of the effect of an electric current on a magnetic needle (Section 3.4.1) was the first indication that electricity and magnetism are connected. Electric currents affect magnets and, according to the Newtonian axiom *action = reaction*, magnets in turn affect currents.

---

#### Experiment 3.8    Current swing

A current-carrying conductor located between the poles of a magnet experiences a force perpendicular to the magnetic field and to the direction of the current (Figure 3.26). We have already become acquainted with this force as the Lorentz force, because the current $I$ flowing in a conductor is the result of the motion of conduction electrons. Let $e$ be the elementary charge, $n$ the electron density in the conductor, $v$ be the velocity of the electrons, $A$ be the cross sectional area, and $l$ be the length of the conductor. Then $I = -envA$. On the other hand, the Lorentz force acting on all the electrons moving in the conductor sums to $F_L = -enAlvB$, where it is assumed that the directions of the current and the magnetic field are mutually perpendicular. These two conditions yield

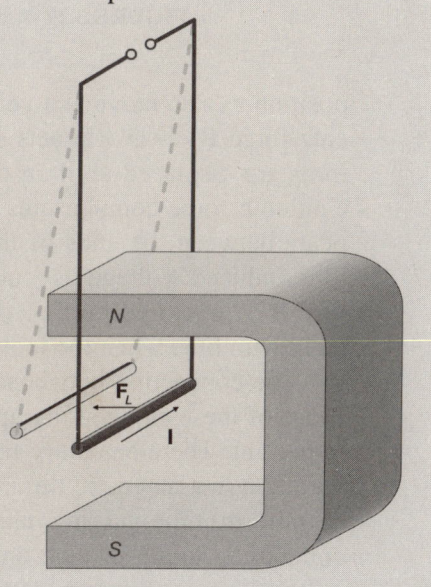

$$F_L = IBl.$$

**FIGURE 3.26**  Current swing.

---

In all *electric motors* the Lorentz force is used as the propelling force. As a simple example, consider a rotatable conducting loop suspended between the poles of a magnet (Figure 3.27). When a current flows, a torque acts on the conducting loop (Section 1.5.1).

Conversely, one could also turn the conducting loop with a crank and rotate it at an angular velocity $\omega$. Then the conduction electrons in the conductor (at

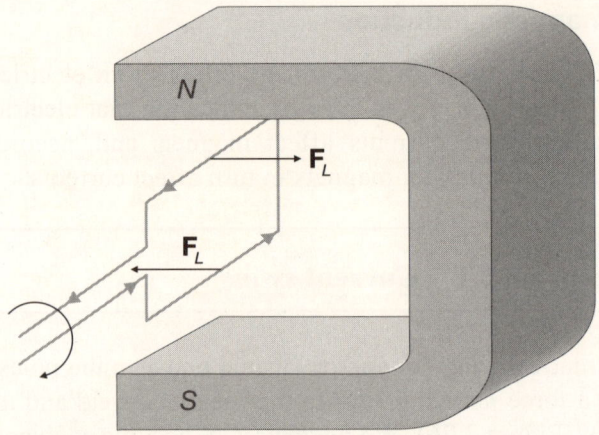

**FIGURE 3.27** A conducting loop between the poles of a magnet.

position $\mathbf{r}$) also move at a velocity $\mathbf{v} = -\boldsymbol{\omega} \times \mathbf{r}$ in the magnetic field. Thus, a Lorentz force $\mathbf{F} = -e\mathbf{v} \times \mathbf{B}$ acts on them. Under the influence of this force the electrons are displaced until an opposing electric field $\mathbf{E} = -\mathbf{v} \times \mathbf{B}$ develops and the Coulomb force compensates the Lorentz force. Then a voltage $U = \int \mathbf{E} \cdot d\mathbf{r}$ appears between the ends of the conducting loop. The rotating loop is a *dynamo*. The induced voltage is caused by the time variation in the magnetic flux $\Phi = \mathbf{B} \cdot \mathbf{A}$ enclosed by the conducting loop of area $\mathbf{A}$. The flux changes periodically with time when the loop is rotating.

The connection between voltage and magnetic flux is not just of interest because of these engineering applications. Its profound scientific aspect is far more important. The elementary relationship between an induced voltage and the time variation in a magnetic flux raises a question of great scientific significance: does a (uniform) time-varying magnetic field $\mathbf{B}(t)$ also induce a voltage in a stationary conducting loop? In fact, any time-varying magnetic field $\mathbf{B}(t)$ does produce a voltage in a conducting loop which encompasses an area $\mathbf{A}$ and thereby encloses magnetic field lines. The induced voltage $U_{\mathrm{ind}}$ is given by the time derivative of the magnetic flux $\Phi = \mathbf{B} \cdot \mathbf{A}$ through the area $\mathbf{A}$:

$$U_{\mathrm{ind}} = \frac{d\Phi}{dt} = \frac{d\mathbf{B}}{dt} \cdot \mathbf{A}. \qquad \text{(Faraday's law of induction)}$$

Faraday's law of induction is the physical basis of transformers and many other engineering devices. A coil (generally with multiple turns) with an alternating current flowing through it produces an alternating magnetic field, which in turn will produce an alternating current in a second, nearby coil. A secondary coil with multiple turns encloses the magnetic field produced by the primary $N$ times more than a single loop, so the induced voltage in the secondary is $N$ times greater than for a single conducting loop.

Unlike the voltages induced in rotating conducting loops, the voltages induced by a variable field cannot be explained by the Lorentz force. Since the loop is at rest, there is no Lorentz force acting on the conduction electrons. Faraday's law of induction thus goes fundamentally beyond the elementary laws dealing with the interaction of static fields with electric fields and currents. So far, the fields have been treated as auxiliary quantities for the description of long-range electric and magnetic forces. With Faraday's law of induction, these fields acquire an independent physical significance (Section 3.5.2).

---

**Problem 3.14**

Calculate the voltage induced in a conducting loop for the case in which the area enclosed by the loop is $A = 0.1\,\text{m}^2$ and a magnetic field **B** perpendicular to this area rises linearly from 0 to 1 T in $10^{-2}$ s.

---

**Notes**    A time-dependent magnetic field $\mathbf{B}(t)$ creates a voltage $U_{\text{ind}}$ in a coil which encloses the magnetic field. A spatially uniform alternating field $\mathbf{B}(t) = \mathbf{B}_0 \exp(-i\omega t)$ with frequency $\omega$ induces a voltage

$$U_{\text{ind}} = i\omega(\mathbf{B}_0 \cdot \mathbf{A})\exp(i\omega t)N$$

in a coil with $N$ turns and cross-sectional area **A**.

## 3.4.4 Electromagnetic oscillator circuit

In the ideal case, an electromagnetic oscillator circuit is a current loop with a capacitor of capacitance $C$ and a coil with inductance $L$ (Figure 3.28). If an oscillator circuit of this sort is excited to electromagnetic oscillations, it behaves as a harmonic oscillator (Section 1.6.1). An oscillator circuit can be excited to forced oscillations when, for example, an alternating magnetic field from an exciter coil acts on the coil in the circuit, in a way similar to the excitation of a spring pendulum with a periodic exciter (Figure 1.6.3). The characteristic frequency of an oscillator circuit (resonant circuit) of this sort is given by

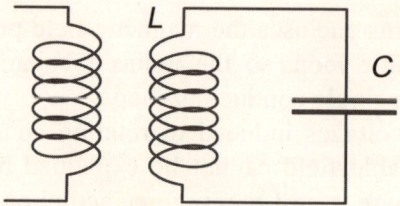

**FIGURE 3.28** An electromagnetic oscillator circuit with an inductive exciter.

$$\omega_0 = \sqrt{\frac{1}{LC}}.$$

In the simplest case, a *capacitor* consists of two flat electrodes with area $A$ opposite each other separated by a distance $d$, to which a voltage $U_K$ is applied (Figure 3.20). An electric field $E = U_K/d$ then exists between the plates. The lines of force of the field begin at the positive and end at the negative electrode. Hence there is, respectively, a positive or negative charge density $q/A = E/\varepsilon_0$ on the inward-facing surfaces of the capacitor plates. The *capacitance* $C$ of the capacitor is the ratio of the charge and voltage:

$$C = \frac{q}{U_K}.$$

A plate capacitor with area $A$ and distance $d$ between the plates therefore has a capacitance $C = \varepsilon_0 A/d$. The unit of capacitance is $1\,\text{F} = 1\,\text{A}\cdot\text{s/V}$ (Farad).

The self-induction or *inductance* $L$ of a coil follows from Faraday's law of induction (Section 3.4.3). The time-dependent magnetic field $B(t)$ produced by a current $I$ in the coil induces a voltage $U_{\text{ind}}$ in the coil itself that opposes the voltage driving the current. The inductance $L$ is the ratio of the induced voltage $U_{\text{ind}}$ to the time derivative $dI/dt$ of the current in the coil, so that

$$U_{\text{ind}} = -L\frac{dI(t)}{dt}.$$

The inductance of a long cylindrical coil with $N$ windings, cross-sectional area $A$, and length $l$ is $L = N^2\mu_0 A/l$. The unit of induction $L$ is $1\,\text{H} = 1\,\text{V}\cdot\text{s/A}$ (Henry).

Besides the capacitance $C$ of a capacitor and the inductance $L$ of a coil, the Ohmic *resistance* $R$ of the wires determines the oscillatory behavior of an oscillator circuit. In particular, the coil in an oscillator circuit has a (internal) resistance.

If a time-independent current $I$ flows through a resistance, the Ohmic resistance $R$ is given by the ratio of the applied voltage $U_R$ to the current $I$:

$$R = \frac{U_R}{I}.$$

The unit of resistance is $1\,\Omega = 1\,\text{V/A}$ (Ohm).

The equation for the oscillations of an oscillator circuit follows from the voltage balance

$$U_K + U_{\text{ind}} = U_R.$$

Since the current in the coil $I = -dq/dt$ of the oscillator circuit is equal to the time variation in the charge $q$ on the capacitor, the voltage balance yields the following differential equation for $q$:

$$L\frac{d^2q}{dt^2} + R\frac{dq}{dt} + \frac{1}{C}q = 0.$$

This is the equation for the oscillations of a harmonic oscillator (Section 1.6.2). The oscillator circuit has an initially specified characteristic frequency $\omega_0 = \sqrt{1/LC}$ and damping constant $2\delta = R/L$.

The forced oscillations produced by an exciter coil (Figure 3.28) are, like the forced oscillations of a spring pendulum (Section 1.6.3), solutions of an inhomogeneous differential equation for the oscillations. They manifest the same resonant behavior in the amplitudes and phases near the characteristic frequency $\omega_0$ found previously for the oscillations of a spring pendulum.

---

### Problem 3.15

The inductance of a simple conducting loop with radius $R$ is on the order of $L \sim \mu_0 R$. Calculate the order of magnitude of the characteristic frequency of an oscillator circuit consisting of a conducting loop with a radius $R = 1\,\text{m}$ and a plate capacitor with $A = 1\,\text{cm}^2$ and $d = 1\,\text{mm}$.

---

> | **Notes** | An oscillator circuit with inductance $L$ and capacitance $C$ has the characteristic frequency |
>
> $$\omega_0 = \sqrt{\frac{1}{LC}}.$$

## 3.5 THE ELECTROMAGNETIC FIELD

The experiments of Oersted (Section 3.4.2) and Faraday (Section 3.4.3) provide clear indications that electric and magnetic fields are closely related to one another. Electric fields are produced by charges at rest and magnetic fields by moving charges. The same holds for the effects of the fields: the Coulomb force of the electric field acts on charges at rest, while the Lorentz force of the magnetic field only acts on moving charges. Both fields serve as intermediaries for the forces operating between electrically charged objects. In accordance with the Newtonian axiom *action = reaction*, the force acts in both directions. Every one of a set of interacting objects contributes to the creation of the fields and, at the same time, each is accelerated by the forces owing to those fields.

Faraday's law of induction requires extensive examination. Electric and magnetic fields are more than just intermediaries of long-range forces. Faraday's law stood at the beginning of a stormy revolution in physics which finally ended with Heinrich Hertz's proof of the existence of electromagnetic waves in 1888. The physics of electricity and magnetism led the way to a unified theory of the electromagnetic field and of the interaction of this field with electrical charges and currents.

In the first section of this lecture we discuss the properties of matter in electric and magnetic fields. For that we shall assume that the fields are time independent. The significance of the electromagnetic field as an independent medium that can be used to transfer energy and information over large distances in the form of electromagnetic waves only becomes evident when the properties of time-dependent fields are studied. We shall discuss time-dependent electromagnetic fields in the following sections while limiting ourselves to the fields in a vacuum.

### 3.5.1 Polarization and magnetization

The electric field strength **E** and the magnetic induction **B** were defined in Section 3.4.2 based on the force experienced by a test charge $e$ in an electric or magnetic field. Alternatively, electric and magnetic fields can be characterized in terms of the electrical charges and currents which produce them. Especially for the description of electromagnetic fields in matter, for both electric and magnetic fields, there is some advantage in introducing two field vectors for each, one for the dynamic effects and one that is involved in creation of the field.

Let us first consider the field of a point charge $q$. The electric field strength $\mathbf{E}$ surrounding the charge is directed radially and falls off as the inverse square of the distance $r$. If the charge is in a vacuum, $|\mathbf{E}| = (4\pi\varepsilon_0)^{-1} \cdot q/r^2$ (Section 3.4.2). A different value is obtained if the charge is in an insulator. Even if the insulator consists exclusively of neutral atoms or molecules, so it is electrically neutral overall, it has an influence on the electric field emanating from the charge $q$. Here the insulator is a *dielectric*. The atoms (or molecules) of the insulator become polarized in an electric field; that is, the center of charge of the electron sheath (Section 5.3) is displaced slightly relative to the atomic nucleus. The atoms are thereby converted into tiny electric dipoles, which together lead to surface charges and reduce the field of the charge $q$ (Figure 3.29). Thus, the electric field is not produced by the charge $q$ alone, but also by the charges bound in the atoms or molecules of the dielectric. In order to describe the fields produced in this way quantitatively, we use a measure for the atomic electric dipoles.

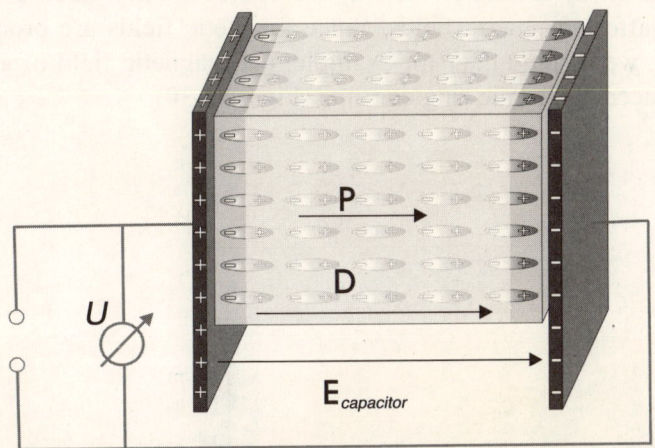

**FIGURE 3.29** Polarization of a dielectric in the field of a parallel plate capacitor.

As an elementary model for an electric dipole, let us consider two point charges $+e$ and $-e$ at neighboring points $\mathbf{r}_+$ and $\mathbf{r}_-$. In an external field $\mathbf{E}$ a torque $\mathbf{T} = e(\Delta\mathbf{r} \times \mathbf{E})$ acts on a dipole of this sort, where $\Delta\mathbf{r} = \mathbf{r}_+ - \mathbf{r}_-$. The vector

$$\mathbf{d} = e\,\Delta\mathbf{r} \qquad \text{(electric dipole moment)}$$

which shows up here is the *electric dipole moment* of the dipole. The (*displacement or electric*) *polarization* **P** of a dielectric is the result of the combined dipole moments of the individual atoms. It is defined as the dipole moment of the dielectric per unit volume:

$$\mathbf{P} = n\mathbf{d}. \qquad \text{(electric polarization)}$$

Here $n$ is the number of atoms per unit volume. The polarization produces an electric field $\mathbf{E}_P = -\mathbf{P}/\varepsilon_0$ inside the dielectric. The electric field in a dielectric is, therefore, the sum of two parts: external charges produce a field **E** and the polarization of the atoms and molecules of the dielectric create a field $\mathbf{E}_P$. In order to emphasize the relationship to the charges which produce it, the electric field is described by the vector **D**, the *electric displacement*. The displacement has the same dimensions $[\mathrm{A \cdot s/m^2}]$ as the electric polarization **P**. In a dielectric

$$\mathbf{D} = \varepsilon_0 \mathbf{E} + \mathbf{P}. \qquad \text{(electric displacement)}$$

In a way similar to the way dielectrics modify electric fields, magnetic materials affect magnetic fields. Since magnetic fields are produced by electric currents, we assume that, for example, the magnetic field of a permanent magnet is produced by atomic ring currents (Figure 3.30).

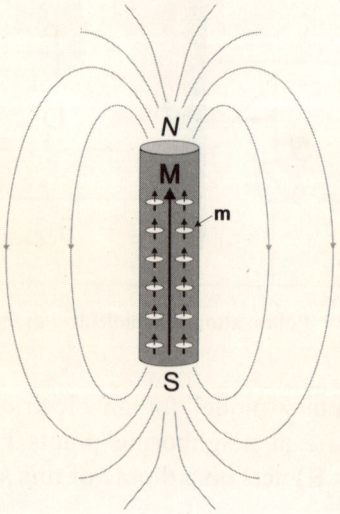

**FIGURE 3.30** Atomic ring currents in a permanent magnet.

Each ring current is a small magnetic dipole **m** on which an external magnetic field **B** exerts a torque $\mathbf{T} = \mathbf{m} \times \mathbf{B}$. As a simple model for a magnetic dipole let us consider an atomic electron (with charge $-e$) moving in a trajectory $\mathbf{r}(t)$ around an atomic nucleus. In the magnetic field **B** the electron experiences a Lorentz force $\mathbf{F}_L$ (Section 3.4.2) and, thereby, a time-averaged torque

$$\mathbf{T} = \langle \mathbf{r} \times \mathbf{F}_L \rangle = -\frac{1}{2} e[(\mathbf{r} \times \mathbf{v}) \times \mathbf{B}].$$

(The calculation of the time average indicated by the angle brackets $\langle \rangle$ requires some vector algebra.) The atomic magnetic dipole has, therefore, a dipole moment $\mathbf{m} = -e(\mathbf{r} \times \mathbf{v})/2$. Since $\mathbf{L} = m_e(\mathbf{r} \times \mathbf{v})$ is the orbital angular momentum of the electron (with mass $m_e$), we obtain

$$\mathbf{m} = -\frac{e}{2m_e} \mathbf{L} \qquad \text{(magnetic dipole moment)}$$

for the *magnetic dipole moment* of an atomic ring current.

By analogy with the electric polarization in dielectrics, in magnetic materials the magnetic dipole moment **m** of the atomic ring currents yields the *magnetization* **M**:

$$\mathbf{M} = n\mathbf{m}. \qquad \text{(magnetization)}$$

The magnetization produces a magnetic field $\mathbf{B}_M = \mu_0 \mathbf{M}$ inside a magnet.

---

### Experiment 3.9    Einstein-de Haas effect

The angular momentum of the electron motion associated with magnetization can be demonstrated experimentally (Einstein-de Haas effect). This is done by suspending a soft iron rod, which is easily demagnetized, on a quartz filament in a vertically oriented coil (Figure 3.31). The iron rod is demagnetized by changing the direction of current flow in the magnetic field coil, thereby changing the angular momentum of the electrons.

*continued*

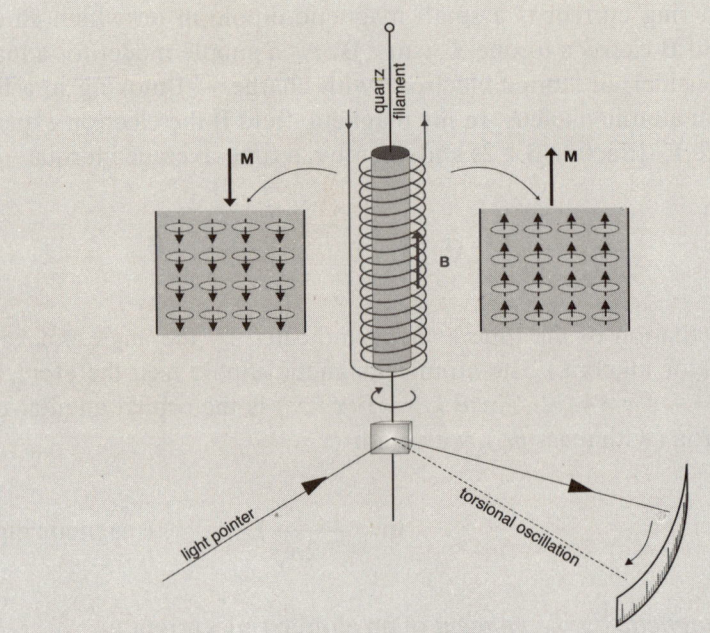

**FIGURE 3.31**  The Einstein-de Haas effect.

As in the experiments with a rotating table (Section 1.4.4), the *z* component of the angular momentum of the rod is conserved upon demagnetization. Thus, the rod will be set to rotating when it is demagnetized. Figure 3.32 is a schematic illustration of a mechanical model for the Einstein-de Haas effect. If the axes of two rotating gyroscopes on the ends of the dumbbell beam are flipped (by a suitable mechanism attached to the beam) by 180°, the beam will start to rotate immediately.

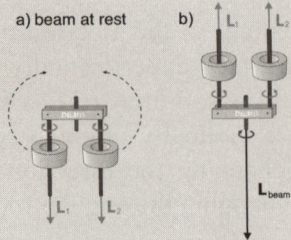

**FIGURE 3.32**  Model experiment for the Einstein-de Haas effect.

## Problem 3.16

Calculate the rotation frequency $\omega$ at which a circular iron rod of radius $r = 1$ mm suspended at one end will rotate after being magnetized once.

Like the electric field, the magnetic field is described by two field vectors. The magnetic induction $\mathbf{B}$ is involved in dynamic effects on a moving test charge. With reference to the creation of the field, one defines the *magnetic intensity* $\mathbf{H}$ (often referred to as the magnetic field). It has the same dimensions [A/m] as the magnetization. In magnetic materials

$$\mathbf{H} = \frac{\mathbf{B}}{\mu_0} - \mathbf{M}. \qquad \text{(magnetic intensity)}$$

> **Notes** Electric fields are described by the field vectors $\mathbf{E}$ [V/m] and $\mathbf{D}$ [$\mathrm{A \cdot s/m^2}$] and magnetic fields, by the field vectors $\mathbf{B}$ [$\mathrm{V \cdot s/m^2}$] and $\mathbf{H}$ [A/m]. The field vectors $\mathbf{E}$ and $\mathbf{B}$ are defined with reference to forces acting on charges and the field vectors $\mathbf{D}$ and $\mathbf{H}$, with reference to the electrical charges and currents that generate the field.
>
> The following relationships hold in a dielectric with electric polarization $\mathbf{P}$ and in a magnetic material with magnetization $\mathbf{M}$:
>
> $$\varepsilon_0 \mathbf{E} = \mathbf{D} - \mathbf{P} \quad \text{and} \quad \frac{\mathbf{B}}{\mu_0} = \mathbf{H} + \mathbf{M}.$$

### 3.5.2  Electric field in a vacuum

The displacement $\mathbf{D}$ was defined in Section 3.5.1 with reference to the creation of electric fields by charges. Electric fields can, however, also be produced by time-dependent magnetic fields (Section 3.4.3). The dynamics of electric and magnetic fields can, therefore, be quite complicated. On the other hand, the behavior of electromagnetic fields in a vacuum is mathematically amazingly simple. This behavior is the foundation of *electrodynamics* or *electromagnetic theory*. In order to formulate the laws associated with this behavior, we first consider the electric and magnetic fields separately in vacuum once again.

In a vacuum, the two field vectors $\mathbf{E}$ and $\mathbf{D}$ of the electric field are related by

$$\mathbf{D} = \varepsilon_0 \mathbf{E}.$$

When only charges at rest contribute to the production of a field, the lines of force of the field begin and end at charges. This simple rule is no longer valid, however, if a time-dependent magnetic field $\mathbf{B}(t)$ also contributes to the production of the electric field. This contribution follows from Faraday's law of induction (Section 3.4.3). The voltage $U_{ind}$ induced in a conducting ring is the result of the electric field strength $\mathbf{E}$ along the conducting ring (which encompasses an area $A$):

$$U_{ind} = -\oint_A \mathbf{E}\, d\mathbf{r}.$$

Thus, Faraday's law of induction can also be interpreted so that the time-dependent magnetic field $\mathbf{B}(t)$ induces an electric field whose lines of force surround the magnetic field (Figure 3.33). Then the lines of force of this field form closed curves without beginning or end. Proceeding from this interpretation yields the following relationship between the time derivative of the magnetic field $\mathbf{B}(\mathbf{r},t)$ and the induced electric field $\mathbf{E}(\mathbf{r},t)$:

$$\oint_A \mathbf{E}(\mathbf{r},t)\cdot d\mathbf{r} = -\iint_A \frac{d\mathbf{B}(\mathbf{r},t)}{dt}\cdot d\mathbf{A}. \qquad \text{(integral form of Maxwell's second time-dependent equation)}$$

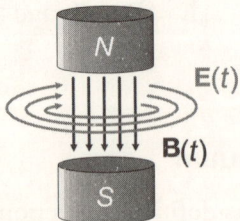

**FIGURE 3.33** The electric field induced by a time-dependent magnetic field.

---

### Problem 3.17

Estimate the electric field strength in the neighborhood of a circular ($R = 0.1\,\text{m}$) electromagnet supplied by a 50 Hz alternating current that reaches a field strength $B = 1\,\text{T}$.

In fact, this field can be demonstrated directly using test charges. In the neighborhood of a time-dependent magnetic field electrons will be accelerated. With a suitable rotationally symmetric magnetic field configuration (a *betatron*) it is even possible to have the electrons move in circular orbits around the magnetic field and be accelerated as long as the magnetic field increases. Electrons can reach energies of many MeV ($1\,\mathrm{MeV} = 10^6\,\mathrm{eV}$) in this way.

Maxwell's equations make it clear that the change from Newton's concept of action at a distance to the Faraday-Maxwell field concept is physically necessary. Thus, electric and magnetic fields acquire an independent physical reality. This paradigm shift also requires that the concept of potential energy, which is based on the idea of action at a distance, be replaced. Rather than the potential energy of a charged object in an electric field, the energy stored in the electric field itself is to be considered. Since the field extends continuously over space, its energy is described by a position and time-dependent energy density $u_{el}(\mathbf{r},t)$. For example, the energy density of the electric field in a parallel plate capacitor can, for example, be derived from the energy $E_C = q^2/2C$ required to charge the capacitor. In general,

$$u_{el}(\mathbf{r},t) = \frac{1}{2}\mathbf{E}(\mathbf{r},t)\cdot\mathbf{D}(\mathbf{r},t).$$

In a vacuum, therefore, the energy density of an electric field is proportional to the square of the field strength; that is, $u_{el} = \frac{1}{2}\varepsilon_0 E^2$.

**Notes**
Stationary charges $q$ produce electric fields with lines of force that begin at positive charges and end at negative charges. A time-dependent magnetic field $\mathbf{B}(\mathbf{r},t)$ induces an electric field $\mathbf{E}(\mathbf{r},t)$ with field lines that close on themselves.

The energy density of an electric field (in a vacuum) is proportional to the square of the field strength:

$$u_{el} = \frac{1}{2}\varepsilon_0 E^2.$$

### 3.5.3 Magnetic field in a vacuum

In a vacuum the field vectors $\mathbf{B}$ and $\mathbf{H}$ of the magnetic field are proportional to one another, as are the field vectors of the electric field:

$$\mathbf{B} = \mu_0 \mathbf{H}.$$

So far we have proceeded under the assumption that magnetic fields are produced by electrical currents. The lines of force of these fields are closed rings surrounding the current carrying conductor (Figure 3.22). This assumption is adequate for determining the magnetic fields of stationary electrical currents. But when we calculate the magnetic fields owing to time-dependent currents, for example, during charging of a capacitor, this assumption leads to contradictions. In order to avoid them, in 1864 James Clerk Maxwell postulated that, just as a time-dependent magnetic field induces an electric field, a time-dependent electric field induces a magnetic field; thus,

$$\oint_A \mathbf{H}(\mathbf{r},t) \cdot d\mathbf{r} = \oiint_A \frac{d\mathbf{D}(\mathbf{r},t)}{dt} \cdot d\mathbf{A}. \qquad \text{(integral form of Maxwell's first time-dependent equation)}$$

According to this equation, the lines of force of a magnetic field induced by a time-dependent electric field are also closed on themselves. Electric and magnetic fields differ in this regard: while the lines of force of an electric field may begin and end at electrical charges, the lines of force of a magnetic field always close on themselves. No magnetic monopoles are known which might be the beginning or end point of a magnetic line of force. Permanent magnets always have a north pole and a south pole of equal strength. Thus, these magnets can be regarded as an ensemble of atomic ring currents (Figure 3.30).

According to Maxwell's first time-dependent equation, a magnetic field develops during the charging of a parallel plate capacitor with lines of force that both encircle the charging current and exist in the volume surrounding the capacitor. There they encircle the time varying electric field (Figure 3.34). In practice, magnetic fields induced in this fashion are, of course, extremely weak compared to the magnetic fields in the vicinity of an electric current I.

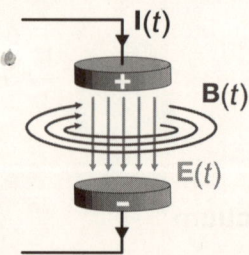

**FIGURE 3.34** The magnetic field that surrounds the electric field during charging of a parallel plate capacitor.

## Problem 3.18

Estimate the magnetic field in the vicinity of a circular ($R = 0.1\,\mathrm{m}$) capacitor in which a 50 Hz alternating current generates alternating electric fields of up to $\mathbf{E} = 1\,\mathrm{MV/m}$.

In an oscillator circuit (Section 3.4.4) a capacitor with capacitance $C$ is charged through a coil with inductance $L$. Then the electric field in the capacitor, with energy density $u_{\mathrm{el}}$, decays and, in its place, a magnetic field with energy density $u_{\mathrm{m}}$ builds up in the coil. Subsequently, the magnetic field decays and an electric field with the opposite polarity builds up in the capacitor. This interplay between electric and magnetic fields shows that magnetic fields, as well as electric fields, have an energy density. By analogy with the energy density of an electric field, the field vectors $\mathbf{B}$ and $\mathbf{H}$ give

$$u_{\mathrm{m}} = \frac{1}{2}\mathbf{H} \cdot \mathbf{B}.$$

In a vacuum $u_{\mathrm{m}} = \frac{1}{2}B^2/\mu_0$. Magnetic fields on the order of 1 T can be produced with an electromagnet. A magnetic field of that strength has an energy density $u_{\mathrm{m}} = 0.4\,\mathrm{MJ/m^3}$. Compared to this, the energy densities attainable in stationary electric fields are very small. An electric field of strength $E = 1\,\mathrm{MV/m}$, for example, has an energy density of about $u_{\mathrm{el}} = 4\,\mathrm{J/m^3}$. The difference in these energy densities explains why heavy pieces of iron can be lifted with magnets while only small shreds of paper can be picked up with electric fields.

| **Notes** | A time-dependent electric field produces a magnetic field whose lines of force surround the electric field and a time dependent magnetic field produces an electric field whose |

lines of force surround the magnetic field. This reciprocity of electric and magnetic fields is implicit in the time-dependent Maxwell equations (in a vacuum):

$$\oint_A \mathbf{H}(\mathbf{r},t) \cdot d\mathbf{r} = \iint_A \frac{d\mathbf{D}(\mathbf{r},t)}{dt} \cdot d\mathbf{A} \qquad \text{(integral form of Maxwell's first time-dependent equation)}$$

$$\oint_A \mathbf{E}(\mathbf{r},t) \cdot d\mathbf{r} = -\iint_A \frac{d\mathbf{B}(\mathbf{r},t)}{dt} \cdot d\mathbf{A}. \qquad \text{(integral form of Maxwell's second time-dependent equation)}$$

### 3.5.4 Maxwell's equations

The formulation of Maxwell's equations in terms of line and surface integrals relies closely on the experimental data presented in Figures 3.33 and 3.34. Physically important quantities can often be estimated rapidly using the integral equations. For exact calculations and more penetrating theoretical studies, however, there is some advantage to a mathematical formulation of the interaction between electric and magnetic fields in terms of differential equations.

A mathematical reformulation of the integral equations into differential equations is made possible by means of *Stokes' theorem*. This theorem states that the integral of a vector field $\mathbf{B}(\mathbf{r})$ over a closed path can be transformed into a surface integral as follows:

$$\oint_A \mathbf{B}(\mathbf{r})\,d\mathbf{r} = \iint_A \nabla \times \mathbf{B}(\mathbf{r})\,d\mathbf{A}. \qquad \text{(Stokes' theorem)}$$

Here the vector $\nabla \times \mathbf{B}$ ($= \operatorname{curl}\mathbf{B} = \operatorname{rot}\mathbf{B}$) has the components

$$(\nabla \times \mathbf{B})_x = \partial B_z/\partial y - \partial B_y/\partial z$$
$$(\nabla \times \mathbf{B})_y = \partial B_x/\partial z - \partial B_z/\partial x$$
$$(\nabla \times \mathbf{B})_z = \partial B_y/\partial x - \partial B_x/\partial y.$$

It is referred to as the *curl* (*rotation*) of the vector field and is a measure of the vorticity of the field at position $\mathbf{r}$.

Because of Stokes' theorem the time-dependent Maxwell integral equations can also be written in differential form:

---

**Maxwell's (Time-Dependent) Equations (in a Vacuum)**

$$\nabla \times \mathbf{H}(\mathbf{r},t) = \frac{\partial \mathbf{D}(\mathbf{r},t)}{\partial t} \qquad \text{(differential form of first time-dependent Maxwell equation)}$$

$$\nabla \times \mathbf{E}(\mathbf{r},t) = -\frac{\partial \mathbf{B}(\mathbf{r},t)}{\partial t}. \qquad \text{(differential form of second time-dependent Maxwell equation)}$$

---

In this form, Maxwell's equations couple *local* properties of electric and magnetic fields. The time variation in one of these fields at position $\mathbf{r}$ at time $t$ is related to the curl, that is, to a spatial derivative, of the other field. This local coupling of the two fields makes emphatically clear the departure from the Newtonian concept of action at a distance. Forces which act on objects separated by large distances are replaced by electric and magnetic fields which mediate the forces and interact with each other *locally*.

> **Notes**
>
> Forces which act between objects that are far apart are mediated by fields. In particular, electric and magnetic forces are mediated by the electromagnetic field. The interaction between the electric and magnetic field is *local*. These fields obey Maxwell's differential equations at every point in space-time $(\mathbf{r},t)$.

## 3.6 ELECTROMAGNETIC WAVES

Maxwell's equations imply that electromagnetic fields can also exist independently of electrical charges and currents and are able to propagate as waves even in a vacuum. Thus, along with matter, the electromagnetic field is to be treated as a self-contained medium that can not only store energy, but also transport it (Sections 3.5.2 and 3.5.3). Electromagnetic waves were first produced and identified by Hertz in 1888 using an electrical oscillator circuit. Today we use them for all kinds of information transfer, such as radio, television, and cell telephones. All electromagnetic waves propagate in vacuum at the speed of light, $c = 3 \times 10^8$ m/s. Furthermore, light, which is familiar to all of us, is also an electromagnetic wave.

### 3.6.1 Wave propagation in a vacuum

According to Maxwell's equations, a time-dependent electric field produces a magnetic field and a time-dependent magnetic field produces an electric field. This interplay leads to the propagation of electromagnetic waves. In order to examine this interplay quantitatively let us consider the propagation of a plane electromagnetic wave in vacuum as a simple example. In a plane wave the field vectors $\mathbf{E}(z,t)$ and $\mathbf{B}(z,t)$ depend only on one position coordinate (here $z$) and time $t$. With this restriction and the assumption that the densities of electrical charge $\rho$ and current $\mathbf{j}$ are identically zero everywhere in space, the time-dependent Maxwell equations yield the following differential equations for the fields $\mathbf{E}(z,t)$ and $\mathbf{B}(z,t)$:

$$\frac{\partial E_y(z,t)}{\partial z} = \frac{\partial B_x(z,t)}{\partial t}$$

$$\frac{\partial B_x(z,t)}{\partial z} = \mu_0 \varepsilon_0 \frac{\partial E_y(z,t)}{\partial t}.$$

To simplify this system of equations here it was assumed that only the $y$ component $E_y$ of the electric field and only the $x$ component $B_x$ of the magnetic field are nonzero. In general, it follows from Maxwell's theory of the electromagnetic field that the field vectors in a plane electromagnetic wave in vacuum are perpendicular to the direction of propagation and perpendicular to one another.

These two partial differential equations, therefore, yield wave equations for both the electric and magnetic fields:

$$\frac{\partial^2 E_y}{\partial t^2} - \frac{1}{\varepsilon_0 \mu_0} \cdot \frac{\partial^2 E_y}{\partial z^2} = 0$$

$$\frac{\partial^2 B_x}{\partial t^2} - \frac{1}{\varepsilon_0 \mu_0} \cdot \frac{\partial^2 B_x}{\partial z^2} = 0.$$

The functions

$$E_y(z,t) = E_y(z \pm ct)$$
$$B_x(z,t) = B_x(z \pm ct)$$

with $c = (\varepsilon_0 \mu_0)^{-1/2}$ are general solutions of these wave equations for the electric and magnetic fields, respectively, which propagate at the speed of light in the $\pm z$ directions. In the following we shall consider the special case of *harmonic* waves propagating in the $z$ direction:

$$E_y(z,t) = E_0 \exp(i(kz - \omega t))$$
$$B_x(z,t) = B_0 \exp(i(kz - \omega t)).$$

These are solutions of the wave equations if the wave number $k$ and frequency $\omega$ of the waves satisfy the relation $k/\omega = c$. If these solutions are substituted into the Maxwell equations for a plane wave, the amplitudes $E_0$ and $B_0$ of the electric and magnetic components of the wave are found to obey

$$E_0 = cB_0.$$

Since $c = (\varepsilon_0 \mu_0)^{-1/2}$, at all times and all places the electric and magnetic fields of a plane electromagnetic wave have equal energy densities; that is,

$$\frac{\varepsilon_0}{2} E_y^2(z,t) = \frac{1}{2\mu_0} B_x^2(z,t).$$

In general, the total energy density $u(z,t) = u_{el} + u_m$ of a plane electromagnetic wave propagating in the $z$ direction with field vectors $\mathbf{E}$ and $\mathbf{B}$ or $\mathbf{D} = \varepsilon_0 \mathbf{E}$ and $\mathbf{H} = \mathbf{B}/\mu_0$ is

$$u = \mathbf{E} \cdot \mathbf{D} = \mathbf{B} \cdot \mathbf{H}.$$

As an example, Figure 3.35 shows an instantaneous picture of an electromagnetic wave propagating in the $z$ direction. Here, as in the above solutions, it is assumed that the electric field vector oscillates everywhere in one and the same direction, in this case in the $x$ direction, while the magnetic field vector oscillates in the $y$ direction. This type of wave is referred to as *linearly polarized*. In general, however, the electric field of an electromagnetic wave can change its direction, since the Maxwell equations merely imply that the fields $\mathbf{E}$ and $\mathbf{B}$ are perpendicular to the direction of propagation, and therefore to the wave vector $\mathbf{k}$, and are perpendicular to one another at every point. Thus, for example, an electromagnetic wave can be *circularly* or *elliptically* polarized. The direction of the field can also fluctuate randomly back and forth in a plane perpendicular to the direction of propagation. In this case, we speak of an *unpolarized* wave.

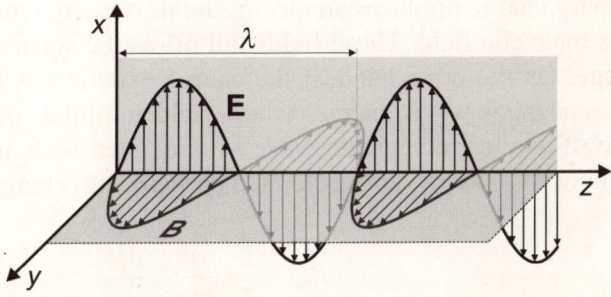

**FIGURE 3.35** A linearly polarized plane electromagnetic wave propagating in the $z$ direction.

Regardless of the polarization, the phase of an electromagnetic wave in a vacuum travels in the direction of the wave vector $\mathbf{k}$ at the speed of light, $c$. An energy flux is associated with the propagation of the wave. The energy flux (energy per unit time per unit area) of an electromagnetic field is given, in general, by the vector product of the electric and magnetic fields, i.e.,

$$\mathbf{S} = \mathbf{E} \times \mathbf{H}. \qquad \text{(Poynting vector)}$$

The direction of the Poynting vector $\mathbf{S}$ for a plane electromagnetic wave is that of the wave vector $\mathbf{k}$ and its magnitude is $|\mathbf{S}| = cu(\mathbf{r}, t)$.

---

**Problem 3.19**

Calculate the field strengths $|\mathbf{E}|$ and $|\mathbf{B}|$ of an electromagnetic wave with intensity $I = |\mathbf{S}| = 1\,\text{W/m}^2$.

---

> **Notes**
>
> A plane electromagnetic wave transports energy in its direction of propagation at the speed of light $c = 3 \times 10^8$ m/s. The energy density $u$ of the wave is proportional to the square of the field vectors:
>
> $$u = \varepsilon_0 E^2 = \frac{B^2}{\mu_0}.$$
>
> Both field vectors are perpendicular to the direction of propagation.

## 3.6.2 Radiation of a Hertzian dipole

A stationary charge produces an electric field. A charge moving at velocity $v$ also creates a magnetic field. These fields fall off as the square of the distance $r$ from the charge. On the other hand, if the charge experiences an acceleration $\mathbf{a}$, then an electromagnetic wave emerges whose field amplitude only falls off as $1/r$. The emission of an electromagnetic wave is associated with an energy loss. According to Maxwell's theory, an accelerated particle with charge $q$ radiates a power

$$P_{\text{rad}} = \frac{2}{3} \cdot \frac{q^2}{4\pi\varepsilon_0} \cdot \frac{a^2}{c^3}. \qquad \text{(Larmor formula)}$$

We now discuss the relationship between acceleration and radiation with the aid of a simple example.

Atoms consist of a positively charged nucleus and negatively charged electrons which orbit the nucleus. Based on the laws of Newtonian mechanics, one would expect that the electrons would orbit the atomic nucleus because of Coulomb attraction, much as the planets orbit the sun. For simplicity, let us consider a hydrogen atom, in which one electron moves in an orbit around a proton (Section 5.1.2). A hydrogen atom like this is a rotating electrical dipole. The rotation frequency $\omega$ of the dipole is, as for the planets, dependent on the radius $r$ of the orbit. Taking $2r \approx 10^{-10}$ m as the order of magnitude of the diameter of the atom gives $\omega \approx 4.4 \times 10^{16}$ s$^{-1}$. Even without a detailed calculation it is clear that a dipole rotating in this way will emit electromagnetic waves. This happens because the dipole first creates a time-varying electric field which, in turn, creates a magnetic field in accordance with the first Maxwell equation. The latter also varies in time, so it again creates an electric field in accordance with the second Maxwell equation. And this interplay continues. Figure 3.36 illustrates the distribution of the fields in the neighborhood of a dipole oscillating harmonically back and forth along a straight line, a so-called Hertzian dipole. Because the emission of electromagnetic waves is associated with energy loss, the oscillations of a Hertzian dipole are damped. For an electron with mass $m_e$ moving in a circular orbit with frequency $\omega$, the Larmor formula gives a damping factor of

$$\delta = \frac{1}{3} \cdot \frac{e^2 \omega^2}{4\pi\varepsilon_0 m_e c^3}.$$

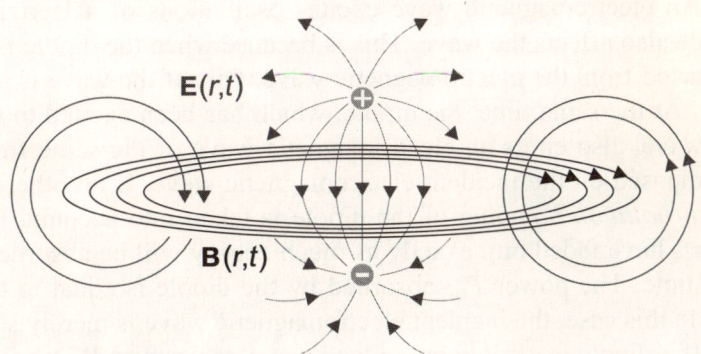

**FIGURE 3.36** The electromagnetic field in the vicinity of a Hertzian dipole.

## Problem 3.20

Calculate the power radiated by an electron moving in a circular orbit of diameter $2r = 10^{-10}$ m around the nucleus of an atom. What is the wavelength of the emitted light? (Hint: the factor $r_e = e^2/4\pi\varepsilon_0 m_e c^2$ is referred to as the *classical electron radius*. Its value is $r_e = 2.8 \times 10^{-15}$ m.)

**Notes**

According to Maxwell's theory of electromagnetic waves, every charge emits electromagnetic waves when it is accelerated and thereby loses energy. Thus, the motion of an electron which oscillates harmonically at a frequency $\omega$ is damped. The damping factor $\delta$ for the oscillations is proportional to the square of the frequency $\omega$.

### 3.6.3 Absorption and scattering

On one hand, a harmonically oscillating electric dipole emits an electromagnetic wave. On the other, oscillations of this Hertzian dipole can also be excited by an electromagnetic wave with frequency $\omega$. This is because the dipole will be polarized in an electric field. The electric field of a wave oscillating at frequency $\omega$ at the position of the dipole then excites the dipole to forced oscillations (Section 1.6.3). As with a mechanical oscillator, at resonance, when the frequency $\omega$ of the electromagnetic wave is equal to the characteristic frequency $\omega_0$ of the Hertzian dipole, the amplitude of the forced oscillations is maximal and the phase of the dipole oscillations is shifted by $\Delta\varphi = \pi/2$ relative to the phase of the wave.

An electromagnetic wave excites oscillations of a Hertzian dipole, but the dipole also affects the wave. This is because when the dipole is excited, energy is extracted from the electromagnetic wave. Part of the wave is absorbed by the dipole. At the same time, the dipole, which has been excited to the frequency $\omega$ of the wave, also emits an electromagnetic wave of the same frequency that is superimposed on the incident electromagnetic wave. If no other damping than that from *radiation damping* of the dipole is taken into account, then after transient effects have faded out, exactly as much energy will be absorbed as is emitted per unit time. The power $P_{abs}$ absorbed by the dipole is equal to the power it emits, $P_{em}$. In this case, the incident electromagnetic wave is merely scattered.

If a dipole is excited by a plane wave, the power $P_{abs}$ absorbed by the dipole is proportional to the intensity (energy flux) $S = cu$ of the wave. The proportionality factor has the dimensions of area. By analogy with the collision cross section for atomic collisions in a gas (Section 3.2.2), a *scattering cross section* $\sigma$ can be defined for the scattering of an electromagnetic wave on a Hertzian dipole. In effect, a power $cu\sigma$, given by the product of the energy flux $cu$ of the plane wave and the cross section $\sigma$, is scattered by the dipole. Thus,

$$P_{abs} = P_{em} = cu\sigma.$$

The magnitude of the scattering cross section $\sigma$ at resonance is of some interest. If the frequency of the electromagnetic wave is equal to the characteristic frequency of the Hertzian dipole, then the scattering cross section $\sigma = \sigma_{res}$ is at its maximum. Its magnitude is independent of the parameters of the dipole and depends only on the wavelength $\lambda$ of the resonant wave:

$$\sigma_{res} = \frac{3\lambda^2}{2\pi}.$$

## Problem 3.21

Calculate the resonance cross section for scattering of an electromagnetic wave with a frequency of 100 MHz on a Hertzian dipole.

Up to now we have assumed that an electromagnetic wave excites oscillations of a dipole which is then damped only by reradiation, so that exactly as much energy is emitted as is absorbed. A receiver antenna should, however, divert part of the absorbed energy to a load (a user). Thus, in the case of an antenna, an additional damping owing to the load must be taken into account. The energy delivered to the load is then calculated from the *effective antenna area* $\sigma_w$. With optimum matching (load impedance = radiation impedance), we have

$$\sigma_w = \frac{\sigma_{res}}{4}.$$

## Problem 3.22

A 2 W lamp is to be lit with a 100 MHz antenna. What is the minimum required intensity of the received electromagnetic wave?

---

### Experiment 3.10   Transmitting antennas

---

A 450 MHz oscillator circuit is coupled inductively to a rod antenna of length $l = \lambda/2$. The electromagnetic wave emitted by the antenna ($\lambda = 67$ cm) is received by a second rod antenna with an incandescent lamp as load. If the two antennas are not too far apart ($d < 1$ m) and parallel, then the incandescent bulb lights up. If, however, the antennas are perpendicular to one another, then oscillations are not excited in the receiver antenna and the bulb remains dark.

Explanation: the electromagnetic wave emitted by the transmitting antenna is linearly polarized. The electric field vector oscillates in the direction of the transmitting antenna. The electric field of the wave cannot excite oscillations in a rod antenna that is perpendicular to the direction of the field.

---

**Notes**

An antenna tuned to resonance absorbs a power from an electromagnetic wave with wavelength $\lambda$ equal to the energy flux flowing through an area on the order of $\lambda^2$; that is,

$$P_{\text{abs}} \sim cu\lambda^2.$$

### 3.6.4 Waveguides

Electromagnetic waves not only propagate in a vacuum independent of any matter, they can also propagate in dielectrics and be guided along electrical conductors. In order to calculate the distribution of the field of these waves, the electric polarization **P** and magnetization **M** of the dielectrics, as well as the charges and currents in the conductors, must be included in Maxwell's equations. Here we avoid a detailed discussion of these waves, and merely describe the propagation of electromagnetic waves along electrical conductors. We shall consider waves with wavelengths $\lambda$ in the cm range and frequencies $\nu$ in the GHz range. *Coaxial conductors*, in which the wave is guided between an inner and an outer conductor (Figure 3.37), and hollow *waveguides,* in which the electromagnetic field propagates in a hollow space (Figure 3.38), are especially suited for guiding electromagnetic waves along electrical conductors. The latter is possible only if the width of the hollow space is greater than half the wavelength. Coaxial conductors and waveguides have the advantage that the electromagnetic fields are completely enclosed in conductors, so that no waves can be emitted into the surrounding space.

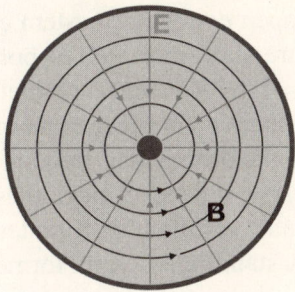

**FIGURE 3.37**  A cross section of a coaxial cable showing electric and magnetic field lines.

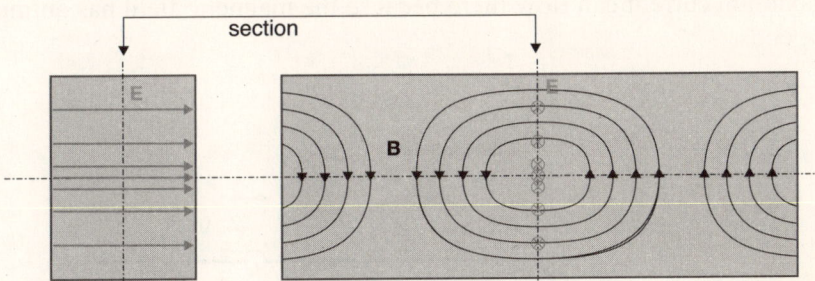

**FIGURE 3.38**  Cross sections of a waveguide showing electric and magnetic field lines.

The electric and magnetic fields in coaxial conductors and waveguides, unlike those of electromagnetic waves propagating freely in space where the electric and magnetic fields are mutually interdependent, are to a significant extent determined by the currents flowing on the surfaces of the conductors and the resulting surface charges. Thus, in a coaxial conductor the magnetic field lines form concentric rings around the inner conductor, while the electric field lines are directed radially.

In a waveguide, waves can propagate in different oscillatory modes. Figure 3.38 shows the field lines of a $TE_{10}$-wave in a waveguide with a rectangular cross section. In the cross section of the waveguide, the electric field lies parallel to the narrow side and the magnetic field has annular field lines in the plane perpendicular to the electric field. It also has a component in the direction of propagation, parallel to the longitudinal axis of the waveguide.

## Experiment 3.11    Lecher lines

We illustrate the propagation of electromagnetic waves along electrical conductors in an open conductor system consisting of two parallel conducting wires which are short-circuited at both ends (Figure 3.39). The *Lecher line* has length $L = 1\,\text{m}$. An electromagnetic wave is produced in the Lecher line through an inductive coupling to an oscillator circuit at one of its short-circuited ends. For a transmitter frequency $\nu = 450\,\text{MHz}$ the wave has a wavelength $\lambda = c/\nu = 0.67\,\text{m}$. Under these conditions the Lecher line can be excited to resonant oscillations if its length is an integral multiple of half the wavelength. A standing wave is formed on the Lecher line. Unlike in a travelling wave, in these standing waves the nodes of the electric and magnetic field oscillations alternate along the line. The nodes of the electric field lie where there is no voltage between the wires and therefore, in particular, at the short circuits at the ends of the Lecher line. Nevertheless, a maximum current can flow there because the magnetic field has antinodes at the ends.

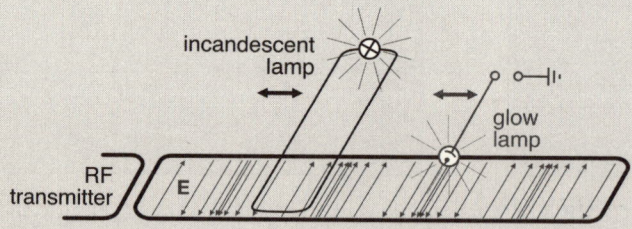

**FIGURE 3.39**  A Lecher line coupled to an oscillator circuit.

The distributions of the field along Lecher lines can be made visible with a simple glow lamp or an incandescent bulb. In a glow lamp a weak gaseous discharge burns between two electrodes. If it is brought near a voltage antinode, it will light up brightly. Then the electrons in the gaseous discharge will be strongly accelerated in the radio frequency (RF) electric field, so that when they collide with other atoms, more atoms will be excited and ionized, and the discharge will grow stronger. At the current antinodes, enough current can be coupled to another wire loop tuned to resonance that an incandescent bulb in that loop will light up.

**Notes**    Electromagnetic waves can be guided in waveguides. The field distribution inside waveguides depends essentially on the current and charge distributions on the surfaces of the waveguide.

# Chapter 4

# Electromagnetic Radiation

## Summary

- Geometrical optics
- Wave optics
- Photons
- Thermal radiation

Sound waves are coupled to a carrier medium, such as gas filling a space, but electromagnetic waves can also propagate in a vacuum. This difference is of practical significance for applications of waves, but it also raises some fundamental questions. Since electromagnetic waves can propagate in a vacuum, a distant object in the skies, such as the sun, can warm the earth and make life on earth possible, and we can see the sun and more distant stars and galaxies and observe the processes taking place in them with telescopes and radio telescopes. But we cannot hear the stars, since sound waves do not propagate in the interstellar medium (which has only a few atoms per $cm^3$).

One of the consequences of the discrete structure of matter is that no sound waves with wavelengths $\lambda < l$ (the mean free path of the atoms) exist (Section 3.3.1). But since Maxwell's theory is a continuum theory, electromagnetic waves can have wavelengths over many orders of magnitude, ranging from subatomic to cosmic (Figure 4.1). In addition, according to Maxwell's theory they are not attenuated in a vacuum by dissipation.

The contrast between the discrete structure of matter and the continuous structure of the electromagnetic field seemed to be a serious obstacle in 1900 when Max Planck was trying to explain the spectrum of thermal radiation. This is because the classical theory of heat is based on the *atomic hypothesis* and the laws of chance for the thermal motion of the atoms. In terms of Maxwell's continuum theory of the electromagnetic field, there was no way to use the laws of chance for solving the thermal radiation problem. Thus, Planck postulated the *quantum hypothesis* named after him and thereby ascribed a discrete structure to the electromagnetic field. The quantum hypothesis can be regarded as an important step in the direction of a unified physical world view. In fact, it provided the impetus for the fantastic development of science and technology during the 20th century. This chapter leads to the quantum hypothesis. In the next two chapters we shall describe the development of physics based on the quantum hypothesis.

## 4.1 GEOMETRICAL OPTICS

We are familiar with electromagnetic radiation as light, that is, electromagnetic waves with wavelengths in the range $0.4\,\mu m < \lambda < 0.7\,\mu m$ (Figure 4.1). With increasing wavelength $\lambda$ the color of the light shifts from violet, through blue, green, yellow, and orange, to red. These are the colors of the rainbow. Since the wavelength $\lambda$ of visible light is very small compared to the typical distances $d$ of the objects surrounding us in daily life, light propagates approximately in the form of rays, despite its wave nature. Thus, under illumination, opaque objects cast a shadow. The laws of geometric (ray) optics will be discussed in this section. They are valid when the objects affecting the propagation of light have very large sizes $d \gg \lambda$.

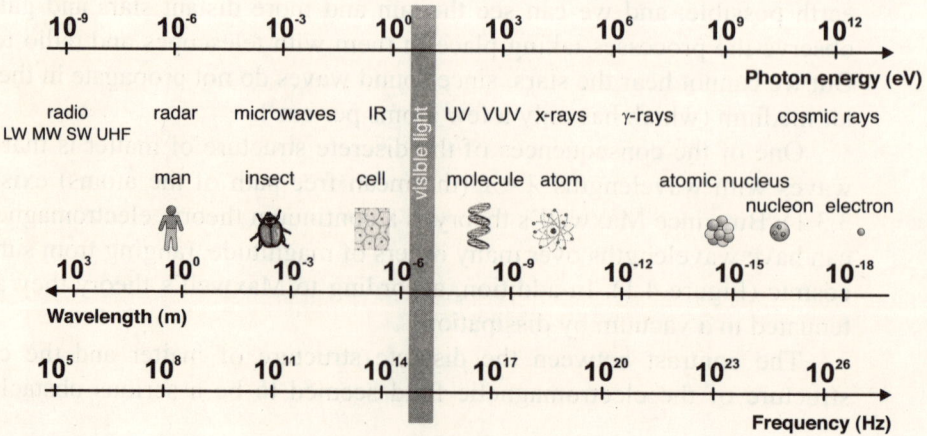

**FIGURE 4.1** The spectrum of electromagnetic waves.

## 4.1.1 Reflection and refraction

As daily experience teaches us, light propagates in a straight line in space (a vacuum). Thus, we speak of *rays of light*. More precisely, we mean beams of light with diameters $\phi \gg \lambda$. A laser beam with a length of a few meters has a diameter of at least 1 mm. If such a ray of light is incident on the surface of a transparent medium, it will generally split into a *reflected* and a *refracted* ray (Figure 4.2). The directions of the rays are customarily specified relative to the *surface normal*. All three rays usually lie in a single plane with the surface normal, and the angle of incidence $\alpha$, the angle of reflection $\alpha'$, and the angle of refraction $\beta$ satisfy the following equations:

$$\alpha = \alpha' \qquad \text{(law of reflection)}$$

$$n_\alpha \sin \alpha = n_\beta \sin \beta. \qquad \text{(Snell's law of refraction)}$$

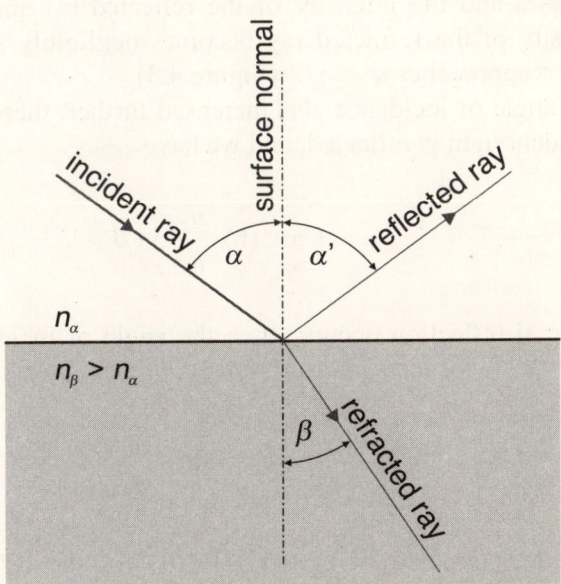

**FIGURE 4.2** Refraction and reflection of a ray of light at a boundary surface.

Here the *refractive indices n* are properties of the materials of the two media $\alpha$ and $\beta$ above and below the boundary surface. The refractive index of the vacuum is arbitrarily set at $n_{\text{vac}} = 1$. Then, for example, for air (under normal conditions) and water we have approximately $n_{\text{air}} = 1.0003$ and $n_{\text{w}} = 1.33$, respectively.

Since water has a higher refractive index, it is said to be *optically thicker* than air and, correspondingly, air is *optically thinner* than water.

In general, on passing from an optically thinner to an optically thicker medium a ray of light will be refracted toward the surface normal ($\beta < \alpha$), and in the opposite case, it will be refracted away from the surface normal. If a ray of light moves from an optically thicker medium $\beta$, e.g., water, across the surface to an optically thinner medium $\alpha$, e.g., air, then the following two cases are to be distinguished with respect to refraction:

$$\text{Case (a)} \quad \sin\alpha = \frac{n_\beta}{n_\alpha}\sin\beta < 1.$$

In this case there is one refracted and one reflected ray. Of course, the intensity ratio of the two rays falls off if the angle of incidence changes. For near perpendicular incidence, the refracted ray is substantially more intense than the reflected ray. As the angle of incidence is increased, however, the intensity of the refracted ray decreases and the intensity of the reflected ray increases correspondingly. The intensity of the refracted ray becomes negligibly small when the angle of refraction $\alpha$ approaches $\alpha = \pi/2$ (Figure 4.3).

If the angle of incidence $\beta$ is increased further, there is no refracted ray. All of the incident light is reflected, and we have

$$\text{Case (b)} \quad \frac{n_\beta}{n_\alpha}\sin\beta > 1. \qquad \text{(total internal reflection)}$$

Total internal reflection occurs when the angle of incidence is greater than the *critical angle* $\beta_T$ for *total internal reflection*, that is, when $\beta > \beta_T$, where

$$\beta_T = \sin^{-1}\frac{n_\alpha}{n_\beta}.$$

Since the deviation $\Delta n_{air} = n_{air} - 1 \approx 3 \times 10^{-4}$ of the refractive index $n_{air}$ of air from the vacuum refractive index is proportional to the density $\rho = nm$ of the air, total internal reflection also occurs where layers of air with different densities meet. Air reflection of this type, or a mirage, happens, for example, on hot summer days over asphalt roads when the layer of air over the black asphalt is substantially hotter than the layers of air above it. Because of the mirage, from far away the road appears to be wet. In the desert this phenomenon is known as a *fata morgana*.

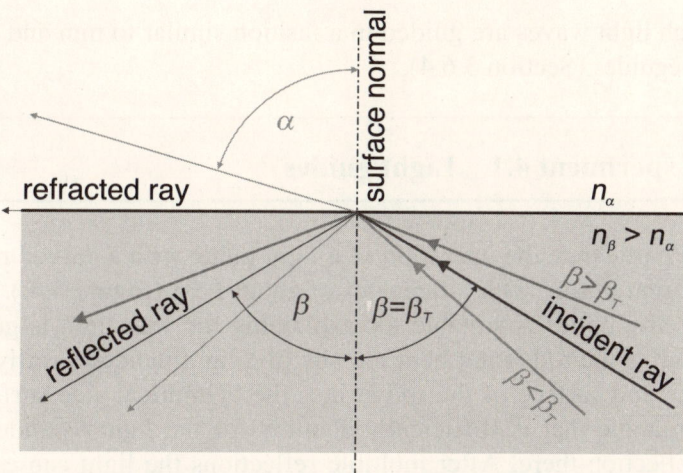

**FIGURE 4.3** Illustrating the definition of the critical angle $\beta_T$ for total internal reflection.

## Problem 4.1

Calculate the critical angle for total internal reflection for the case in which adjacent layers of air have a temperature difference $\Delta T = 30$ K. At what distance is a level street reflected if it is observed at a height of 1 m?

> **Notes**  At the interface between two media with different refractive indices $n_1 \neq n_2$ rays of light are usually broken up into a refracted ray and a reflected ray. The directions of the refracted and reflected rays obey the Snell's law of refraction and the reflection law. On passing from an optically thicker medium to an optically thinner medium, an incident ray is totally reflected if the angle of incidence is greater than the critical angle for total internal reflection.

### 4.1.2 Optical components

The reflection and refraction laws are the basis of an elementary understanding of many optical components. Here we consider a few examples.

**Light guides.** In modern optoelectronics, light guides have many applications. These are glass fibers with a diameter of a few tens of micrometers in which light waves are guided in a fashion similar to mm and cm waves in hollow waveguides (Section 3.6.4).

---

## Experiment 4.1    Light guides

---

We illustrate the operation of a light guide with a curved rod made of transparent plastic with a diameter of about 1 cm (Figure 4.4.). In this case, geometric optics is suitable for explaining the way light is guided through the rod. If the rod is not bent too sharply, light incident nearly perpendicular to the end surface of the rod strikes the cylindrical side surfaces of the rod at an angle that is sufficiently shallow for the light to undergo total internal reflection there. After multiple reflections the light can escape through the other end of the rod. This demonstration also shows, however, that a lot of the light is scattered on impurities and air bubbles in the rod material, and thereby emerges through the sides. Commercial light guides are, however, made of extremely pure material, so these losses are greatly reduced.

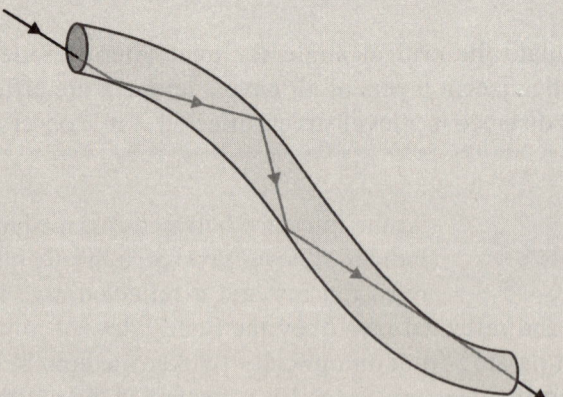

**FIGURE 4.4** A curved rod of transparent plastic operating as a light guide.

**Prisms.** A prism is an object with a triangular cross section. If a beam of white light is sent through a transparent prism, the light will be decomposed into the colors of the rainbow (Figure 4.5). Since the prism is usually optically thicker than the surrounding medium, when the light beam enters the prism, it is refracted toward the surface normal and as it leaves the prism, the beam is refracted away from the surface normal. Since the surfaces at which the beam enters and leaves the prism are not parallel, the light beam is deflected. The spectral decomposition of the incident light beam is a consequence of the *dispersion* of light waves (Section 3.1.4). The phase velocity $v_{ph}$ of the light waves in the material is given by the refractive index $n$ and the speed of light $c$ in a vacuum:

$$v_{ph} = \frac{c}{n}.$$

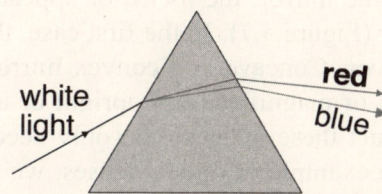

white light

red

blue

**FIGURE 4.5** Passage of light through a prism.

Only in a vacuum do all electromagnetic waves propagate at one and the same phase velocity $c$ independently of their wavelength. The propagation speed of light in matter depends on the refractive index, which usually varies with wavelength. The refractive index is usually higher at shorter wavelengths. Thus, violet light will be more strongly refracted than red light.

**Mirrors.** A plane mirror projects a *virtual* mirror image behind the plane of the mirror. The path of the light rays is determined strictly by the reflection law (Figure 4.6). The rays of light emerging from a luminous point in front of the mirror will be reflected by the mirror and then appear to come from a mirror image point behind the mirror.

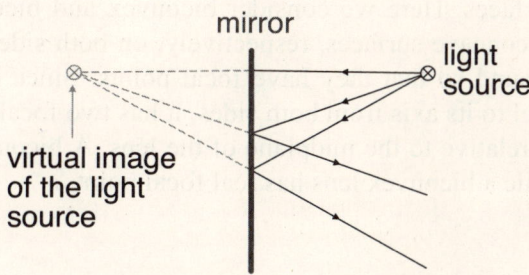

mirror

light source

virtual image of the light source

**FIGURE 4.6** Ray paths at a flat mirror.

### Problem 4.2

Show that the mirror image of a righthanded screw is a lefthanded screw.

It is more difficult to construct the path of the rays for curved mirror surfaces. The mirrors employed in optics are, however, usually rotationally symmetric and curved so that all light rays which are parallel to the axis of symmetry, the *optical axis* of the mirror, and are incident on the mirror will either converge at a point in front of the mirror, the *focus*, or appear to emerge from a virtual focus behind the mirror (Figure 4.7). In the first case, the mirror surface is *concave* and in the latter, *convex*. Concave and convex mirrors can produce either virtual or real, demagnified or magnified, and upright or upside down images. In order to be able to construct these images, you only need to know the focal point of the mirror. Using the example of optical lenses, we shall show how the image of an object can be determined graphically by simple *ray tracing*.

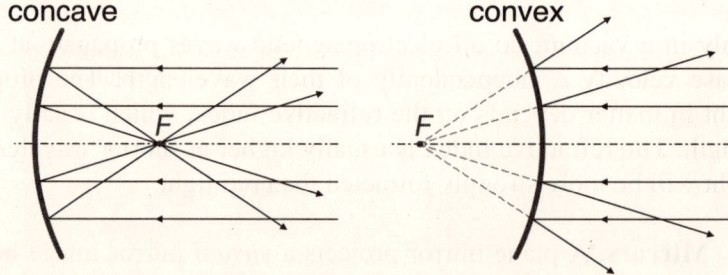

**FIGURE 4.7** Reflection of paraxial rays by concave and convex mirrors (convergence of the rays at the focus $F$).

**Lenses.** Optical lenses are typically thin, usually rotationally symmetric objects made of glass or other transparent material. Lens can also have concave or convex surfaces. Here we consider biconvex and biconcave lenses which have convex or concave surfaces, respectively, on both sides (Figure 4.8). Lenses can also be ground so that they have focal points. Since light can be incident on a lens parallel to its axis from both sides, it has two focal points, which have mirror symmetry relative to the midplane of the lens. A biconcave lens has virtual focal points, while a biconvex lens has real focal points.

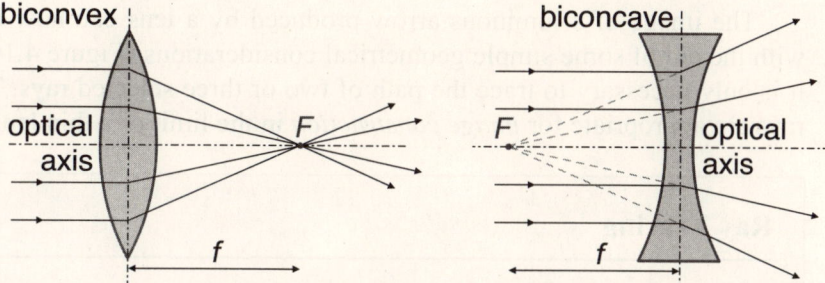

**FIGURE 4.8** Refraction of paraxial rays by biconvex and biconcave lenses.

A ray passing through the midpoint of the lens travels in a straight line, unrefracted, through the lens. Except at the midpoint, however, the lens acts as a prism and deflects the rays toward the focus. In the following we shall neglect the many small corrections to this model, known as *lens aberrations*. In particular, we neglect the dispersion of light in the lens material and the thickness of the lens and proceed as if the rays were refracted only at the midplane of the lens. Thus, in design drawings, convex (converging) and concave (diverging) lenses are often simply represented by double pointed arrows whose points are directed outward or inward, respectively.

In addition, if a beam of parallel rays that are not parallel to the optical axis is incident on a lens, in this approximation the beam will be focused onto a point (Figure 4.9) which lies in the focal plane, but away from the focal point (focus). The position of the focal point (in the focal plane) is indicated by the principal ray, which passes unrefracted through the center of the lens.

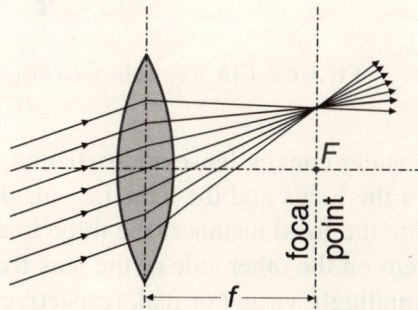

**FIGURE 4.9** Focusing of a parallel bundle of rays onto a point in the focal plane away from the focus.

The image of a luminous arrow produced by a lens can now be determined with the aid of some simple geometrical considerations (Figure 4.10). To do this, it is only necessary to trace the path of two or three selected rays. The following rays are appropriate for *image construction* in the limit of a thin lens:

---

## Ray Tracing

- **Paraxial ray**: it is refracted in a way such that it passes through the focal point on the opposite (emerging) side.
- **Principal (chief) ray**: it passes unrefracted and along a straight line through the lens.
- **Focal ray**: it passes through the focus and parallel to the axis on the other side of the lens.

---

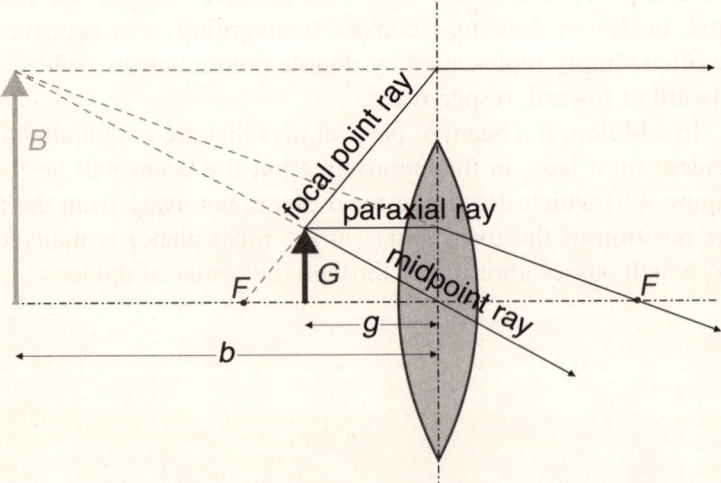

**FIGURE 4.10**  Ray paths for a magnifying glass.

We shall consider convex lenses in particular. Depending on whether the arrow lies between the focus and the lens, i.e., inside the first *focal distance* (*focal length*), or outside the focal distance, the three image construction rays are divergent or convergent on the other side of the lens from the arrow. The image of the arrow is correspondingly virtual or real, respectively.

In order to use a convex lens as a *magnifying glass*, the object must lie inside the first focal spot of the lens (Figure 4.10). Behind the lens one sees a magnified, upright virtual image $B$ of the object $G$. Of course, rays emerge from every point of the illuminated object in all directions, and all the rays which pass through the lens and enter the eye of the observer contribute to forming the image seen by the observer. As for the rays used to construct the image, it does not matter whether they intersect the lens or not.

A convex lens projects an inverted real image of an object lying outside its first focus (Figure 4.11). In a camera such an image is retained on film. If the object $G$ lies within a distance equal to twice the focal length, then the object is magnified, otherwise it is demagnified. The magnification $V = B/G$ (image size to object size) and the distance $b$ of the image from the lens are given in terms of the focal length $f$ of the lens and the object distance (distance of the object from the lens) $g$ by

$$\frac{1}{f} = \frac{1}{g} + \frac{1}{b} \qquad \text{(thin lens formula)}$$

$$V = \frac{b}{g}. \qquad \text{(linear magnification)}$$

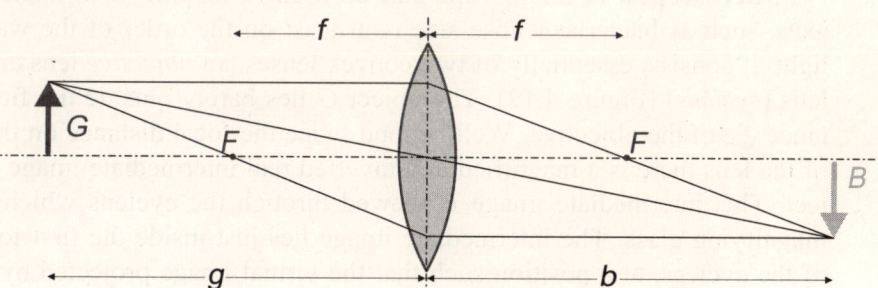

**FIGURE 4.11** Ray paths through a convex lens yielding a real image of an object.

## Problem 4.3

Prove these equations using drawings with image construction rays. Show that the thin lens equation can also be used with magnifying glasses if the image distance $b$ is taken to be negative.

**Problem 4.4**

Make a sketch of the ray paths for convex lenses. Under what condition does a virtual image result? Under what condition does a demagnified (magnified) real image result? Are the images right side up or upside down?

---

**Notes**   Optical lenses and mirrors are usually ground so that paraxial rays will be focused to a real or virtual focus. Lenses and mirrors can produce real or virtual, demagnified or magnified, and right side up or upside down images of objects. In each case the images can be determined geometrically using image construction rays.

### 4.1.3  Microscopes and telescopes

Many optical instruments can be built from the components described in Section 4.1.2. We shall consider microscopes and telescopes as examples.

**Microscopes.** A microscope can be used to magnify and make visible objects, such as bacteria, whose size is at least on the order of the wavelength of light. It consists essentially of two convex lenses, an *objective* lens and an *ocular* lens (*eyelens*) (Figure 4.12). The object $G$ lies barely outside the first focal distance $f_{ob}$ of the objective. Well beyond twice the focal distance on the other side of the lens there is a magnified and inverted real intermediate image $Z$ of the object. This intermediate image is viewed through the eyelens which serves as a magnifying glass. The intermediate image lies just inside the first focal distance of the eyelens, at a position such that the virtual image projected by the eyelens will lie within the clear seeing distance $s = 0.25$ m.

**Telescopes.** With a telescope one can see Jupiter, for example, as a luminous disk in the sky like our moon and see the planet's moons orbiting around it. In order to see objects such as Jupiter and its moons, which are so close, distinctly, the angular separation $\Delta\alpha$ at which we see these objects from the earth must be magnified. This goal is also reached with the aid of two convex lenses arranged in a row (Figure 4.13). The objective lens has a large diameter. It projects an intermediate image of a distant object approximately into the focal plane, or a distance $f_{ob}$ from the lens. An essentially infinitely distant star will be imaged as a

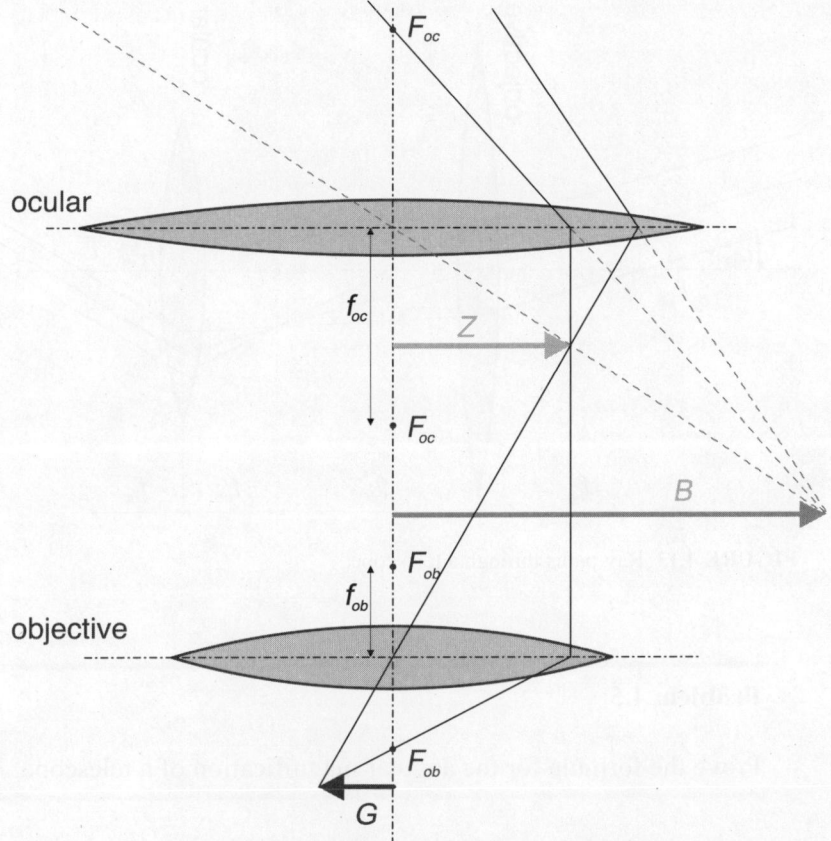

**FIGURE 4.12** Ray paths through a microscope.

point in the focal plane. In a telescope this focal plane of the objective coincides with the focal plane of the eyelens (ocular). Through the eyelens, therefore, one again sees an intermediate image of a star in the form of a point as an infinitely distant point. But the angular separation of the star from the optical axis of the telescope has been magnified by a factor $V$, which is given by

$$V = \frac{\alpha_B}{\alpha_G} \approx \frac{\tan \alpha_B}{\tan \alpha_G} = \frac{f_{ob}}{f_{oc}}. \quad \text{(angular magnification of a telescope)}$$

Likewise, two neighboring stars in the sky will be seen through a telescope with an angular separation that is magnified by a factor $V$.

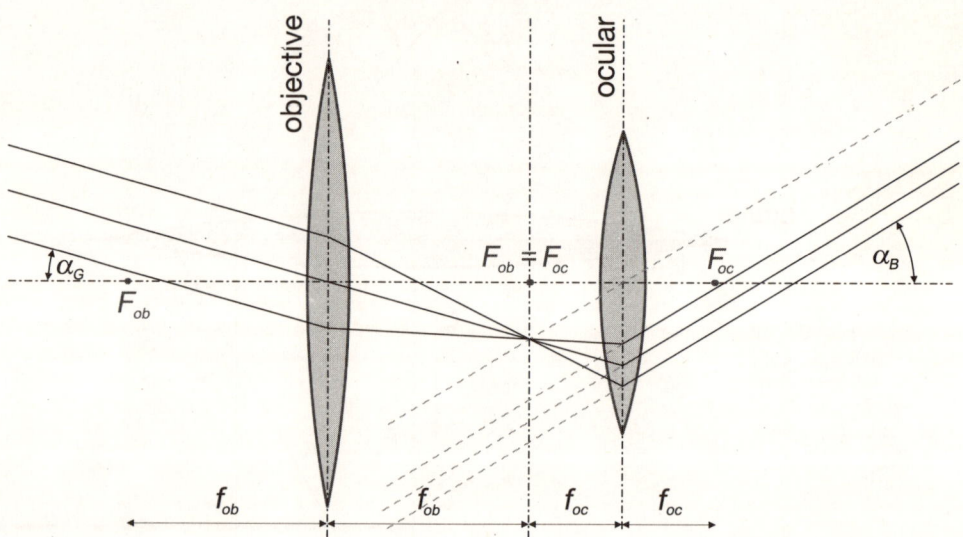

**FIGURE 4.13** Ray paths through a telescope.

---

### Problem 4.5

Prove the formula for the angular magnification of a telescope.

---

| **Notes** | Microscopes and telescopes are optical instruments that essentially consist of two lenses, an objective lens and an eyelens. In a microscope the objective produces a magnified |

real image of the object, which is observed through the eyelens as through a magnifying glass. In a telescope, the focal points of the two lenses coincide between the objective lens and the eyelens. The angular magnification is equal to the ratio of the focal lengths of the two lenses. In a telescope, as opposed to a microscope, $f_{ob} \gg f_{oc}$.

## 4.1.4 Measuring the speed of light

With observations of the moons of Jupiter it became possible for the first time to determine the propagation speed of light. Light takes about 8 minutes to reach the earth from the sun. Hence, the time taken for light from Jupiter to reach the earth varies over about 16 minutes. It reaches us about 16 minutes earlier when the earth and Jupiter are on the same side of the sun, than when the two planets are on opposite sides of the sun. This time interval affects the observed orbital periods of Jupiter's moons and leads to deviations from the theoretical values. Olaus Römer analyzed these deviations and in 1676 determined the speed of light.

Since the speed of light in a vacuum is a constant of nature that is independent of the relative velocity of a light source and observer and also does not depend on a relative velocity with respect to any medium (Section 1.2.4), it is of fundamental significance for physics and, since 1983, for metrology (Section 1.1.1). In a lecture demonstration we now show how the speed of light (in air) can be measured using the methods of geometrical optics, specifically Foucault's three-mirror method (1862) (Figure 4.14).

---

### Experiment 4.2   Measuring the speed of light

---

A laser beam is reflected by a mirror (plane and mirror reflecting on both sides) rotating at frequency $\nu$, travels a distance $3f$, and is then reflected back onto the rotating mirror by a fixed (flat) mirror. Before the light has returned to the rotating mirror, the latter has turned a bit further and, therefore, no longer reflects the laser beam back in its original direction, but in a slightly different direction. The propagation speed of the light can be determined from the change in direction, $\Delta\varphi$, in a way similar to the velocity measurement in Section 1.2.2. In order to measure $\Delta\varphi$ as accurately as possible, the laser beam is focused using a lens with a focal length $f = 5\,\text{m}$. The lens images the stop in a proportion of $1\!:\!1$ onto the fixed mirror. The distance travelled by the light from the stop to the mirror is thus equal to $4f$. The rotating mirror lies at the focal length of the lens. On its return, the reflected light passes the lens once again, so that another image of the stop is produced at the screen. The beam splitter, which in each passage transmits 50% of the light and reflects the remaining 50%, makes it possible for the laser and screen to be mounted in different spatial positions. The displacement $\Delta s$ of the light spot on the screen is measured as a function of the rotation frequency $\nu$ of the rotating mirror. Here $\Delta\varphi = \Delta s / f$. The speed of light, $c$, is then calculated using the formula

$$c = \frac{24\pi\nu f}{\Delta\varphi}.$$

*continued*

$\Delta s$ is measured by determining the position of the light spot when the rotating mirror is rotating and when it is stopped. The displacement of the light spot gives the familiar value $c = 3 \times 10^8$ m/s (with reasonable accuracy).

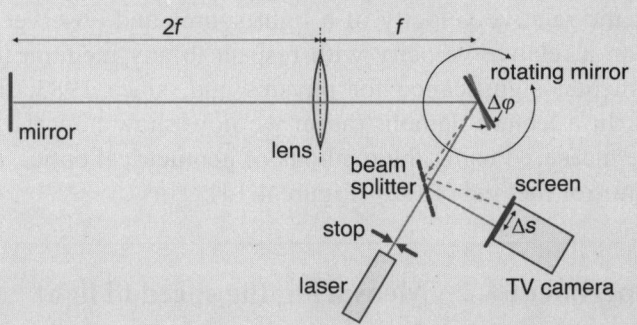

**FIGURE 4.14** Experimental apparatus for measuring the speed of light.

**Problem 4.6**

Prove the formula for calculating the speed of light. What rotation frequency $\nu$ of the rotating mirror is required for the light spot to be displaced by $\Delta s = 3$ mm on the screen?

**Notes**   Light travels a distance of 300 km (Berlin-Hamburg) in 1 ms. Light reaches the moon from the earth in somewhat over 1 s. The sun's light reaches us in about 8 min. The universal constant of nature $c$ is approximately equal to $c = 3 \times 10^8$ m/s.

## 4.2  WAVE OPTICS

Light and electromagnetic waves propagate in vacuum at the same speed,

$$c = \sqrt{\frac{1}{\varepsilon_0 \mu_0}} \approx 3 \times 10^8 \text{ m/s}.$$

For this reason Maxwell assumed that they are physically equivalent and differ only in frequency and wavelength. The typical radio waves and microwaves of electrical engineering have wavelengths in the km, m, and mm ranges. Visible light, on the other hand, has wavelengths less than $1\,\mu m$ (Figure 4.1). Like all waves, light waves also obey the superposition principle (Section 3.1.3). Under suitable conditions, the superposition of light waves produces interference and diffraction patterns. Since about 1800 *interference* and *diffraction* of light have been demonstrated and studied experimentally, first by **Thomas Young (1773–1829)** and later by **Augustin Jean Fresnel (1788–1827)**, in particular. As opposed to the idea of a ray of light, electromagnetic waves are, in principle, extended in space. An exactly plane wave stretches out over all space (Section 3.3.1). But light propagating in a finite volume $V$ (whose extent is very large compared to the wavelength of the light) can often be described approximately as a plane wave of amplitude $A$ with a wave vector $\mathbf{k}$. The wave vector points in the direction of the *wave normal*, so it is perpendicular to the wave front. Thus, in the framework of wave optics, the elementary picture of light rays must be replaced by a picture of plane waves propagating like rays in the direction of the wave normal.

## 4.2.1  The colors of thin films

The display of colors produced by soap bubbles and oil spots in the sunlight offers a beautiful and impressive example of the phenomenon of interference. If the wave nature of light is taken into account, this display of colors can be explained by the laws of reflection and refraction. Here it is important that the light is not only reflected and refracted at an interface, but that the two interfaces are separated by only a few wavelengths or are just a few $\mu m$ apart (Figure 4.15). The outer and inner surfaces of a soap bubble or the upper and lower surfaces of an oil film on an asphalt street are appropriate examples.

A plane light wave with wavelength $\lambda$ incident on a thin film with a thickness of a few $\mu m$, such as a soap bubble or an oil film, is initially decomposed into a reflected component and a refracted component on striking the thin film. The refracted wave is, however, reflected and refracted as it emerges from the film. The reflected wave is once again refracted and reflected at the entry surface, and this process continues onward. In this way, many partial waves are produced, which move out in the direction of the reflection from the entry surface and are superimposed. Depending on the relative phase $\varphi$ of the partial waves, *constructive* or *destructive interference* take place with the superposition of the waves. The phase $\varphi$ is determined by the *optical path difference* $\Delta$ of the partial waves and the wavelength $\lambda$ of the light. If the phase discontinuity $\Delta\varphi = \pi$ which

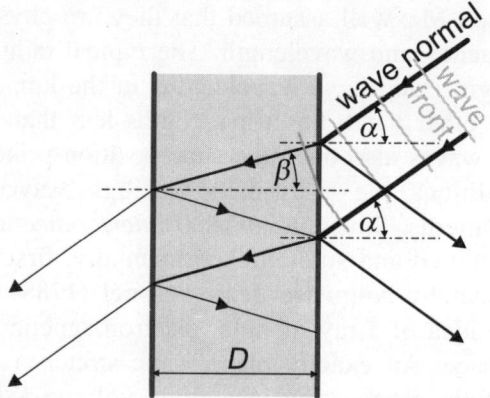

**FIGURE 4.15** Reflection and refraction of a light wave at a thin film.

occurs upon reflection at an optically thicker medium (as during the reflection of sound waves from the closed end of a flute, Section 3.1.2) is taken into account, then we obtain the

## Conditions for interference

- **Constructive interference**, when $\varphi = 2n\pi$ is an integral multiple of $2\pi$, and
- **Destructive interference**, when $\varphi = (2n+1)\pi$ is a half-integer multiple of $2\pi$.

Here $\varphi = \Delta/\lambda + \pi$ and $\Delta$ is obtained from the thickness $D$ of the layer, the angle of incidence $\alpha$ of the wave, and the refractive index of the medium. As a crude approximation we neglect the difference in the refractive index of the different media in calculating $\Delta$, so that we can set $\alpha = \beta$. Then the wave normals move along a straight line across the interfaces and the geometry of the ray path implies that

$$\Delta = 2D\cos\alpha.$$

---

### Problem 4.7

Prove the formula for $\Delta$. In doing this, keep in mind that the wave fronts (loci of equal phase) are perpendicular to the wave normals.

---

Along with the optical path difference $\Delta$, the phase difference $\varphi$ of the partial waves arising from multiple reflections is also dependent on the film thickness $D$ and the angle of incidence $\alpha$. The condition for constructive interference is, therefore, satisfied for different colors at different angles of incidence. An observer thus sees different parts of the film in different colors, depending on the film thickness and the angle of incidence of the light.

> **Notes**  Visible light is an electromagnetic wave. The wavelength increases on going from violet to red from about 0.4 µm to about 0.7 µm. Two light waves can interfere if they have a constant phase relation. They interfere constructively where the electromagnetic field of the light oscillates in phase and destructively, where it oscillates out of phase.

## 4.2.2 Coherence

Up to now we have assumed tacitly that a light wave is a strictly periodic occurrence. Under this assumption interference structures should be expected when light is reflected at thick glass plates, such as window panes. In fact, sunlight in particular is made up of all colors of the spectrum and, therefore, represents the superposition of a broad spectrum of monochromatic waves. A superposition of this sort is generally an utterly nonperiodic phenomenon. This is because, according to the Fourier theorem, an arbitrary function $f(x - ct)$ of space and time can be decomposed into strictly periodic waves $\exp(i(kx - \omega t))$ with $\omega/k = c$. Thus, if light is represented as a wave, we find that, as with water waves, the distances of neighboring wave peaks fluctuate irregularly with time and from place to place. These irregular fluctuations mean that interference structures can only be observed under appropriate *interference conditions*.

For a strictly periodic plane electromagnetic wave the electric and magnetic field strengths $\mathbf{E}(\mathbf{r},t)$ and $\mathbf{B}(\mathbf{r},t)$ are determined everywhere in space and at all times, if only the wave vector $\mathbf{k}$ and the frequency $\omega$, as well as the field strengths at *one* place and *one* time, are known. In this case we speak of a spatially and temporally fully *coherent* wave. For real light waves, however, these conditions can only apply for limited intervals of space and time. Thus, the coherence of a light wave is characterized in terms of its *coherence length* and *coherence width* and in terms of its *coherence time*. We shall illustrate the significance of the coherence length for the case of measurements with a Michelson interferometer.

**Michelson interferometer.** A Michelson interferometer basically consists of a beam splitter and two mirrors (Figure 4.16). The light from a light source (laser) is initially incident on the beam splitter, e.g., a partially transmitting mirrored glass plate, and is then broken up into a transmitted (path 1) and a reflected component (path 2). The transmitted light is reflected back toward the beam splitter by a fixed mirror and the reflected light, by a mirror that can be moved with micrometer precision. Both parts of the beam are again split by the beam splitter into transmitted and reflected components. Thus, two light waves which have travelled along different path lengths are superimposed on the screen after the beam splitter. The path difference $\Delta$ can be varied since the moveable mirror can be displaced. Depending on the magnitude of $\Delta$, intensity maxima and minima appear at the screen $S$ as a result of constructive and destructive interference.

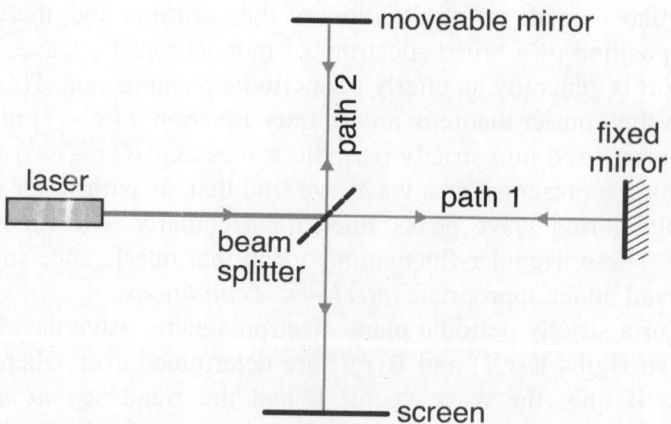

**FIGURE 4.16** Schematic representation of a Michelson interferometer.

## Experiment 4.3   Coherence length

We use a Michelson interferometer to measure the coherence length of light from various light sources. If light from a (approximately) point source of monochromatic light is sent through the interferometer, an image of the light source, i.e., a small spot of light, will be seen on the screen (when imaging optics are used). The brightness of the image varies periodically with the path difference $\Delta$. In this experiment let us use an extended light source which emits light with a spectral bandwidth $\Delta \nu$ (Figure 4.32). In this case one sees an interference pattern on the screen which consists of concentric bright and dark rings, provided the optical path lengths of the two partial waves are somewhat similar. When the path difference is increased or decreased by displacing the moveable mirror, interference rings initially appear or disappear, respectively, in the center of the interference pattern. But as the path difference is increased further, the interference structure fades away until finally no interference structure is recognizable.

Let us study the interference patterns of an incandescent lamp and an Hg spectrum lamp. In the incandescent lamp a hot tungsten wire glows. It emits light, similarly to the sun, over the entire spectral range of visible light (Section 4.4.3). In the Hg spectrum lamp, mercury vapor is excited to emission in an electrical discharge. Then the mercury atoms radiate a yellow spectrum line that is characteristic of mercury (Section 5.3.4). A color filter absorbs the other visible spectrum lines of mercury. Interference patterns can be observed with both light sources. But in both cases the path difference $\Delta$ must not be made too large. With the incandescent lamp, concentric interference rings can only be seen if $\Delta$ is, at most, about 2 μm. The path difference $\Delta$ can be varied over a substantially wider range when the spectrum lamp is used as a light source. The interference rings disappear only when $\Delta$ is on the order of 1 cm.

The experimental results show that interference patterns only develop if the path difference $\Delta$ does not exceed the order of magnitude of the *coherence length* $L_{\text{coh}}$ of the light. The coherence length is a measure of the *spectral width* $\Delta \nu$ of the light (Figure 4.17). It characterizes the spectral intensity distribution $I(\nu)$ of the light being analyzed. The spectral width of the spectrum lines of the Hg lamp is mainly determined by the thermal motion of the Hg atoms. The broadening of the spectrum line is, therefore, a consequence of the Doppler effect (Section 3.3.4). For this reason, $\Delta \nu$ for the yellow Hg spectrum line with a frequency

$v_0 \approx c/\lambda_0 \approx 5\times10^{14}$ Hz follows from the ratio of the magnitude of the average thermal velocity $v_{th}$ of the atoms to the velocity of light, i.e.,

$$\Delta v \approx 2\frac{v_{th}}{c}v_0.$$

The coherence length $L_{coh}$ of a spectrum line is inversely proportional to the width of its spectral distribution, with

$$L_{coh} = \frac{c}{\Delta v}.$$

The coherence length for the yellow Hg spectrum line is on the order of 1 cm, in agreement with the experiment. The light from an incandescent bulb, however, has a much greater spectral width, so that its coherence length is on the order of the wavelength of visible light.

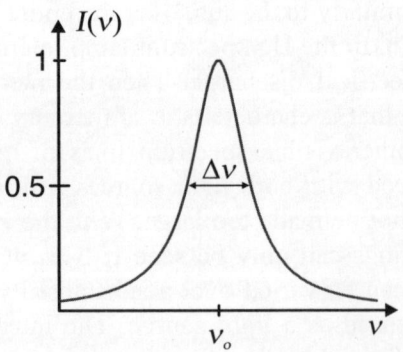

**FIGURE 4.17** Spectral intensity distribution $I(v)$ of a spectrum line as a function of frequency $v = c/\lambda$.

## Problem 4.8

Calculate the coherence length of light whose spectral width is $\Delta\lambda = 1$ nm on a wavelength scale.

| **Notes** | The coherence length of a light source is inversely proportional to the spectral width of the light. |
|---|---|

### 4.2.3 Diffraction

Geometrical optics is based on the seemingly obvious assumption that light propagates along a straight line in a homogeneous medium. As an electromagnetic wave, however, a "ray of light" also extends outward perpendicular to the direction of propagation. The question therefore arises of what happens when an extended light ray of this type is drastically limited by a small aperture diaphragm. How does the light propagate beyond such a diaphragm?

This question was answered in 1690 by **Christian Huygens (1629–1695)**, a contemporary of Newton's. According to the *Huygens-Fresnel principle* an *elementary wave* propagates as a sphere away from the hole in the stop into the entire space behind the stop (Figure 4.18) provided the aperture is small enough, i.e., its diameter is less than or equal to the wavelength of the light. Thus, the propagation direction can change when the waves pass through an aperture stop. The wave is said to be diffracted. *Diffraction* can be observed directly for water

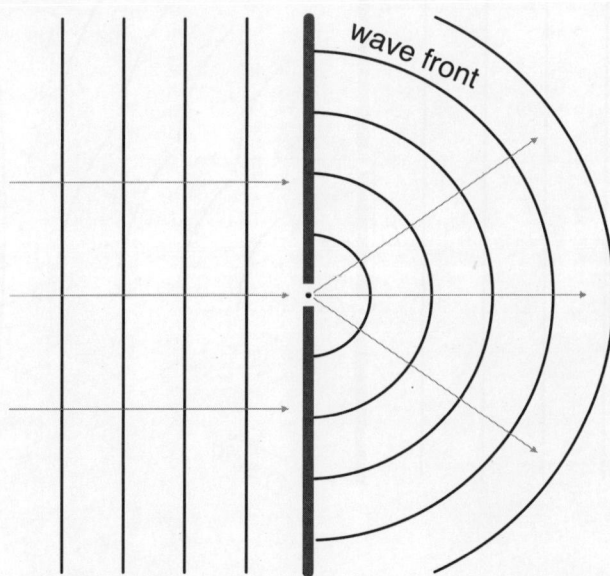

**FIGURE 4.18** Propagation of an elementary wave behind an aperture stop.

waves travelling through a slot. In the following we describe some simple experiments with light which demonstrate that light waves are also diffracted.

**Diffraction at a double slit.** Light from a sufficiently monochromatic and plane (and, therefore; coherent) light source (these days it is best to use a laser) is incident on a double slit (Figure 4.19). According to the Huygens-Fresnel principle, two elementary waves move out from the double slit and interfere with each other. Depending on the diffraction angle $\alpha$, the two elementary waves interfere constructively or destructively. If a plane wave is normally incident on a double slit with a distance $D$ between the slits, the two elementary waves will be excited in phase. Let us consider the two elementary waves in the direction of the wave normals with a diffraction angle $\alpha$. Because of the path difference $D\sin\alpha$, in this direction the elementary waves have a phase difference $\Delta\varphi$ given by

$$\Delta\varphi = 2\pi\frac{D\sin\alpha}{\lambda}.$$

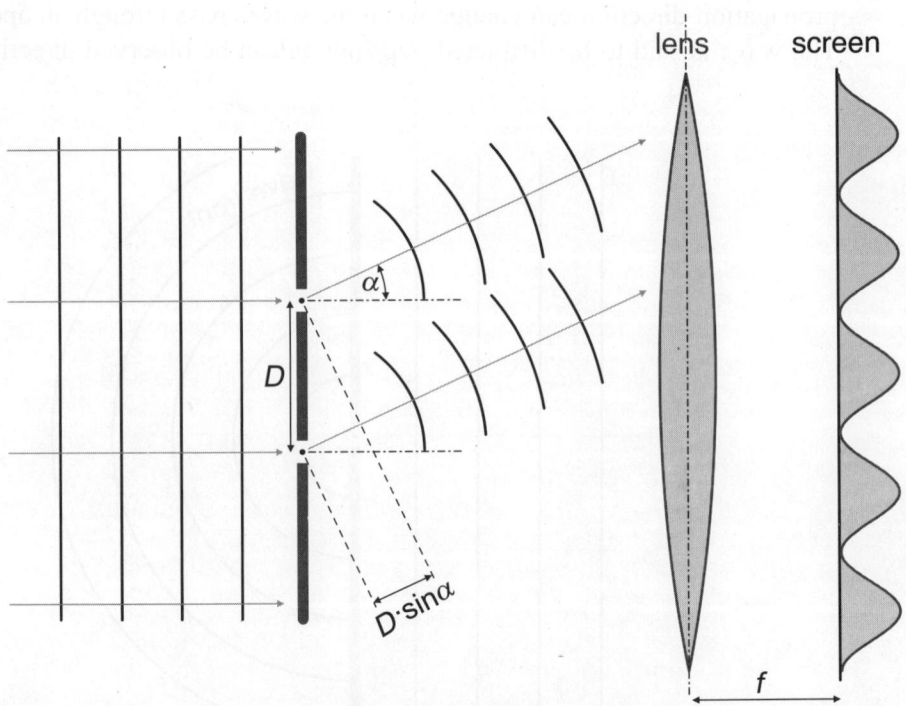

**FIGURE 4.19** Diffraction (interference) at a double slit.

The waves propagating along the parallel wave normals are focused by a collecting lens $L$ to a point in the focal plane of this lens and interfere constructively or destructively there, depending on the phase difference $\Delta\varphi$ and whether the path difference $D\sin\alpha$ is an integral or half-integral multiple of the wavelength $\lambda$. On a screen mounted in the focal plane, therefore, one can see an interference pattern with periodically varying intensity. Since $\alpha \ll 1$ for these experimental conditions, $\sin\alpha \approx \alpha$. The angular separation of neighboring diffraction maxima is given in this case by the ratio of the wavelength to the distance between the slits; i.e., $\Delta\alpha \approx \lambda/D$.

**Diffraction at a single slit:** In the discussion of diffraction at a double slit we have tacitly assumed that the widths of the slits are no greater than the wavelength $\lambda$ of the light, so that only *one* elementary wave emerges from each slit. In an experiment, the slits widths are usually several times the wavelength. Thus, it is important to study how a slit width $b \gg \lambda$ affects the interference pattern.

Here we consider a plane wave that is normally incident on a single slit of width $b$ and determine the wave field behind the slit. In the limiting cases of very small and very large slit widths, we obtain, respectively, the wave field of an elementary wave or an (almost) sharply bounded wave field with light and shadow regions determined by the laws of geometrical optics. We want to investigate how one limiting case goes to the other as the slit width is increased. To do this, we imagine dividing the slit into many small segments, out of which an elementary wave emerges. We combine these elementary waves into pairs of waves which interfere destructively with one another (Figure 4.20). This is possible if the path difference $b\sin\alpha$ of the two waves emerging from the slit boundaries is an integral multiple of the wavelength $\lambda$; that is,

$$b\sin\alpha = N\lambda, \text{ with } N \geq 1. \qquad \text{(condition for destructive interference at an individual slit)}$$

To prove this statement, let us divide the slit into $2N$ equal segments and number them in order with numbers from 1 to $2N$. In the direction of the diffraction angle $\alpha$ the elementary waves emerging from the odd segments cancel out those emerging from the even segments owing to mutual destructive interference, since obviously only wave pairs with a path difference of $\lambda/2$ can be formed.

These considerations do not apply for $N = 0$, since $\alpha = 0$ in this case. In the forward direction all the elementary waves are in phase and therefore interfere constructively. Thus, there is an intense maximum in the forward direction. Besides this zero order maximum, weaker side maxima appear between the minima formed by destructive interference. The intensities of the side maxima fall off rapidly with increasing order.

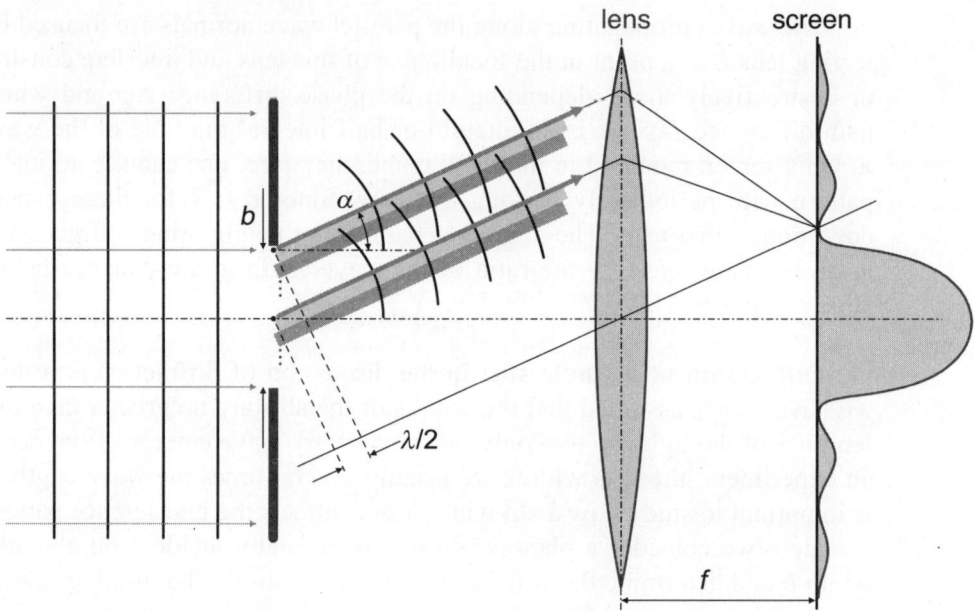

**FIGURE 4.20** Diffraction at a single slit of width $b$.

Judging from the example of diffraction at a single slit, it is clear that light behind a stop is never a plane wave in the strict sense. Rather, the wave field behind a stop is always somewhat divergent, even when a strictly plane wave is incident on the stop. The aperture angle $\delta\alpha$ of the wave field behind a stop is usually specified by the half width of the zeroth order diffraction maximum. For a slit stop of width $b$, this gives

$$\delta\varphi \approx \frac{\lambda}{b}.$$

A (diffraction limited) laser beam with $\lambda = 1\,\mu\text{m}$ and a diameter $D = 0.1\,\text{m}$ which is aimed at the moon from the earth will have a diameter of about $D^* \approx \lambda s/D = 4\,\text{km}$ when it reaches the moon, a distance of about $s \approx 400{,}000\,\text{km}$ away.

---

### Problem 4.9

Calculate the angular width of the zeroth order diffraction maximum for red light incident on a slit with width $b = 1\,\text{mm}$.

---

| Notes | A plane wave incident on an aperture stop of radius $r$ will travel as a spherical wave beyond the stop. The aperture angle of the zeroth order diffraction maximum is on the order of $\lambda/2r$. |
|---|---|

### 4.2.4 Resolution of optical instruments

It is impossible to make atoms visible with an (optical) microscope or to resolve small structures on distant stars with a telescope. The diffraction of light at the *aperture stops* of these optical instruments limits their *resolution* (*resolving power*).

According to geometrical optics (Section 4.1.3) the objective of a microscope images every point of an object as a point in the plane of the intermediate image. According to wave optics, however, a point object appears in the plane of the intermediate image as a small, but extended spot of light, a *diffraction disk*. If the diffraction disks of two neighboring light points overlap, these points of light cannot be observed as separate with the microscope. The size of the light spot is a result of the diffraction of light on the entrance aperture determined by the objective lens of the microscope, the aperture stop. With a microscope the object lies near the focal point of the objective lens whose focal length is usually of the same order of magnitude as its diameter. In this case, points on objects can only be observed as separate if the distance between them is at least on the order of the wavelength of the light, or not much less than $1\,\mu m$.

For a telescope the resolution is obtained directly from the angular width $\Delta\varphi$ of the zeroth order diffraction maximum produced by diffraction at the aperture stop. Two stars with an angular separation of $\delta$ can be resolved by a telescope and seen as separate only if $\delta > \Delta\varphi = \lambda/D$. Here $D$ is the diameter of the aperture and $\lambda$ is the wavelength of the light.

---

**Problem 4.10**

Estimate the minimum angular separation of two stars such that they can be perceived as separate by the naked eye.

---

> **Notes**   The resolution of a telescope is limited by diffraction at the aperture stop and is, thereby, determined by the diameter $D$ of the objective lens. The smallest measurable angular separation is on the order of $\delta\varphi = \lambda/D$.
>
> The smallest spatial structures measurable by an optical microscope have sizes on the order of the wavelength $\lambda$ of the light.

## 4.3 PHOTONS

In 1905, his *annus mirabilis*, Albert Einstein published four important papers with which he decisively determined the development of modern physics. The first of these papers was entitled "On a heuristic viewpoint regarding the generation and conversion of light" and began with this sentence: "There is a profound formal difference between the theoretical ideas which physicists have developed about gases and other ponderable bodies and the Maxwellian theory of electromagnetic processes in so-called empty space." While ponderable bodies were considered to be discretely structured systems made up of many, but a finite number of atoms, according to Maxwell's theory electromagnetic processes involve variations in a field that stretches continuously over space. This structural difference leads to difficulties in the interpretation of processes where matter and electromagnetic fields interact, especially the emission and absorption of electromagnetic waves. The problems that arise in the course of interpreting these processes can only be solved if a discrete structure is ascribed to electromagnetic waves. The *photon hypothesis* discussed in this chapter is a counterpart, applicable to the electromagnetic field, of the atomic hypothesis for the description of matter.

### 4.3.1 Photoelectric effect

When light is incident on a metallic surface part of the light is absorbed. Then the energy delivered by the light wave is primarily taken up by the conduction electrons in the metal. Some electrons gain so much energy in this way that they can leave the metal. The energy of the electrons and the electrical current produced by this electron flux can be studied with a simple experimental setup (Figure 4.21).

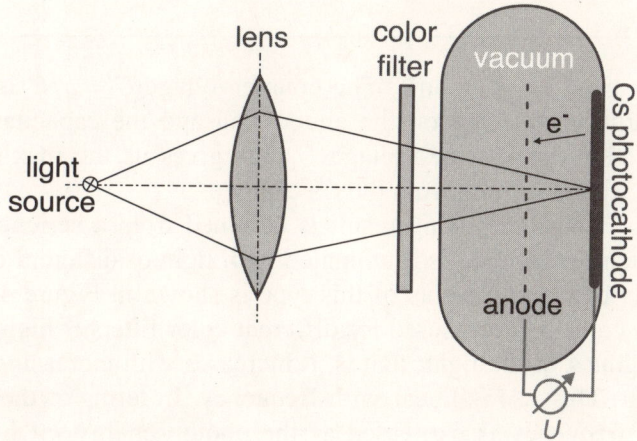

**FIGURE 4.21** Experimental setup for demonstrating the photoelectric effect.

## Experiment 4.4    Photoelectric effect

A glass vessel evacuated to about $10^{-4}$ Pa serves as an electron tube with a cathode and an anode. A device of this type is known as a *diode*. The cathode is a vapor-deposited metallic layer of the alkali metal cesium (Cs) on the glass wall and the anode is a wire ring. The cathode and anode are attached to external electrical leads to which voltage or current measurement apparatus can be connected. In particular, we measure the voltage which builds up between the cathode and anode when the Cs cathode is illuminated with light. For this we connect a high-impedance voltmeter (i.e., a voltmeter with a high internal resistance, $R_{in} \approx 10^9 \ \Omega$) between the anode and cathode.

The measurements show, first of all, that when the Cs cathode is illuminated a voltage $U$ on the order of a few volts builds up between the anode and cathode. Evidently some electrons gain enough kinetic energy when the cathode is illuminated to reach the anode ring and to charge it negatively. This produces an opposing voltage between the anode and cathode, which the electrons must overcome before they can reach the anode. The charging process can continue like this only so long as the kinetic energy $E_{kin} > eU$ of at least a few liberated electrons still exceeds the potential energy $eU$

*continued*

that must be overcome. The countervoltage $U = q/C$ is determined by the charge $q$ collected on the anode ring and the capacitance $C$ of the diode (Section 3.4.4). The voltage $U$ thus gives us the maximum energy of the electrons escaping from the Cs cathode.

A most surprising result is obtained from a series of measurements in which the cathode is illuminated with light of different colors. The result of a set of measurements of this type is shown in Figure 4.22. The photoelectric voltage $U$ measured for different color filters is higher for shorter wavelengths $\lambda$ of the light; that is, it increases with increasing frequency $\nu$ of the light. This rise is linear with frequency. In terms of the photon hypothesis, the frequency is expressed as the photon energy $h\nu$ in energy units [eV] (Section 3.4.1). (Note: in measuring the photoelectric voltage it is necessary to make sure that the photoelectric current is high enough compared to the input current of the voltmeter, $I = U/R_{in}$ .)

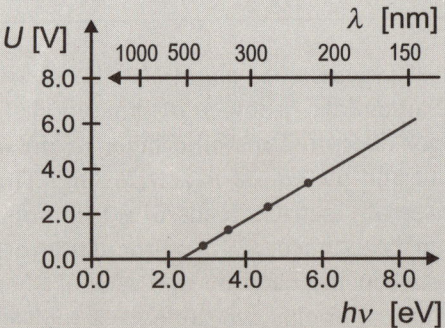

**FIGURE 4.22** Photoelectric voltage as a function of the frequency of light.

The experiment yields the following simple relation between the energy $eU$ required for the electrons to reach the anode from the cathode and the frequency $\nu$ of the light:

$$eU = h\nu - W_A. \quad \text{(Einstein equation for the photoelectric effect)}$$

Here $h$ is Planck's quantum of action (Planck's constant) $h = 6.63 \times 10^{-34}$ J·s, a constant that is independent of the cathode material and of the intensity of the light. The constant $W_A$ is obviously the energy required for conduction electrons to be able to leave the cathode. For a Cs cathode the measurements yield a value for the work function of $W_A = 2.14$ eV. In terms of a simple model we assume that the conduction electrons in the metal move as if in a potential well with a depth equal to the work function $W_A$ (Figure 4.23). Only when an electron gains an energy greater than $W_A$ upon illumination of the cathode can the electron escape the cathode. The experiment shows that on absorbing light of frequency $\nu$ the electrons absorb a maximum energy of $h\nu$. Only if $h\nu > W_A$ can they leave the cathode.

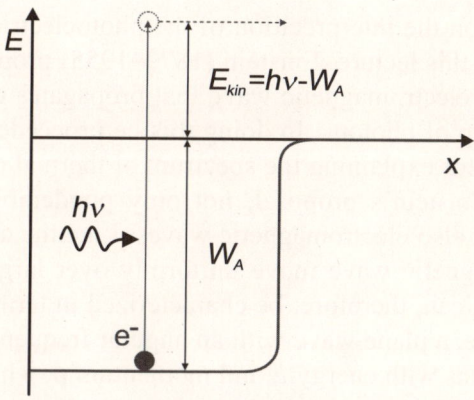

**FIGURE 4.23** The potential well of the conduction electrons in a metal and a schematic representation of the photoelectric effect.

An entirely different result is to be expected from Maxwell's electromagnetic theory, according to which the electrons should be continuously accelerated in the field of a light wave. Thus, the kinetic energy gained by the electrons under the influence of light should depend on the electric field strength of the light wave and thereby on the intensity of the light. The measurements show, however, that the *photoelectric voltage* is independent of the intensity of the light and only depends on the frequency. With increasing intensity only the *photoelectric current*, or the number of electrons liberated per second, increases.

---

**Problem 4.11**

Most metals have work functions ranging from 4 to 5 eV. Calculate the wavelength of electromagnetic radiation which can just barely liberate photoelectrons from these materials. To which spectral range of the electromagnetic spectrum does this radiation belong?

---

| Notes | The maximum energy of photoelectrons is determined solely by the frequency or wavelength of the light, but is independent of its intensity. With rising intensity only the number of photoelectrons liberated per second increases. |
|---|---|

### 4.3.2 Photon hypothesis

In his paper on the interpretation of the photoelectric effect mentioned in the introduction of this lecture, Einstein (1879–1955) proposed that light should not be treated as an electromagnetic wave that propagates continuously in space, but as a gas made up of photons. In doing this he proceeded from Planck's idea of five years earlier for explaining the spectrum of thermal radiation (Section 4.4.4). According to Einstein's proposal, not only ponderable bodies have an atomistic structure, but also electromagnetic waves. Like the atoms in a gas, the photons in an electromagnetic wave move uniformly over large distances in straight lines. Their motion can, therefore, be characterized in terms of energy and momentum. In this picture, a plane wave with an angular frequency $\omega$ and wave vector $\mathbf{k}$ consists of photons with energy $E$ and momentum $\mathbf{p}$, where

$$E = \hbar\omega = h\nu \quad \text{and}$$

$$\mathbf{p} = \hbar\mathbf{k}, \text{ with } \hbar = h/2\pi = 1.0546 \times 10^{-34} \text{ J} \cdot \text{s}.$$

On one hand, the photon hypothesis is a major step toward a unified physical world view, but on the other, it represents a radical renunciation of the classical picture of light as an electromagnetic wave. At this point we shall not elaborate further on the dichotomy between the particle picture and the wave picture of light. At the start it is more important to point out that, besides the photoelectric effect, other phenomena which are involved in the absorption, emission, and scattering of light, can be explained easily in terms of the photon hypothesis.

If we assume that light consists of photons, the photoelectric effect is immediately understandable. Under this assumption the absorption of light can no longer be treated as a continuous process, but is to be seen as a discrete sequence of elementary absorption events in each of which a photon transfers its energy $E = h\nu$, say, to a conduction electron in a metal. Assuming that under the given experimental conditions an individual electron can absorb at most one photon, the particle picture of light then implies that after an electron is ejected from the metal the kinetic energy of the electron is at most $E_{\text{kin}} = h\nu - W_A$. It has exactly this value if no energy is lost in collisions with other electrons or atoms as it leaves the metal.

---

**Problem 4.12**

Calculate the energies of photons in the visible spectrum and compare these energies with the value of the thermal energy $kT$ at $T = 300$ K. What is the energy of radio wave photons with a frequency $\nu = 100$ MHz?

---

| Notes | Electromagnetic waves with a frequency $\nu$ are absorbed in quanta of energy $E = h\nu$. Here Planck's constant $h = 6.6 \times 10^{-34}$ J·s is a universal constant of nature. |
|---|---|

## 4.3.3 X-ray *bremsstrahlung* (slowing-down radiation) spectrum

In the same way as the absorption of light, the emission process must also be treated as a discrete sequence of elementary events, each involving a photon with energy $E = h\nu$. An experimental example of this statement is the x-ray bremsstrahlung spectrum of an *x-ray tube*.

An x-ray tube (Figure 4.24) is an electron tube with an incandescent cathode (a tungsten wire) and a massive anode made, for example, of molybdenum (Mo). Conduction electrons emerge from the incandescent cathode by *thermionic emission*. This is done by heating the cathode electrically to a temperature of about 3000 K (white heat). The electrons than have an average thermal energy of about $kT \approx 0.25$ eV. At thermal energies of this magnitude the fastest electrons can overcome the work function $W_A = 4.5$ eV of tungsten and thereby contribute to the formation of an electron current in the mA range from the cathode to the anode.

---

**Problem 4.13**

Estimate the probability that a conduction electron at a temperature of 3000 K will have a kinetic energy $E_{kin} > 4.5$ eV. How many conduction electrons are there in 1 mol of tungsten?

---

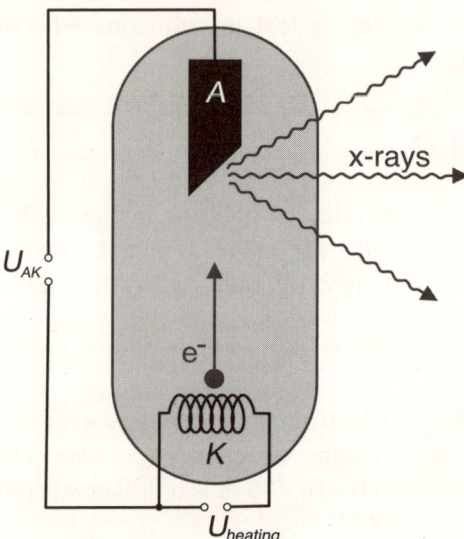

**FIGURE 4.24** Structure of an x-ray tube.

In the diode, the thermally emitted electrons are accelerated to energies of a few tens of keV, corresponding to a high voltage $U_{AK}$ of a few tens of kV between the anode $A$ and cathode $K$, and strike the anode with this energy. There, if they pass close enough to an atomic nucleus, they will be sharply deflected in the Coulomb field of the nucleus; that is, they experience an acceleration and therefore emit electromagnetic waves (Section 3.6.2). The closer the electrons come to a nucleus, the greater the deflection and, therefore, the higher the frequency of the emitted electromagnetic waves will be. The electrons lose energy in amounts corresponding to the radiated energy. Thus, they are slowed down, as well as deflected.

---

**Problem 4.14**

Under the assumption that the electrons travel in circular orbits of radius $r$ around the atomic nucleus, you can calculate the frequency and wavelength of the emitted electromagnetic waves. What frequency do you obtain for electrons orbiting at a distance $r = 10^{-12}$ m from a Mo nucleus (nuclear charge $Z = 42$)? What is the energy of the corresponding photons?

---

## Problem 4.15

Calculate the power with which an electron current $I \sim 1\,\text{mA}$ heats the anode of an x-ray tube. Why are molybdenum and tungsten suitable anode materials?

The x-ray bremsstrahlung spectrum (bremsstrahlung = slowing-down radiation) produced as the electrons are slowed down in the anode has a maximum at wavelengths below 0.1 nm, or wavelengths which are shorter than the distances between atoms in crystalline solids. X-ray bremsstrahlung spectra obtained by diffraction and interference on crystal lattices can therefore be analyzed (Section 5.6.1). Some x-ray spectra produced by an x-ray tube with a Mo anode are shown in Figure 4.25.

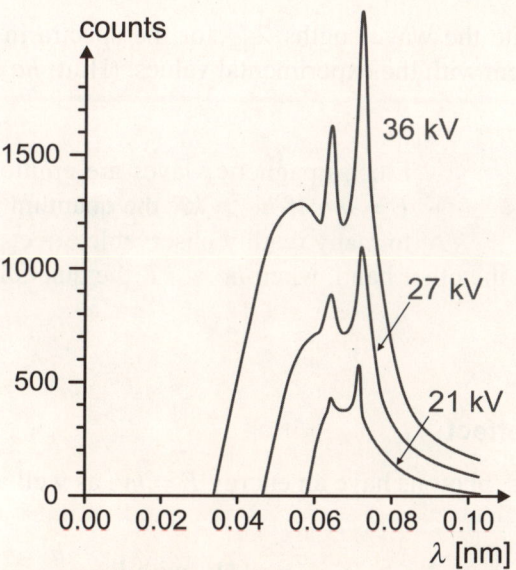

**FIGURE 4.25** X-ray spectra of a molybdenum anode for different voltages $U_{AK}$.

The sharpest intensity maxima, which always appear at the same wavelengths regardless of the accelerating voltage $U_{AK}$, are characteristic of the anode material Mo. They will be explained in a later lecture (Section 5.3.2). Here we are mainly interested in the broad x-ray bremsstrahlung spectrum underneath the peaks. It is noteworthy that this spectrum extends to ever shorter wavelengths with increasing voltage $U_{AK}$, but that there it cuts off suddenly at a wavelength

$\lambda_{min}$. (In interpreting these spectra, the resolution of the crystal spectrometer must be taken into account.)

This sudden cutoff of the x-ray bremsstrahlung spectrum cannot be explained in terms of Maxwell's electromagnetic theory, but it can be explained by Einstein's photon hypothesis. There each electron has a kinetic energy $E_{kin} = eU_{AK}$ when it strikes the anode. An individual electron can thus only produce photons with energies this high. The inequality $hv < eU_{AK}$ yields the following condition for the wavelengths of the x-ray spectrum:

$$\lambda > \lambda_{max} = \frac{hc}{eU_{AK}}.$$

---

**Problem 4.16**

Calculate the wavelengths $\lambda_{min}$ for the spectra in Figure 4.25 and compare them with the experimental values. (Hint: $hc = 1.237 \times 10^{-6}$ eV·m.)

---

| **Notes** | Electromagnetic waves are emitted and absorbed in quanta $E = hv$. If $hv \gg kT$ the quantum nature of the waves leads to many readily observable effects, such as the photoelectric |

effect. On the other hand, when $hv \ll kT$ the quantum steps are lost in thermal vibrations.

## 4.3.4 Compton effect

As particles, photons have an energy $E = hv$, as well as a momentum

$$\mathbf{p} = \hbar\mathbf{k}, \text{ with } \hbar = \frac{h}{2\pi}.$$

Here $\mathbf{k}$ is the wave vector of the light wave associated with the photon. Since $|\mathbf{k}| = 2\pi/\lambda$, the magnitude of the momentum of a photon is $p = h/\lambda$. The momentum has an effect in experiments where photons with sufficiently high energies are scattered on free electrons. Then collision processes of this sort are to be treated in the particle picture as elastic collisions of photons and electrons. Here energy and momentum are conserved (Section 1.3.4). When a photon is scattered on an electron initially at rest, the momentum of the photon changes by $\Delta \mathbf{p} = \mathbf{p}_i - \mathbf{p}_f$ and the electron experiences a corresponding recoil and acquires a corresponding recoil energy $E_R = (\Delta \mathbf{p})^2 / 2m$. Here $m$ is the electron mass. This energy is lost by the photon. The scattered photon thus has a lower energy than initially. The energy loss rises with increasing scattering angle $\theta$ and is greatest for backward scattering ($\theta = \pi$). In this case $|\Delta \mathbf{p}| = h(v_i - v_f)/c$. Then the energy lost by the electron by scattering is

$$E_R = h(v_i - v_f) = \frac{h^2 (v_i + v_f)^2}{2mc^2}.$$

Since $2mc^2 \approx 1\,\text{MeV}$, the energy loss is negligibly small for scattering of visible light photons, but is clearly observable (Compton effect) in experiments involving the scattering of photons in the x-ray region of the electromagnetic spectrum (Figure 4.1).

---

### Problem 4.17

Calculate the change $\Delta E$ in energy of a photon in the visible range ($\lambda = 500\,\text{nm}$) during backscattering on an electron. How precisely does the energy of the photon have to be measured in order to detect the change in energy? How big is the change in energy for an x-ray photon with energy $E = 50\,\text{keV}$?

---

| **Notes** | The photons of an electromagnetic wave with frequency $v$ and wave vector $\mathbf{k}$ behave during collisions as particles with energy $E = hv$ and momentum $\mathbf{p} = h\mathbf{k}/2\pi$. |
|---|---|

## 4.4 THERMAL RADIATION

Incandescent objects emit light. The higher its temperature, the more brightly an object emits and, as the brightness increases, the color of the light changes from a dark red glow at about 1000 K, through a bright red glow, to a white glow. But, even at temperatures below 1000 K, electromagnetic waves are emitted by all objects. Though we cannot see these waves, we can feel them as warmth or detect them with infrared detectors. The radiation emerging from objects with temperatures $T > 0\,\text{K}$ is referred to as *thermal radiation*.

Besides heat conduction and convection (Section 3.2.3), thermal radiation also contributes to temperature equilibration. Unlike heat conduction and convection, thermal radiation can transfer heat between objects in a vacuum. The sun's radiation which warms us reaches the earth through interplanetary space, which is almost free of matter. The behavior of thermal radiation will be discussed in this lecture. It is of surprisingly fundamental importance for physics. In the course of his efforts to understand the behavior of thermal radiation theoretically, in 1900 **Max Planck (1858–1947)** found it necessary to postulate the quantum hypothesis and, thereby, became the first to cast doubt on the continuity axioms of classical physics.

### 4.4.1 Emittance and absorptance

The characteristics of thermal radiation are mainly studied using measurement devices that can determine its intensity and spectrum quantitatively. With intensity measurements it is important to be sure that the measurement apparatus is able to detect the entire spectrum of the thermal radiation ranging from the far infrared to, as the temperature increases, shorter wavelength ranges, initially the visible and into the ultraviolet, with equal sensitivity. A detector of this sort is referred to as a *bolometer*. A *thermopile* (Figure 4.26) is quite suitable as a bolometer. It consists of a chain of thermocouples (Section 2.1.3) connected in series, with the junctions alternating between a blackened layer exposed to the radiation and a location that is shielded from the radiation. When it is irradiated the blackened layer (which is a good absorber over the entire spectral range) is heated to a slightly higher temperature. The temperature difference $\Delta T$ relative to the ambient, which is often less than $1\,\mathrm{K}$ and is measured as the thermocouple voltage, is an indicator of the absorbed radiant power.

The thermal radiation emitted by hot objects can be studied with a thermopile of this type and, in particular, the total and the spectral *emittance* of the object can be determined. The total emittance $E_K(T)$ is the overall radiant power emitted per unit area from the surface of the object;

$$E_K(T) = \text{radiated power/area} \quad (\text{units: W/m}^2).$$

In addition to the total radiated power, we are also naturally interested in how the radiated power is distributed over different spectral intervals. This can refer, as desired, either to the frequency $\nu$ or the wavelength $\lambda$ of the radiation. With respect to frequency, we have the *spectral emittance* $E(\nu, T)$, which is defined as the power radiated per unit area per unit frequency interval $\Delta \nu$:

absorber

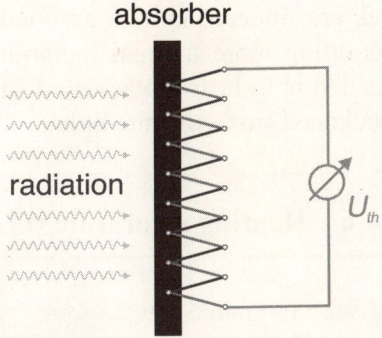

**FIGURE 4.26** Circuit diagram of a thermopile.

$$E(\nu,T) = \text{radiated power}/(\text{area} \times \Delta\nu) \quad (\text{units: J/m}^2).$$

With respect to wavelength, it becomes, correspondingly,

$$E(\lambda,T) = \text{radiated power}/(\text{area} \times \Delta\lambda)$$

$$= E(\nu,T)\left|\frac{d\nu}{d\lambda}\right| = E(\nu,T)\frac{c}{\lambda^2} \quad (\text{units: W/m}^3).$$

The emittance of an object usually depends strongly on the condition of its surface, in particular, its color. This is shown by the following experiment:

---

### Experiment 4.5   Leslie cube

A cubic metal container with one polished side and one soot-coated side is filled with boiling water and the thermal radiation of these two sides is measured with a thermopile (Figure 4.27). The thermal emission from the soot-coated side is about 10 times greater than that from the polished side.

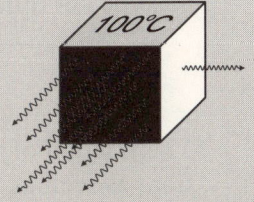

**FIGURE 4.27** Leslie cube.

Black surfaces are, indeed, known as good absorbers of light, but are less well known for emitting more thermal radiation, as this experiment teaches. In fact, all that is needed is to heat a blackened surface sufficiently in order to *see* how brightly a blackened surface emits light.

---

**Experiment 4.6   Heating a ceramic tube with black writing on it**

---

A ceramic tube with the name "Pythagoras" written on it in black is heated with a hot flame until it glows red (Figure 4.28). In its cold state and under normal illumination, the letters are black and the ceramic is white, since the black letters absorb the light and the white ceramic reflects it, while thermal radiation is emitted only in the invisible infrared portion of the spectrum. But when the ceramic pipe is heated with a hot flame to a temperature over 1000 K do we see (at least when the lights are turned off) that the black letters are much brighter than the ceramic tube.

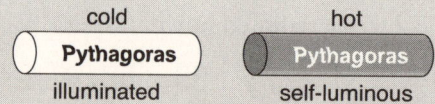

**FIGURE 4.28**  A ceramic tube with black lettering (left) illuminated in its cold state and (right) emitting light on its own when heated.

---

The clear relationship between absorption and emission that showed up here is worth investigating more precisely. To do this, we define the *absorptance* of the surface of an object. Assuming that the object is not transparent, electromagnetic waves incident on it with a spectral intensity distribution $I(\nu)$ are either absorbed or reflected, so that

$$I(\nu) = I_{abs}(\nu) + I_{refl}(\nu).$$

The fraction of the light that is absorbed depends on the frequency of the radiation, as well as on many other factors, such as the surface condition, color, and temperature of the object. As with the emittance, here we emphasize the dependence $\alpha(\nu, T)$ on just the frequency and temperature in defining the absorptance:

$$\alpha(\nu, T) = \frac{I_{abs}(\nu)}{I(\nu)}.$$

Since $I_{\text{abs}}(\nu) < I(\nu)$, the absorptance can only take values between 0 and 1. Black bodies have an absorptance $\alpha = 1$ in the visible range, while totally reflecting (white) objects have $\alpha \approx 0$. In the next section we shall show that the absorptance of an object is closely related to its emittance.

---

### Problem 4.18

Estimate the approximate spectral emittance $E(\lambda)$ of a hotplate heated with $1\,\text{kW}$ near the peak of the distribution (at $\lambda \approx 3\,\mu\text{m}$). Assume that the effective width of the distribution is $\Delta\lambda = \lambda$. What is the magnitude of the corresponding $E(\nu)$?

---

**Notes**    The spectral emittance $E(\nu, T)$ is the radiated power per unit area per unit frequency interval. On the other hand, $E(\lambda, T)$ is the radiated power per unit area per unit wavelength interval. Accordingly,

$$E(\nu, T)\,d\nu = E(\lambda, T)\,d\lambda.$$

Here the (positive) intervals $d\nu$ and $d\lambda$ refer to the same infinitesimal spectral range.

## 4.4.2 Kirchhoff's radiation law

Although the emittance $E(\nu, T)$ and the absorptance $\alpha(\nu, T)$ are functions that depend on the material and surface properties of particular objects, the ratio $E(\nu, T)/\alpha(\nu, T)$ is a universal function that is the same for all objects and, therefore, independent of any material or surface properties of objects. This, *Kirchhoff's radiation law*, is a consequence of the second law of thermodynamics. In order to derive it, let us consider two neighboring objects in a vacuum. Heat exchange between the two objects is possible only through emission and absorption of thermal radiation. According to the second law, net heat always flows from the warmer of the two objects. If both objects have the same temperature, they are in thermal equilibrium, and the heat fluxes flowing back and forth between the two must balance exactly.

In order to obtain a graphic illustration of the behavior of the heat flux, let one of the two objects be an ideal *black body* (BB). An object of this sort is distinctive in that it absorbs all the radiation incident on it, i.e., its absorptance

$\alpha_{\mathrm{BB}}(v,T)$ has the maximum value at all temperatures over the entire spectral range:

$$\alpha_{\mathrm{BB}}(v,T)=1. \qquad \text{(definition of a black body)}$$

Actually, no surface is black in this idealized sense, even if it seems black. The idea of a black body is very closely approached by a small hole in a large box made of black material (Figure 4.29). Radiation entering the hole is multiply reflected inside the box and, therefore, so strongly attenuated during absorption that essentially none of the incident radiation can reach the outside again.

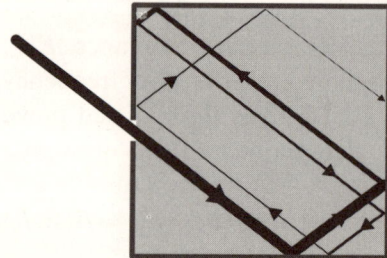

**FIGURE 4.29** Realization of an (almost) black body.

We now consider the radiation flux between an arbitrary object $X$ and a black body BB (Figure 4.30).

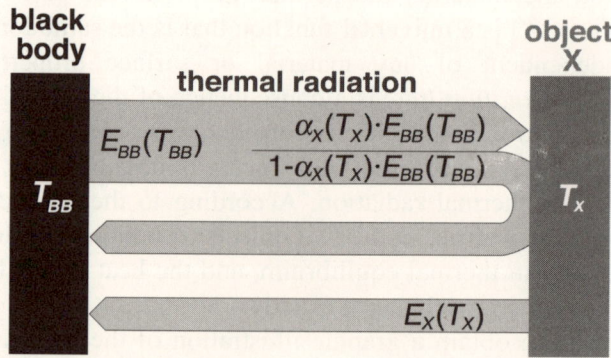

**FIGURE 4.30** Radiative flux between a black body and an arbitrary object $X$.

Both objects emit thermal radiation. But while object $X$ reflects part of the radiation incident on it, all of the radiation incident on the black body is absorbed. Suppose we have an (idealized) color filter that only transmits radiation between the two objects within a narrow frequency interval $\Delta \nu$. Then heat $E_{BB}(\nu,T)\Delta \nu \alpha_X(\nu,T)$ flows from the black body to object $X$ and heat $E_X(\nu,T)\Delta \nu$ flows from object $X$ to the black body. In thermal equilibrium, therefore, the balance equation $E_{BB}(\nu,T)\Delta \nu \alpha_X(\nu,T) = E_X(\nu,T)\Delta \nu$ holds, so that

$$\frac{E_X(\nu,T)}{\alpha_X(\nu,T)} = E_{BB}(\nu,T). \quad \text{(Kirchhoff's radiation law)}$$

Therefore, according to the second law of thermodynamics, for all objects the ratio of the emittance and the absorptance is equal to the emittance of a black body. This function is of fundamental importance in physics and by end of the 19th century physicists were aware of its significance. Efforts were made, on one hand, to determine it as exactly as possible experimentally and, on the other, to interpret it theoretically. In the course of these efforts, in 1900 Planck became convinced that the spectral emittance of black bodies could not be interpreted in terms of the physical theories of that time, the so-called classical physics. Only by postulating the then revolutionary quantum hypothesis could he explain the black-body spectrum.

## Problem 4.19

Draw a diagram of the radiative flux between two black bodies at different temperatures $T_1 > T_2$. In which direction does more radiant energy flow? Also consider the radiative flux for the case in which a color filter that only transmits light optimally within a narrow spectral interval $d\lambda$ and otherwise reflects it completely is placed between the two black bodies. What does this imply about the spectral emittance of black bodies with different temperatures?

| Notes | Of all objects at a given temperature $T$, a black body has the highest spectral emittance $E(\nu,T)$ over the entire spectrum. |
|---|---|

### 4.4.3 Emittance of a black body

An experimental setup suitable for measuring the emittance $E_{\mathrm{BB}}$ of a black body is shown in Figure 4.31. Here an electrically heated hollow object whose temperature is measured by a thermocouple serves as the black body. The thermal radiation emitted by a hole in the cavity is measured with a thermopile. In order to block radiation from the region near the hole, the hole is surrounded by a water cooled stop.

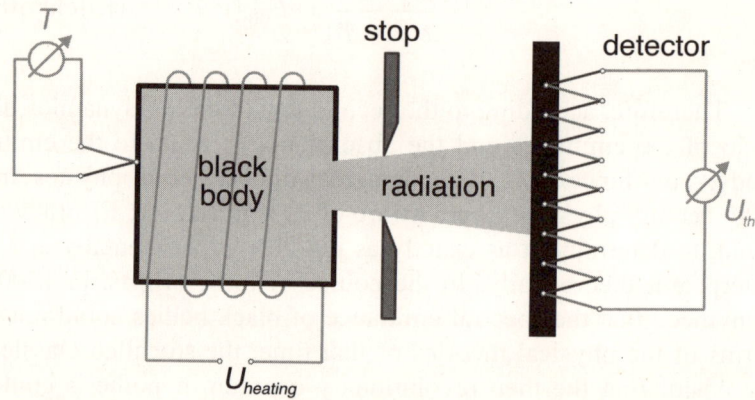

**FIGURE 4.31** Experimental setup for measuring the emittance of a black body.

It is easiest to determine the total intensity of the thermal radiation emerging from the hole and thereby the total emittance $E_{\mathrm{BB}}(T)$ of the black body. By definition,

$$E_{\mathrm{BB}}(T) = \int_0^\infty E_{\mathrm{BB}}(v,T)\,dv.$$

Measurements and theoretical considerations based on the second law show that the intensity of thermal radiation is proportional to the fourth power of the absolute temperature:

$$E_{\mathrm{BB}}(T) = \sigma T^4. \qquad \text{(Stefan-Boltzmann law)}$$

The proportionality constant is given by $\sigma = 5.67 \times 10^{-8}\ \mathrm{W \cdot m^{-2} \cdot K^{-4}}$. Thus, a black surface of area $1\ \mathrm{cm^2}$ at 1000 K radiates heat at a power of 5.67 W.

## Problem 4.20

The sun is also an almost ideal black body. Its surface has a temperature of about 5800 K. Calculate the power radiated by the sun. (The earth-sun distance, $1\,\mathrm{AU} = 150 \times 10^9$ m and the apparent size of the sun imply that $R_S = 0.7 \times 10^9$ m.) Calculate the intensity of the sun's light at the earth, or the *solar constant* $S = 1.4 \times 10^3$ W/m$^2$.

In order to investigate the spectrum of thermal radiation, it must be decomposed spectrally with, for example, a prism and then the intensities within the spectral intervals $\Delta\lambda$ must be measured. Measurements of this kind show that, as the temperature increases, the intensity of thermal radiation increases sharply while the spectrum also shifts to shorter wavelengths (Figure 4.32). The wavelength $\lambda_{\max}$ of the peak in the spectral distribution $E_{\mathrm{BB}}(\lambda, T)$ is inversely proportional to the temperature $T$, with

$$\lambda_{\max} T = const = 2.9 \times 10^{-3} \text{ m} \cdot \text{K}. \qquad \text{(Wien displacement law)}$$

At temperatures below 1000 K the spectrum of thermal radiation lies almost exclusively in the infrared (invisible to humans). At the temperature of the sun, by contrast, the peak in the spectral distribution is in the middle of the visible range.

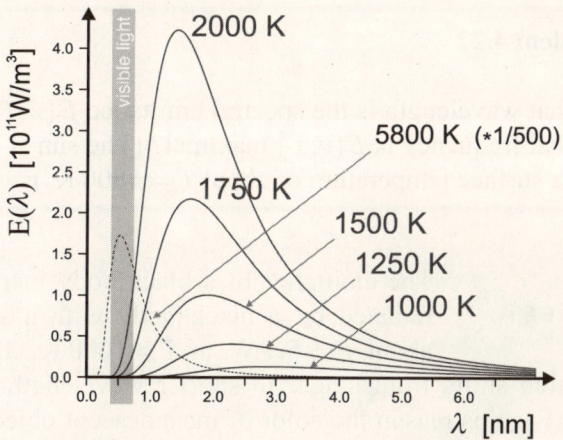

**FIGURE 4.32** Spectral emittance of a black body.

To within the experimental accuracy, the spectral emittance of a black body is given by the Planck's formula of 1900. As a function of frequency $\nu$ and temperature $T$ it has the form

$$E_{\mathrm{BB}}(\nu,T) = \frac{2\pi h}{c^2} \cdot \frac{\nu^3}{\exp(h\nu/kT) - 1}. \qquad \text{(Planck radiation law)}$$

Besides the already known constants $c$ (the speed of light) and $k$ (Boltzmann's constant), this equation contains *Planck's constant h* (the quantum of action) as a parameter. Integrating Planck's radiation formula over $\nu$ yields the Stefan-Boltzmann law with the constant

$$\sigma = \frac{2\pi^5}{15} \cdot \frac{k^4}{c^2 h^3}.$$

The numerical value of $\sigma$ is, as noted above, $\sigma = 5.67 \times 10^{-8} \ \mathrm{W \cdot m^{-2} \cdot K^{-4}}$.

---

### Problem 4.21

Calculate the power radiated by a completely black cube with edges of length 0.4 m and a temperature $T = 1000 \ \mathrm{K}$.

---

### Problem 4.22

At what wavelength is the spectral emittance $E(\lambda, T)$ of the sun greatest? At what frequency is $E(\nu, T)$ maximal? (The sun is an almost black body with a surface temperature of about $T = 5800 \ \mathrm{K}$.)

---

**Notes**   The emittance of a black body increases as $T^4$. The power radiated by a black body with a surface area of $1 \ \mathrm{cm^2}$ is about $P = 5.7 \ \mathrm{W}$ at $T = 1000 \ \mathrm{K}$. The peak in the spectral distribution shifts from longer to shorter wavelengths as the temperature increases. For this reason the color of incandescent objects goes from a dark red glow to a bright white glow as their temperature rises.

## 4.4.4  The quantum hypothesis

The exponent $h\nu/kT$ in the denominator of Planck's radiation formula makes it clear that the spectral distribution of black body radiation depends on the energy $h\nu$. In his efforts to interpret the energy distribution of this radiation physically, Planck postulated the quantum hypothesis. Under the assumption that electromagnetic waves are only emitted and absorbed in quanta $h\nu$, he was able to derive the radiation formula theoretically. Here we shall just point out some of the consequences of the quantum hypothesis.

Once we assume that, in addition to the well-known constants of nature $c$ and $k$ and the elementary charge $e$, Planck's constant $h$ is of fundamental significance for all branches of physics (the constant $G$ in Newton's law of gravity is to be included for cosmological processes), then physical quantities can meaningfully be compared with one another in energy units, even if they are measured in different units. In practice, a comparison based on the energy unit $1\,\text{eV} = 1.6 \times 10^{-19}$ J has proven effective. Thus, for example, temperatures ($kT$), frequencies ($h\nu$), lengths ($hc/l$), voltages ($eU$), and masses ($mc^2$) can be converted into energies and compared with one another. A comparison of temperature and frequency allows us to divide the spectrum of black-body radiation into three distinct spectral ranges (Figure 4.33):

---

### Wave-Particle Duality

- $h\nu \ll kT$    Wave region
- $h\nu \sim kT$    Transition region
- $h\nu \gg kT$    Particle region

---

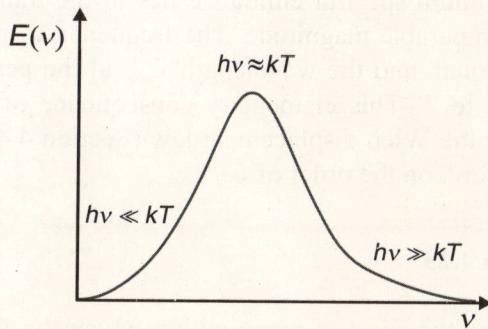

**FIGURE 4.33**  Illustrating the interpretation of Planck's formula in terms of the quantum hypothesis.

In the region $h\nu \ll kT$, the energy of particles in thermal motion changes in such small steps when a photon is absorbed or emitted that absorption and emission can be regarded as quasicontinuous processes. Thus, in this region, the black-body spectrum is consistent with classical physics. The radiation can, therefore, be treated as electromagnetic waves in this region. In this limit the approximation $\exp(h\nu/kT) \approx 1 + h\nu/kT$ holds. Hence, in agreement with classical physics the radiation formula takes the following form in the wave region:

$$E_{\mathrm{BB}}(\nu,T) \approx \frac{2\pi\nu^2}{c^2}kT. \qquad \text{(Rayleigh-Jeans radiation formula)}$$

Planck's constant does not appear in this formula. According to classical physics, this formula should be exact. It implies that the intensity of thermal radiation always rises with increasing frequency and does not fall off after reaching a peak. This result, known as the *ultraviolet catastrophe*, was corrected by the quantum hypothesis.

In the region $h\nu \gg kT$ thermal particles have a very low probability of having sufficient energy to be able to emit a photon with energy $h\nu$. By analogy with the Maxwellian velocity distribution (Section 2.1.4), we should expect that the spectral intensity of thermal radiation would fall off exponentially with the photon energy, just as the Maxwellian distribution function falls off with increasing kinetic energy of the thermal motion of particles. This kind of behavior follows directly from the Planck formula for $h\nu \gg kT$. In this limit $\exp(h\nu/kT) \gg 1$, so that

$$E_{\mathrm{BB}}(\nu,T) = \frac{2\pi h\nu^3}{c^2}\exp\left(-\frac{h\nu}{kT}\right). \qquad \text{(Wien's radiation law)}$$

The maximum spectral emittance lies in the transition region where $h\nu$ and $kT$ are of comparable magnitude. The frequency $\nu_{\max}$ at the peak of $E_{\mathrm{BB}}(\nu,T)$ is thus proportional, and the wavelength $\lambda_{\max}$ at the peak of $E_{\mathrm{BB}}(\lambda,T)$ is inversely proportional, to $T$. This elementary consequence of the quantum hypothesis is equivalent to the Wien displacement law (Section 4.4.3). The constant in Wien's law is, therefore, on the order of $hc/k$.

---

**Problem 4.23**

Determine the spectral range within which the Rayleigh-Jeans radiation formula is valid to within about 1% when $T = 300\,\mathrm{K}$.

**Notes**   The spectral emittance $E(\nu, T)$ has a maximum in the region where $h\nu \sim kT$. If $h\nu \ll kT$, the quantum nature of the electromagnetic waves is insignificant relative to the thermal fluctuations. In this region, therefore, the equations of classical physics are valid. If $h\nu \gg kT$ the quantum nature of the radiation is decisive.

Many quantities in atomic physics, such as frequency and temperature, can be compared with one another in energy units. There is some advantage to using the unit eV. For this reason, take note of the following constants:

- $c \approx 3 \times 10^8$ m/s the speed of light,
- $h \approx 4 \times 10^{-15}$ eV·s Planck's constant,
- $hc \approx 1.2 \times 10^{-6}$ eV·m,
- $k \approx 0.8 \times 10^{-4}$ eV/K Boltzmann's constant, and
- $e^2/4\pi\varepsilon_0 \approx hc/861$ Coulomb interaction constant for elementary charges.

Chapter

# 5 Atomic Structure of Matter

## Summary

- The atom
- Electron waves
- The electron clouds of atoms
- The atomic nucleus
- Chemical bonds
- Lattice structure of crystals

The statistical theory of heat is based on the assumption that matter and radiation have discrete structures. The assumption that matter is made up of atoms, electrons, and ions, and that the radiation field consists of photons, has also stood up to the test in many other areas of physics. In terms of the theory of heat, atoms can be regarded as point masses or small spheres with a diameter of about $10^{-10}$ m. In order to interpret the electrical and mechanical properties of matter, it is assumed that the electrically neutral atoms can be broken up into charged particles, such as electrons and ions (Section 3.4.1). Thus, one proceeds from the idea that atoms not only have a finite size, but also an internal structure. In this chapter we shall first be concerned with explaining the structure of atoms. This creates a foundation for explaining the structure of macroscopic particles made of atoms, electrons, and ions, as well as many properties of macroscopic objects.

Planck's constant $h$ turns out to be a key to atomic physics. The quantum hypothesis, according to which electromagnetic radiation can only be absorbed in discrete quanta $h\nu$, is akin to the assumption that a harmonic oscillator can only be excited to oscillate in discrete steps of magnitude $h\nu$. This assumption obviously contradicts the concepts of Newtonian mechanics and Maxwell's electromagnetic theory, for according to these theories the excitation of an oscillator by an exciter is a continuous process (Section 1.6.3). Nevertheless, this sort of assumption is consistent with every experiment on forced oscillations in classical physics. That is because frequencies from a few Hz to, at most, a few kHz are involved in mechanical oscillations, or frequencies up to the MHz range in the case of electromagnetic oscillations. In all these cases, the steps $h\nu$ are very much smaller than the thermal energy of the vibrations of the oscillator. Because of the experimental uncertainties, the energy steps in these experiments seem as tiny as the steps on a miniature stairway for which the height $H$ of an individual step is much smaller than the size of the shoes with we walk on it.

Rather than trying to understand atomic physics in terms of classical mechanics and electromagnetic theory, it is more reasonable to develop new concepts of the nature of matter, in which the discrete structures perceptible in all branches of nature form a unifying, universal basic idea. Historically, quantum mechanics, which was formulated in the 1920s by Werner Heisenberg, Erwin Schrödinger, Paul Dirac, and other physicists, arose from such an effort.

## 5.1 THE ATOM

The discovery of the elementary laws of chemistry, such as the law of constant and multiple proportions, according to which chemical elements combine with each other only in certain ratios, led around 1800 to the assumption that all matter consists of smallest, indivisible particles, or atoms. The chemical theory of atoms reached its crowning conclusion in the periodic system of the elements (Section 5.3.3) advanced by Dmitri Mendeleev. In that way, by assuming that there are only about 92 different atoms, which can be combined into the most dissimilar molecules in accord with the laws of chemistry, the variety of chemical substances found in nature could be explained. Chemically, the atoms of an element behave exactly similarly. This similarity would be understandable in terms of classical physics, if the atoms, as originally assumed, actually were the smallest, indivisible and unchangeable particles. In fact, they can be broken up into ions and electrons, so it must be assumed that the atoms of an element also have an internal structure. Given a structure of this sort, the similarity of atoms is a most surprising property. How can atoms be dissociated into ions and electrons in gaseous discharges or electrolytes (Section 3.4.1), but subsequently recombine into the original atoms and become indistinguishable from all other atoms of the element? This astonishing stability of atoms is explained by the quantum hypothesis.

## 5.1.1 The structure of atoms

The structure of an atom cannot be studied with a microscope, since the resolution of a microscope is limited by the wavelength of light, $\lambda \sim 1\,\mu m$ (Section 4.2.4) and so is not adequate for making atoms, which are four orders of magnitude smaller, visible. The size and structure of atoms must, therefore, largely be deduced indirectly. The density of liquids and solids and the mean free paths of atoms in gases indicate that all atoms have diameters on the order of $10^{-10}$ m. More precise studies show, however, that the size of atoms varies from element to element. The smallest atoms are the noble gases and the largest are the alkali metals.

The mass of atoms follows from the mass of 1 mol of the material and the Avogadro number $N_A \approx 6 \times 10^{23}$. The lightest atom is hydrogen, with a mass number $A = 1$. Thus, the mass of a H atom is $m_{\mathrm{H}} \approx N_A^{-1} \times 10^{-3}$ kg $\approx 1.6 \times 10^{-27}$ kg. Studies of *gaseous discharges* and the laws of *electrolysis* (the dissociation of chemical bonds by electrical currents in solutions or melts) show that atoms can be broken up into electrons and ions. The masses of the electrons and ions are determined from the acceleration and deflection of these particles in electric and magnetic fields (Section 3.4.2). The mass of the ions is approximately equal to that of the corresponding atoms. But the mass $m_e$ of an electron, on the other hand, is considerably lower:

$$m_e = \frac{m_{\mathrm{H}}}{1837}.$$

In order to ignite gaseous discharges at gas pressures of about 100 Pa, voltages $U \sim 100$ V must be applied to the electrodes. But during the discharge, the applied voltage is a few tens of volts. This indicates that energies on the order of 10 eV are needed in order to ionize an atom, since this is the highest energy to which the charged particles in the electric field between the electrodes can be accelerated.

A first glimpse into the inner structure of atoms was made possible by *scattering experiments* in which the deflection of charged particles with very high energies as they passed through thin foils (so-called *targets*) was studied. High-energy particles of this sort are produced by radioactive decay (Section 5.4.2).

## Experiment 5.1   Absorption of electron (beta) rays

When radioactive strontium atoms ($^{90}$Sr) decay, electrons with kinetic energies of up to 0.55 MeV are emitted. Charged particles with energies this high can initiate pulse discharges in a Geiger counter tube (Section 6.6.1) and can thereby be detected individually and counted. A simple experimental setup (Figure 5.1) can thus be used to study the passage of high-energy electrons through thin foils of, for example, aluminum (Al).

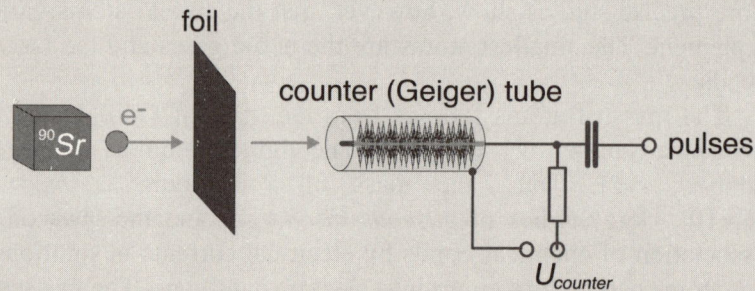

**FIGURE 5.1** Experimental setup for measuring the absorption of high-energy electrons.

The number of electrons detected in a time span of, say, 10 s depends on the activity of the $^{90}$Sr source and the distance between the detector and the source (Section 5.4.2). In order to avoid radiation damage, the measurements are made with low-activity sources. For an average counting rate of about 100 events per second one measures about $N = 1000$ events in 10 s. Since radioactive decay obeys the laws of chance, the number of events counted in an individual measurement fluctuates from one measurement to another. The standard deviation (Section 6.6.2) of the average is $\sqrt{N} \approx 32$. Hence, the measurements only provide values with moderate accuracy. With Al foils as absorbers, the measurements show that the count rate falls off exponentially with increasing thickness of the Al foils, but only with an amazingly low absorption coefficient. About 80% of the incident electrons pass through a 0.1 mm thick Al foil, or about $10^6$ layers of Al atoms piled up on top of one another.

These measurements show that the atoms cannot be represented as impenetrable spheres. Obviously they contain a lot of free space within which fast electrons can move unimpeded.

A more precise glimpse into the structure of atoms was provided by the experiments of Ernest Rutherford (1910). He and his collaborators studied the scattering of $\alpha$ particles in gold foils. The particles produced by $\alpha$ decay (Section 5.4.2) are doubly charged positive He ions with mass number $A = 4$. They have a kinetic energy of about 5 MeV. The experimenters were surprised to find that, despite their high energy, some of these particles simply bounced off the Au atoms, so that after scattering they flew in a backward direction with almost no reduction in energy. In order to interpret these experiments, Rutherford assumed that the atom consists of a very small $Z$-fold positively charged nucleus, in which almost all of the mass of the atom is concentrated, and of $Z$ electrons which orbit the nucleus in a way similar to the way the planets orbit the sun. This *Rutherford planetary model* is the foundation of modern atomic physics. But, in order to explain the stability of the atoms in terms of this model, the universal validity of classical mechanics must be questioned. A new approach to explaining both the stability and the spectrum of atoms was provided by Planck's quantum hypothesis (Section 5.1.3).

**Problem 5.1**

Calculate the distance of approach of $\alpha$ particles to a gold atom during backward scattering. How much energy must the $\alpha$ particles have in order to contact a gold nucleus (with a diameter $2R \approx 10^{-14}$ m and charge $Z = 79$) during backscattering? What is the mass density $\rho$ of the gold nucleus?

**Notes** Atoms contain an atomic nucleus, which carries an electrical charge $+Ze$ and has a radius $R < 10^{-14}$ m. The $Z$ electrons orbiting the nucleus travel on orbits with radii of about $10^{-10}$ m. The atomic numbers (nuclear charges) $Z$ of the natural elements are integers with $1 < Z < 92$. The atomic number $Z$ determines the chemical properties of the elements in the periodic system.

## 5.1.2 The spectrum of hydrogen

The simplest atom is the hydrogen atom. According to Rutherford's planetary model, it consists of a singly charged, positive nucleus and only one electron, which orbits the nucleus. The nucleus and the electron are attracted to one another by the Coulomb force. Since, like the force of gravity, the Coulomb force falls off with the square of the distance $r$ between the interacting objects, the electron should orbit about the nucleus in the same way a single planet orbits the sun. In particular, Kepler's laws (Section 1.14) could also be applied to the motion of the electron. According to Kepler's third law, the square of the orbital frequency $v$ for circular orbits would be inversely proportional to the third power of the radius $r$ of the orbit. Thus, based on classical electromagnetic theory, hydrogen atoms should be able to emit electromagnetic waves of arbitrary frequency.

---

**Problem 5.2**

What is the radius of the orbits of electrons that emit electromagnetic waves in the visible spectrum according to classical physics? What is the orbital angular momentum of these electrons? Compare the calculated value with Planck's constant.

---

Spectroscopic studies of the light emitted by gaseous discharges show, however, that all atoms emit a spectrum with discrete spectral lines. The spectrum of the H atoms is especially simple (Figure 5.2). It has four spectral lines in the visible and other lines in the ultraviolet (UV) and infrared (IR) parts of the spectrum. The wavelengths $\lambda$ of these spectral lines can all be calculated using a simple formula obtained by Johann Jakob Balmer in 1885:

$$\frac{1}{\lambda} = Ry\left(\frac{1}{n^2} - \frac{1}{m^2}\right). \qquad \text{(Balmer formula)}$$

Here $n = 1, 2, 3, \ldots$ and $m > n$ are natural numbers and $Ry = 109737 \text{ cm}^{-1}$ is the *Rydberg constant*. For every $n$ there is a series of lines with $m = n+1, n+2, \ldots$. The *Lyman series* ($n = 1$) lies entirely in the UV and the *Balmer series* ($n = 2$) in the visible and near IR ranges, while the *Paschen series* ($n = 3$), and all the series with higher $n$, lie entirely in the IR.

These lines are characteristic of the hydrogen atom. No matter how hydrogen gas, which consists of diatomic molecules ($H_2$) at ordinary temperatures, is

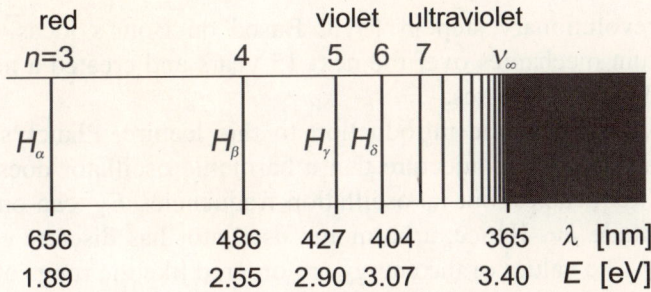

**FIGURE 5.2** The spectrum of hydrogen (Balmer series).

excited, whether in a gaseous discharge or in a flame, precisely these lines are always emitted by hydrogen atoms. The four visible lines of the Balmer series are also observed in the spectrum of the sun and other stars as the *Fraunhofer lines* and lead us to conclude that these heavenly bodies consist essentially of hydrogen.

All the other elements in the periodic system also have characteristic line spectra. The wavelengths of the lines in these spectra cannot, however, be calculated in such a simple way as the spectrum of hydrogen. Nevertheless, knowledge of the spectra offers, just as with hydrogen, the possibility of identifying the elements of which samples of chemical substances consist. This is the basis of *spectral analysis*, developed by **Gustav Robert Kirchhoff (1824–1887)** and **Robert Wilhelm Bunsen (1811–1899)**. For example, if table salt is sprinkled into a flame of a Bunsen burner, the flame emits a yellow light and shows, thereby, that table salt (NaCl) contains sodium. Here the Na atom emits an intense yellow spectral line at $\lambda = 589$ nm in the visible range (Section 5.3.4).

| **Notes** | In hot flames or gaseous discharges, atoms emit a spectrum of discrete spectral lines that is characteristic of the respective chemical elements. For the simplest atom, the hydrogen |

atom, which consists of a singly charged nucleus and an electron, the wavelengths of the spectral lines can be calculated using the Balmer formula.

### 5.1.3 Bohr's model of the atom

From the standpoint of classical physics the stability of the atom and its line spectrum are an unsolvable puzzle. Only by combining the laws of classical physics with other kinds of regularities based on Planck's quantum hypothesis was it possible to solve the puzzle. It was **Niels Bohr (1885–1962)** who ventured this revolutionary step in 1913. Based on Bohr's ideas, physicists developed quantum mechanics over the next 15 years and created a new foundation for the world view of physics.

As noted in the introduction to this lecture, Planck's quantum hypothesis originated in the conjecture that a harmonic oscillator does not oscillate at arbitrary frequencies, but its oscillation frequencies $E_{osc}$ can only change in steps of magnitude $h\nu$. Hence, a harmonic oscillator has discrete energy levels (Section 6.1.1). The values of the energy are ordered like the rungs of a ladder:

$$E_n = E_0 + nh\nu, \text{ with } n = 0, 1, 2, \ldots \qquad \text{(discrete energy levels of the harmonic oscillator)}$$

In a transition to a higher or lower energy level, a photon is absorbed or emitted, respectively.

Bohr generalized this concept and applied it to Rutherford's planet model. In this way he could explain the stability of atoms and the existence of discrete spectral lines. Proceeding from three basic postulates, he was able, in particular, to justify the Balmer formula for the spectral lines of hydrogen theoretically.

Two of Bohr's three postulates are:

---

### Bohr's Postulates

---

- Bohr's first postulate: electrons move in circular orbits around an atomic nucleus. Here the energy of the electrons can only have discrete values $E_n$. During these stationary movements, the electrons emit no electromagnetic waves.
- Bohr's second postulate: the electrons can undergo quantum jumps or leaps. They jump from one stationary state of motion with a discrete energy $E_n$ to another stationary state with energy $E_m$. During these quantum jumps, a photon is emitted if $E_n > E_m$, or absorbed if $E_n < E_m$. The energy $h\nu$ of the photon is determined by the energy difference $|E_m - E_n|$, i.e.,

$$h\nu = |E_m - E_n|. \qquad \text{(Bohr's frequency condition)}$$

---

With these two postulates, Bohr formulated some general axioms that were not consistent with the concepts of classical physics. According to Maxwell's electromagnetic theory, an electron that moves around an atom and is, therefore, continuously accelerated (since the direction of its velocity changes), would continuously emit electromagnetic waves (Section 3.6.2). There would be no stationary electron states. Rather the electrons would constantly lose energy and should ultimately fall into the nucleus. The first postulate is obviously necessary in order to explain the stability of the atom. The second postulate can be used to interpret the line spectrum of the atom. Since stationary states are possible only with discrete energies, according to Bohr's frequency condition only spectral lines with discrete frequencies are emitted and absorbed. In contrast to the classical picture where emission and absorption are continuous processes, emission and absorption processes occur discontinuously in *quantum leaps*.

The question now arises of how to calculate the energies $E_n$ and, thereby, the line spectra of the atom based on this new conceptual foundation. For the hydrogen atom Bohr solved this problem with the following postulate:

---

## Bohr's Third Postulate

---

■ The orbital angular momentum $L$ of an electron orbiting about the atomic nucleus can only have values equal to an integral multiple of $h/2\pi$:

$$L = \frac{nh}{2\pi} = n\hbar. \quad \text{(quantization of angular momentum)}$$

---

Given that the orbital angular momentum of an electron can only have discrete values, the laws of classical mechanics imply that the energy of an electron orbiting about the atomic nucleus can also only have certain values. Then the energies are given as a function of the angular momentum for the circular orbits of the electrons in the Coulomb field of the atomic nucleus.

For the orbit of an electron in the Coulomb field of an atomic nucleus with charge $Ze$, Newton's second axiom $\mathbf{F} = m\mathbf{a}$ gives

$$\frac{Ze^2}{4\pi\varepsilon_0 r^2} = m_e \omega^2 r$$

($m_e$ is the electron mass, $e$ the elementary charge, $r$ the radius of the orbit, and $\omega$ the angular velocity of the electron). The potential energy of an electron in the Coulomb potential of the nucleus is $E_{\text{pot}} = -Ze^2/4\pi\varepsilon_0 r$, and when the above equation is multiplied by $r$, it gives the kinetic energy $E_{kin} = -E_{pot}/2$. Thus, the total energy of the electron is

$$E_{tot} = \frac{1}{2} \cdot \frac{Ze^2}{4\pi\varepsilon_0 r}.$$

Since, in addition, $L = m_e \omega r^2$, the relationship between the centripetal force and the acceleration is found on multiplying by $r^3$, i.e., the radius $r$ is proportional to the square of the angular momentum: $L^2 = (m_e Ze^2/4\pi\varepsilon_0)r$. The total energy of the electron is thus inversely proportional to the square of the angular momentum:

$$E_{tot} = -\frac{m_e}{2} \left( \frac{Ze^2}{4\pi\varepsilon_0} \right)^2 \frac{1}{L^2}.$$

The quantization of angular momentum then implies that the electron can only move in orbits with energies $E_{tot} = E_n$. For the discrete energies $E_n$ we obtain

$$E_n = -\frac{m_e}{2} \left( \frac{e^2}{2\varepsilon_0 h} \right)^2 \frac{Z^2}{n^2} \qquad \text{(discrete energy levels of an electron in a Coulomb field)}$$

For a hydrogen atom with its nuclear charge $Z = 1$, we obtain the following values:

$$E_n = -\frac{13.6 \text{ eV}}{n^2}.$$

The position of the energy levels $E_n$ on an energy scale is shown in the *energy level diagram* of Figure 5.3. The lowest energy level $E_1$ is the *ground state* of the hydrogen atom. In order to remove the electron from the nucleus of a hydrogen atom and, thereby, ionize the atom, an energy of at least 13.6 eV must be delivered to the electron. Thus, in the ground state of the hydrogen atom, the nucleus and the electron are bound to one another with a *binding energy* $E_B = 13.6 \text{ eV}$.

The spectrum of the hydrogen atom can be derived from the energy level diagram using Bohr's second postulate. Since $h\nu = E_m - E_n$ and $\nu = c/\lambda$, the reciprocal wavelength $\lambda^{-1}$ of the spectral line of the H atom emitted during a quantum jump from $E_m$ to $E_n$ is given by

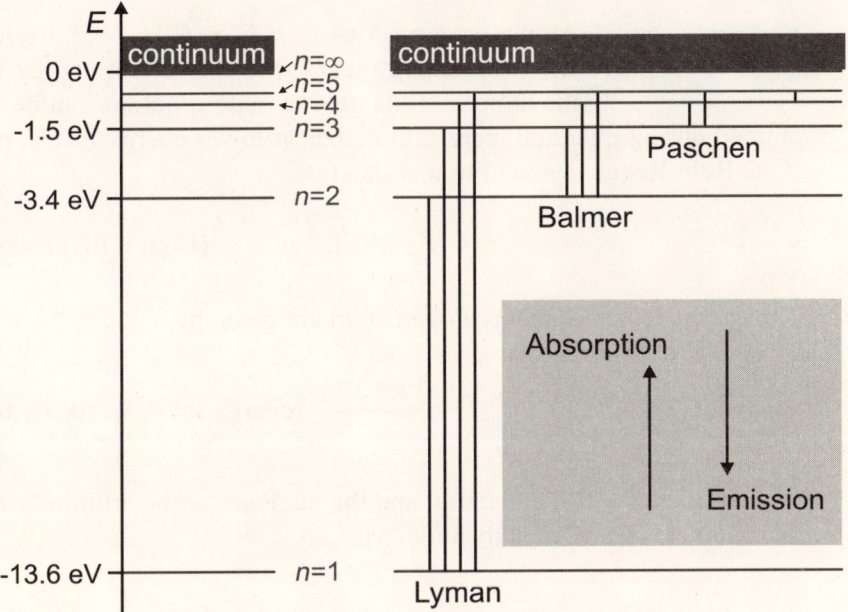

**FIGURE 5.3** Energy level diagram for the hydrogen atom.

$$\frac{1}{\lambda} = \frac{E_m - E_n}{hc} = \frac{m}{2hc}\left(\frac{e^2}{2\varepsilon_0 h}\right)^2\left(\frac{1}{n^2} - \frac{1}{m^2}\right).$$

This is the Balmer formula. The numerical factor $Ry = (m_e/2hc)(e^2/2\varepsilon_0 h)^2 = 1.09737 \times 10^7 \text{ m}^{-1}$ is equal to the Rydberg constant in the Balmer formula (Section 5.1.2).

---

## Problem 5.3

Calculate the minimum energy that has to be delivered to a hydrogen atom in order for the electron to jump from the ground state into another state. Can this energy be transferred in thermal collisions of a gas at room temperature? How much energy is needed to ionize a hydrogen atom?

---

> **Notes**
>
> Atoms have a series $E_1 < E_2 < E_3 < \cdots$ of discrete energies $E_n$. If the energy level with the lowest energy is occupied, the atom is in its ground state. Photons can be absorbed or emitted during quantum leaps into higher or lower energy levels, respectively, if the Bohr frequency condition is satisfied:
>
> $$h\nu = |E_m - E_n|. \qquad \text{(Bohr's frequency condition)}$$
>
> The energy levels of the hydrogen atom are given by
>
> $$E_n = -\frac{13.6\,\text{eV}}{n^2}. \quad \text{(energy levels of the hydrogen atom)}$$
>
> For energies $E > 0$ the electron and the nucleus can be arbitrarily distant from one another. The atom is then ionized.

## 5.1.4 Discrete energy levels

Daily experience seems to teach us that objects move continuously in space and time and waves propagate continuously in space and time. Newtonian mechanics and Maxwell's electromagnetic theory arose in accord with this experience. The quantum hypothesis and the resulting postulates of Bohr are in flagrant conflict with the outlook implanted by daily experience and with the classical continuum theories, but not with daily experience, as such. This is because the state of an object that we are able to perceive changes so little during a quantum jump that, despite discontinuous changes, a macroscopic process seems to be taking place continuously.

Not only our outlook, but also the fundamental theories of classical physics, are on the test bench. These have proven effective in so many ways in natural science and technology that it is hard to believe that the fundamental concepts behind these theories are not of unlimited validity. Besides theoretical considerations, numerous experiments later contributed to the scientific acceptance of Bohr's revolutionary new concept of stationary states and quantum leaps between them. One experiment that revealed the existence of discrete energy levels especially clearly is the experiment of James Franck and Gustav Hertz in 1914.

## Experiment 5.2    The Franck-Hertz experiment

Franck and Hertz studied the excitation of mercury atoms in collisions with electrons and the energy of the electrons after these collisions. Electron collisions with other atoms can be studied in the same way. The experimental apparatus consists of an electron tube with an incandescent cathode, grid, and anode (triode), which contains mercury vapor at low ($\sim 1\,\mathrm{Pa}$) pressure (Figure 5.4). Electrons are liberated by thermionic emission (Section 4.3.3) from the incandescent cathode and accelerated by a voltage $U_{KG}$ of up to 20 V applied between the cathode and the grid to a kinetic energy of up to $eU_{KG}$. Between the grid and anode, the electrons are slowed down by a retarding voltage of about $-0.5$ V between the grid and the anode. Thus, the electrons can only reach the anode if they have an energy of at least 0.5 eV at the grid. The electron current reaching the anode is measured as a function of the voltage $U_{KG}$. The results of such a measurement are shown in Figure 5.5.

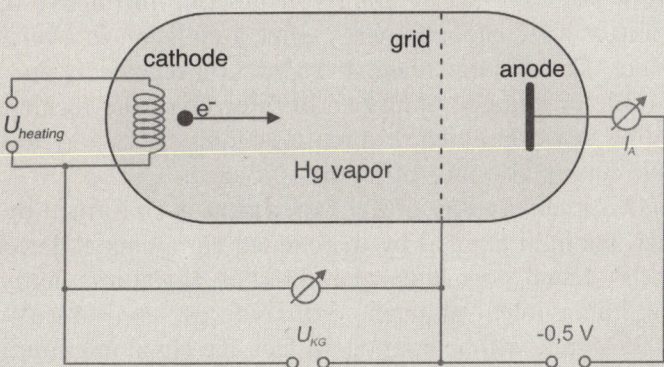

**FIGURE 5.4**  A diagram of the experimental apparatus of Franck and Hertz.

The anode current initially increases with the voltage $U_{KG}$ as more electrons are drawn from the region of the cathode as the voltage rises. As soon as $U_{KG}$ exceeds 4.9 V, the anode current falls off drastically. A similar drop in the anode current is observed at twice and three times this voltage.

*continued*

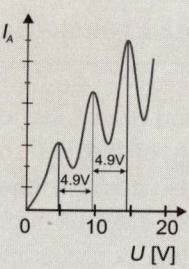

**FIGURE 5.5** Anode current $I_A$ as a function of the voltage $U_{KG}$.

This striking drop in the anode current confirms the existence of discrete energy levels. This is because when they collide with mercury (Hg) atoms, the electrons recoil elastically, or essentially with no energy loss to the atoms, as long as their energy is not high enough to excite the atoms. The lowest excited energy level of the Hg atom lies 4.9 eV above the ground state. This is the minimum energy that the electrons must have in order to collisionally excite an Hg atom and, thereby, lose energy, specifically 4.9 eV. For accelerating voltages $U_{KG}$ just above 4.9 V the electrons no longer have enough energy after a collision to overcome the retarding voltage. The electrons then strike the grid repeatedly and the anode current falls off accordingly. With two or three times the accelerating voltage each electron can excite two or three Hg atoms on its way to the grid. Then the anode current also falls off at these voltages.

This interpretation of the experiment is confirmed by measurements in which the light emitted by the excited Hg atoms is detected. According to Bohr's second postulate, when excited Hg atoms decay into the ground state, ultraviolet photons with energy $h\nu = 4.9$ eV or wavelength $\lambda = 253.7$ nm will be emitted. In fact, the Hg atoms emit no light as long as $U_{KG} < 4.9$ V. As expected, only when $U_{KG} > 4.9$ V can the ultraviolet spectral line of the Hg atom be detected.

Not only electron-atom collisions are completely elastic if the collision energy is too low to excite the atom. Collisions of atoms with one another are likewise completely elastic. This condition is essentially satisfied always for atoms at room temperature moving with a kinetic energy of about $kT \approx 25$ meV. The fundamental assumption of the kinetic theory of gases (Section 2.1.2) that collisions between atoms at room temperature are completely elastic is thus explained with Bohr's first postulate.

**Problem 5.4**

Calculate the probability that an atom in a gas at (a) room temperature $T = 300\,\text{K}$ and (b) the temperature of the sun's surface $T = 6000\,\text{K}$, has enough energy to excite a hydrogen atom.

| Notes | Collisions of atomic particles are completely elastic if the kinetic energy of the relative motion of the two colliding particles is less than the excitation energy for these particles. Most atoms have excitation energies of a few eV. |
|---|---|

## 5.2 ELECTRON WAVES

Bohr's postulates are irreconcilable with the fundamental concepts of classical physics. The existence of discrete energy levels and the changes in atomic systems by quantum jumps are in striking conflict with the idea of the steady motion and evolution of objects and fields. Thus, Bohr's postulates cannot be regarded simply as an addition to Newtonian mechanics or Maxwell's electromagnetic theory. Rather, together with the quantum and photon hypotheses they form a first, tentative step toward a new physics.

This new physics, known as *quantum* or *wave mechanics*, arose in the 1920s. In it, such conflicting ideas of classical physics as the deterministic concept of mechanics and the concept of randomness in thermodynamics, were brought together in a new synthesis. Quantum mechanically, the state of a physical object can change continuously, following deterministic laws, as well as in discrete quantum leaps, following the laws of chance. In this lecture we shall illustrate some of the ideas underlying modern wave mechanics.

Another inherent contradiction in the concepts of classical physics is the contrast between waves and particles. A plane wave $\exp(i\mathbf{k}\cdot\mathbf{r} - i\omega t)$ extends continuously over all space. It is everywhere. But at a given time, a particle is at a given point. It is localizable. In spite of this contrast, we have described electromagnetic radiation both as a wave field and as a particle (Chapter 4). In order to explain interference and diffraction, we treated the radiation as an electromagnetic wave (Section 4.2). To explain the spectrum of thermal radiation and the photoelectric and Compton effects, however, we assumed that electromagnetic radiation consists of photons (Section 4.3). This wave-particle duality cannot be made consistent with the outlook we have derived from classical physics. Nevertheless, we must try to grasp it. That is because it is not just a peculiarity of electromagnetic radiation, but is also a feature of material particles. In fact, the wave-particle duality is consistent with all experiments conducted up to now. Also, there are no conceivable experiments that might bring this duality into question on the basis of quantum physics. And, finally, the wave-particle duality can be described theoretically using quantum mechanics without contradictions. Thus, the problem is to adapt our conceptual representation of nature to the new facts brought to light by many experiments.

### 5.2.1 Electron diffraction

That material particles also can have wave properties was discussed by Louis de Broglie in 1924 and first demonstrated on electron beams in 1927. The same relationship which couples the wave and particle properties of electromagnetic radiation (Section 4.3.4) holds for material particles:

$$\mathbf{p} = \hbar \mathbf{k} \quad \text{or} \quad p = \frac{h}{\lambda}. \qquad \text{(de Broglie equation)}$$

Particles with momentum $\mathbf{p}$ can, therefore, behave as waves with a wavelength $\lambda = h/p$. For electrons with kinetic energy $E_{\text{kin}} = p^2/2m_e$, this implies that

$$\lambda = \frac{hc}{\sqrt{2m_e c^2 E_{\text{kin}}}}.$$

To calculate $\lambda$ it is necessary to know the product $hc$ of Planck's constant and the speed of light [in units of $\text{eV} \cdot \text{m}$] and the rest energy $m_e c^2$ of the electron [in eV] (Section 4.4.4):

$$hc = 1.237 \times 10^{-6} \text{ eV} \cdot \text{m}$$
$$m_e c^2 = 0.511 \times 10^6 \text{ eV}.$$

After being accelerated by a voltage $U = 10\,\text{kV}$, electrons have a wavelength $\lambda = 10^{-11}$ m. Electron beams with this energy thus have a wavelength of the same order of magnitude as x-rays (Section 4.3.3). Therefore, as with x-rays, the wave character of electron beams can be detected by diffraction on crystal lattices.

## Experiment 5.3    Electron diffraction

Electrons are liberated in an electron tube by thermionic emission from an incandescent cathode and accelerated to about 10 keV (Figure 5.6). The electron beam that passes through a hole in the anode is incident on a thin carbon foil (about 10 µm thick) and is diffracted on the graphite lattice (Section 5.6) of the carbon foil. In order to make the points where the electrons strike the glass wall of the tube visible, the wall is coated with a fluorescent layer of ZnS.

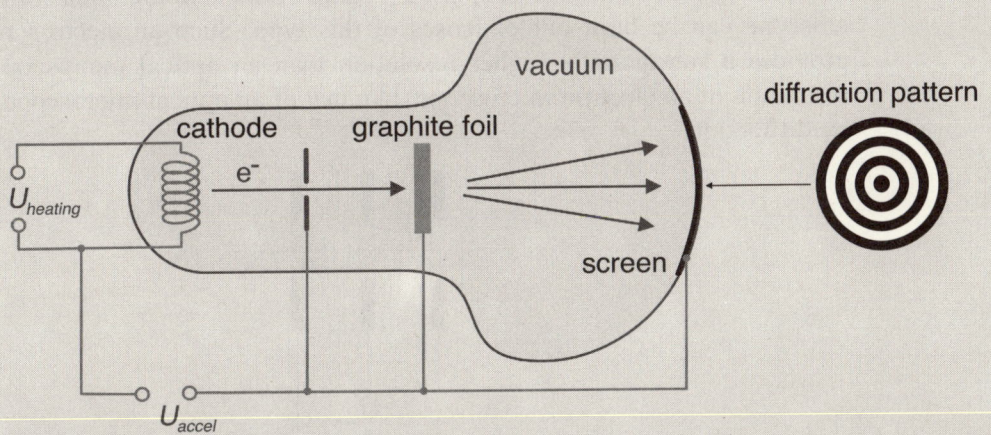

**FIGURE 5.6** An electron tube for demonstrating electron diffraction and a diffraction pattern observed on a fluorescent screen.

On the fluorescent screen the electrons produce a diffraction pattern consisting of concentric rings that is typical of diffraction on microcrystalline crystal lattices. The angular separation $\Delta\alpha$ between neighboring diffraction maxima obeys the laws of diffraction (Section 4.2.3). By analogy with diffraction at a double slit, $\Delta\alpha \approx \lambda/D$. Here $D$ is the distance between neighboring atoms in the crystal. Since $\lambda$ falls off with increasing electron energy, the angular separation between the diffraction maxima also decreases if the accelerating voltage $U_B$ is raised.

## Problem 5.5

Calculate the angular separation $\Delta\alpha$ of neighboring diffraction maxima for the diffraction of an electron beam on a crystal with $D = 0.2\,\text{nm}$ for accelerating voltages $U_B = 10$ and $40\,\text{kV}$.

Electron beams can be focused with electron optical lenses, in a way similar to the focusing of light beams. A simple electrostatic lens, for example, consists of two grounded ring electrodes and an electrode between them to which a (positive or negative) voltage $U$ is applied (Figure 5.7). As in the optics of light, a microscope can be built out of lenses of this type. Such an electron microscope provides a substantially higher resolution than an optical microscope. But the resolution of an electron microscope, like that of an optical microscope, is limited by diffraction.

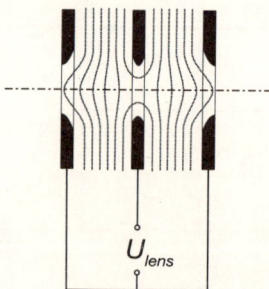

**FIGURE 5.7** An electrostatic electron lens with equipotential lines.

<table>
<tr><td>**Notes**</td><td>Diffraction and interference patterns can be produced with electron beams. These experiments demonstrate the wave character of electrons. The wavelength of electron waves is</td></tr>
</table>

calculated from the momentum of the electrons. The momentum **p** of the particles and the wave vector **k** of the waves obey the fundamental equation

$$\mathbf{p} = \hbar\mathbf{k}.\qquad\text{(de Broglie equation)}$$

This wave-particle duality is not just a characteristic of electrons, but of all atomic particles. Diffraction experiments can also be done, for example, with neutrons, entire atoms, or even with molecules. In doing so, it is important to make sure that the particles undergo no quantum jumps along the path from source to detector.

## 5.2.2 Tunnel effect

In order to quantitatively describe the diffraction of an electron beam on a crystal lattice, let us consider an electron beam as a plane wave with an amplitude $A$, wave vector $\mathbf{k}$, and frequency $\omega$. Accordingly, we describe the electron beam mathematically with a wave function

$$\psi(\mathbf{r},t) = A\exp(i(\mathbf{k}\cdot\mathbf{r} - \omega t)).$$

The wave number $\mathbf{k}$ is obtained from the momentum $\mathbf{p}$ of the electrons using the de Broglie equation. We assume that, as in the case of electromagnetic waves, a corresponding relationship exists between the energy $E$ of the electrons and their frequency $\omega = 2\pi\nu$: $E = h\nu$. As for the energy, in general both the potential energy $E_{\text{pot}}(\mathbf{r})$ of the electrons and their kinetic energy, are to be taken into account; thus, $E = E_{\text{kin}} + E_{\text{pot}}$, or $E = p^2/2m_e + E_{\text{pot}}$. Here $m_e$ is the mass of an electron. The electron energy $E$ and the frequency $\omega$ of the electron waves obey the equation $E = h\omega/2\pi$. Thus, if $E_{\text{pot}}(\mathbf{r}) = const$, then

$$\hbar\omega = \frac{\hbar^2 k^2}{2m_e} + E_{\text{pot}}.$$

Based on this approach for describing the electron waves, it is now possible to search for a wave equation which has the wave function of the electron beam as a solution in the special case $E_{\text{pot}}(\mathbf{r}) = const$. One such wave equation is the *Schrödinger equation*:

$$i\hbar\frac{\partial\psi(\mathbf{r},t)}{\partial t} = -\frac{\hbar^2}{2m}\nabla^2\psi(\mathbf{r},t) + E_{\text{pot}}(\mathbf{r})\psi(\mathbf{r},t). \quad \text{(Schrödinger equation)}$$

Here the *Laplacian operator* $\nabla^2$ is a shorthand notation for the sum of the second partial derivatives with respect to the three position coordinates:

$$\nabla^2\psi(\mathbf{r},t) = \frac{\partial^2\psi(\mathbf{r},t)}{\partial x^2} + \frac{\partial^2\psi(\mathbf{r},t)}{\partial y^2} + \frac{\partial^2\psi(\mathbf{r},t)}{\partial z^2}. \quad \text{(Laplacian operator)}$$

### Problem 5.6

Prove that in the special case $E_{\text{pot}}(\mathbf{r}) = const$ the wave function $\psi(\mathbf{r},t)$ is a solution of the Schrödinger equation. What is the frequency $\omega = 2\pi\nu$ of the electron beam.

Wave mechanics is based on the assumption that the Schrödinger equation also describes the motion of electrons in position-dependent potentials $E_{\text{pot}}(\mathbf{r})$. In order to get a first glimpse of the consequences of this assumption, let us consider the case in which the potential energy has the form of a step function of just one spatial coordinate, say the $x$ coordinate. If $E_{\text{pot}}(x)$ has only one step at $x = 0$ (Figure 5.8), solutions are easily obtained for the left $(x < 0)$ and right $(x > 0)$ half-spaces. Let $E_{\text{pot}}(x) = 0$ for $x < 0$ and $E_{\text{pot}}(x) = E_0$ for $x > 0$.

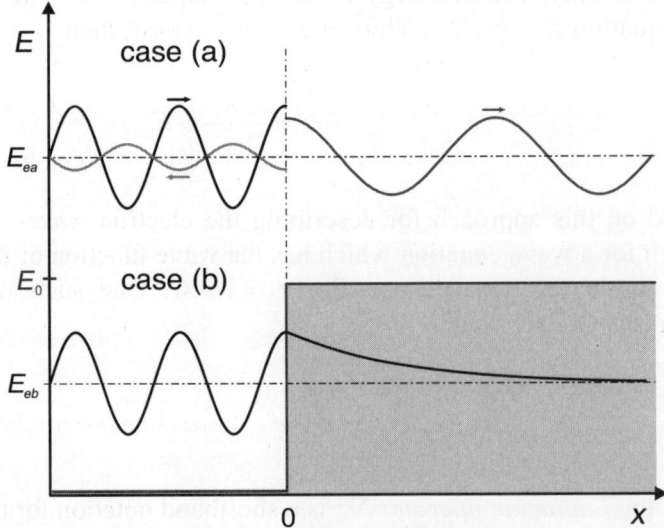

**FIGURE 5.8** Transmission and reflection of an electron wave for electrons with kinetic energy $E_e$ at potential steps $E_0 > E_e$ (a) and $E_0 < E_e$ (b).

A wave incident on the barrier from the left with amplitude $A_0 = 1$ and kinetic energy $E_e$ has the wave function $\psi_0(x,t) = \exp(i(kx - \omega t))$ for $x < 0$, with wave number $k = (2\pi/h)\sqrt{2m_e E_e}$ and frequency $\omega = 2\pi E_e/h$. At the potential step it will be partially reflected and partially transmitted. The reflected wave has the wave function $\psi_R(x,t) = R\exp(i(-kx - \omega t))$ with the (real) amplitude $R$ for $x < 0$. In the half-space $x > 0$ a transmitted wave appears only if $E > E_0$. In this case, the wave is described (for $x > 0$) by the wave function $\psi_T(x,t) = T\exp(i(k'x - \omega t))$ with a (real) amplitude $T$ and wave number

$$k' = \frac{2\pi}{h}\sqrt{2m_e(E_e - E_0)}.$$

The amplitudes $R$ and $T$ are obtained from the condition that the combined function

$$\psi(x,t) = \psi_0(x,t) + \psi_R(x,t) \text{ for } x < 0 \text{ and } \psi(x,t) = \psi_T(x,t) \text{ for } x > 0$$

must be continuous and differentiable at $x = 0$. The continuity condition gives $1 + R = T$ and the differentiability condition, $k(1 - R) = k'T$. Thus, the equations for these conditions yield the amplitudes $R$ and $T$:

$$R = \frac{k - k'}{k + k'}$$

$$T = \frac{2k}{k + k'}.$$

These satisfy the condition $kR^2 + k'T^2 = k$. This condition is an expression of the fact that no electrons are lost during reflection and transmission at a potential barrier (Section 5.2.4).

Let us now consider the interesting case $0 < E < E_0$. In this case the electrons do not have enough energy to reach the potential level in the right half-space. Thus, all the electrons are reflected; that is, the reflected wave has amplitude $R = 1$. Nevertheless, electrons do penetrate the right half-space. This is because the description of the electron beam as a wave implies that in the left half-space a standing wave develops because of the superposition of the incident and reflected waves:

$$\psi(x,t) = \cos(kx + \varphi) \text{ for } x < 0.$$

But the wave equation also has solutions in the right half-space. These are still small wavelike periodic solutions, but they fall off exponentially (Figure 5.8):

$$\psi(x,t) = Te^{-Kx}e^{-i\omega t} \text{ for } x > 0.$$

The decay constant $K$ is given by

$$K = \frac{\sqrt{2m_e(E_0 - E_e)}}{\hbar}.$$

The phase $\varphi$ of the standing wave in the left half-space and the amplitude $T$ of the exponentially decaying solution in the right half-space are again obtained by matching the two partial solutions at $x = 0$. But the most interesting point is that the electron waves also penetrate into the energetically forbidden region $x > 0$.

The penetration of electrons into the energetically forbidden region is physically relevant when the electrons are incident on a potential barrier of finite width $d$ (Figure 5.9). In this case, as well, the electron wave penetrates into the energetically forbidden region, but it then reaches the space beyond the barrier with a certain amplitude and can again propagate there in the form of a wave. Thus, it is capable of *tunneling* through the potential barrier. This *tunnel effect* plays an important role in many areas of physics. One example is the *field emission* of electrons.

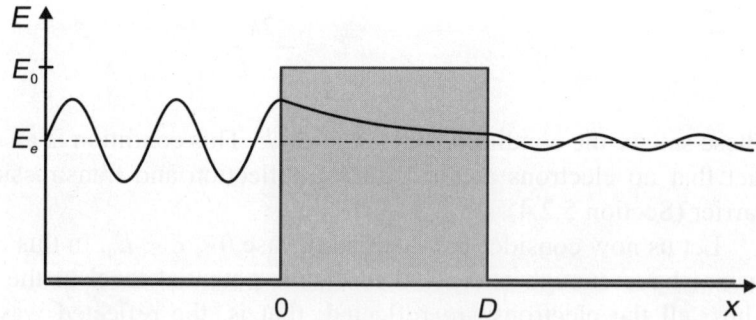

**FIGURE 5.9**   An electron wave at a potential barrier with $E_0 > E_e$.

Usually the conduction electrons of a metal can only enter the space outside the metal if they have energies greater than the work function $W_A$. In photoemission (Section 4.3.1) the required energy is supplied by the absorption of a photon. In thermionic emission (Section 4.3.3) the temperature of the metal is so high that some electrons have energies $E > W_A$ because of the thermal energy distribution. In field emission, on the other hand, conduction electrons with $E < W_A$ are able to emerge from the metal.

Field emission occurs when a very high electric field is set up at the surface of a metal and imposes a negative charge on it. In this case, at the metal's surface a potential barrier develops for the conduction electrons, through which they can tunnel (Figure 5.10).

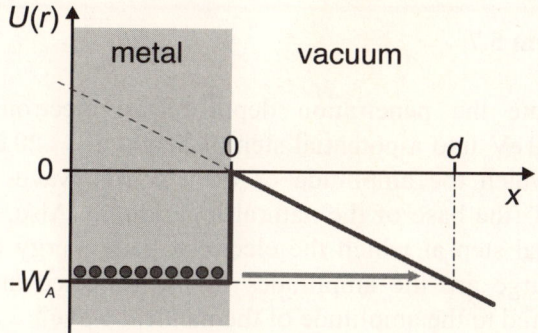

**FIGURE 5.10** A potential barrier at a metal surface at $x = 0$ with a negative charge.

## Experiment 5.4   Field emission

For a potential barrier with a width of only a few atomic diameters to develop at the surface of a metal, an electric field **F** on the order of $F \sim 1\,\text{V/nm} = 10^9\,\text{V/m}$ must be applied. In order to create a field of this magnitude, an electron tube with a cathode made of a sharp metal needle is employed (Figure 5.11). The radius of curvature of the tip is about $R \sim 10\,\mu\text{m}$. A voltage $U$ on the order of $-10\,\text{kV}$ relative to the grounded anode is applied to the cathode so that the metal tip will be sufficiently negatively charged. The field strength near the metal tip is then $F \approx U/R \sim 10^9\,\text{V/m}$.

Under these conditions an electron current actually flows from the cathode to the anode. Here a ZnS layer deposited on the glass wall of the tube serves as the anode and fluoresces when the electrons strike it. In the experiment, structures are seen on the fluorescent screen which can be explained as diffraction patterns corresponding to the atomic structure at the surface of the cathode. This apparatus can, therefore, also be used as an electron microscope (*Field electron microscope*).

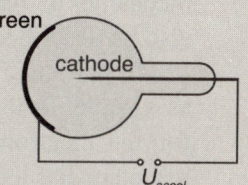

**FIGURE 5.11** Experimental setup for demonstrating field emission.

**Problem 5.7**

Calculate the penetration depth of an electron beam with energy $E_e = 10\,\text{eV}$ into a potential step of height $E_0 = 20\,\text{eV}$. Give the distance over which the amplitude of the electron waves falls by a factor of $e = 2.7$ (the base of the natural logarithms). Also, discuss the case of a potential step at which the electrons gain energy ($E_0 < 0$). In this case how large are the amplitudes of the reflected and transmitted waves compared to the amplitude of the incident wave?

**Notes** Electrons can tunnel through a potential barrier whose height $E_0$ exceeds the potential energy $E_e$ of the electrons. The amplitude $A$ of the electron wave falls off exponentially with increasing width $d$ of the potential barrier. The decay constant $K$ for the exponential decrease $A = A_0 \exp(-Kd)$ is given by

$$K = \frac{\sqrt{2m_e(E_0 - E_e)}}{\hbar}.$$

## 5.2.3 Wave mechanics

Electron diffraction and the tunnel effect show clearly that electron beams not only behave as particle currents, but also as waves under certain experimental conditions. Since both a particle current and a wave have a certain finite size, this dualistic behavior of electron beams is not in such obvious conflict with our customary intuition. The contradiction between the wave and particle representations seemed unbridgeable, however, when Heisenberg and Schrödinger were founding quantum and wave mechanics, respectively, in 1925. That is because wave mechanics is based on the assumption that the motion of a *single* electron can also be described as a wave.

In particular, the motion of the electron in a hydrogen atom can be described as a wave. In terms of wave mechanics the H atom is also treated as a resonator for the electron waves. Like other resonators (Sections 3.1.2 and 3.3.2), this resonator has a discrete series of eigenmodes. Thus, Bohr's first postulate regarding the existence of discrete energy levels follows directly from Schrödinger's assumption that the motion of an individual electron can be described as a wave.

In order to calculate the states of oscillation and the corresponding energies of the H atom and other hydrogen-like systems, it is necessary to solve the

Schrödinger equation (Section 5.2.2) for an electron in the Coulomb potential of a $Z$-fold charged nucleus:

$$i\hbar \frac{\partial \psi(\mathbf{r},t)}{\partial t} = -\frac{\hbar^2}{2m}\nabla^2\psi(\mathbf{r},t) - \frac{Ze^2}{4\pi\varepsilon_0 r}\psi(\mathbf{r},t).$$

(time-dependent Schrödinger equation)

The eigenmodes correspond to the stationary states of Bohr's model for the atom. For these, $\psi(\mathbf{r},t) = \psi(\mathbf{r})\exp(-i\omega t)$ is a periodic function of time. The state functions $\psi(\mathbf{r})$, which now depend only on position, are found by solving the time-independent Schrödinger equation with $E = h\nu$ $(2\pi\nu = \omega)$:

$$-\frac{\hbar^2}{2m}\nabla^2\psi(\mathbf{r},t) - \frac{Ze^2}{4\pi\varepsilon_0 r}\psi(\mathbf{r},t) = E\psi(\mathbf{r}).$$

(time-independent Schrödinger equation)

The eigenvalues $E_n$ of this differential equation are precisely the energies given by Bohr's third postulate (Section 5.1.3):

$$E_n = -13.6\frac{Z^2}{n^2}\ [\text{eV}].$$

The Schrödinger equation yields more than the well-known discrete energy levels of the H atom. Rather, the corresponding stationary states can also be calculated. This shows that each energy level $E_n$ corresponds, not to a single stationary orbit as assumed in Bohr's model of the atom, but to $n^2$ different (mutually orthogonal) oscillation states. Like the acoustic Chladni figures of circular plates (Section 3.3.2), these states of oscillation can be characterized in terms of their antinodes and nodes. Alternatively, they can be characterized by the three quantum numbers $n$, $l$, and $m$:

## Quantum Numbers

- Principal quantum number $n$: $n = 1, 2, 3, \ldots$
- Angular momentum quantum number $l$: $l = 0, 1, \ldots, n-1$
- Directional quantum number $m$: $m = -l, -(l-1), \ldots, +(l-1), +l$

In addition, electrons have an intrinsic angular momentum (spin) which can only have two directions (*up* and *down*). Thus, directional quantum numbers

$m_s = \pm 1/2$ are assigned to the electron spin $s = 1/2$. The number of stationary states corresponding to the energy $E_n$, therefore, increases by a factor of 2 to $2n^2$.

---

**Problem 5.8**

The stationary states with $l = 0$ have wave functions $\psi(r)$ that depend only on the radial coordinate $r$ and not on the polar angles $\theta$ and $\varphi$. The time-independent Schrödinger equation for these states then simplifies to an ordinary differential equation (Section 3.3.1). What does it look like? Calculate the state function $\psi_1(r)$ for an electron in the ground state ($n = 1$) and its binding energy.

---

**Notes** The wave functions $\psi(r)$ for the stationary states of an electron in a Coulomb potential with charge number $Z$ can be uniquely characterized by the three quantum numbers $n$, $l$, and $m$. The states of an electron can also differ in the direction of their spin. The directional quantum number for the electron spin can take the values $m_s = \pm 1/2$. The $2n^2$ electron states with principal quantum number $n$ all have the same energy

$$E_n = -13.6 \frac{Z^2}{n^2} \, [\text{eV}].$$

## 5.2.4 Interpretation of electron waves

Electrons deflected in electron tubes by electric and magnetic fields (Section 3.4.2) follow a particle trajectory that can be calculated using the laws of Newtonian mechanics. In experiments these particle trajectories are made visible since each electron arriving at a fluorescent screen produces a short pulse of light. Thus, it seems that electrons are particles in the sense of Newtonian mechanics and move along a continuous trajectory $\mathbf{r}(t)$.

Diffraction experiments show, however, that electrons propagate in space like waves. The wave motion is not localized as is a particle trajectory, but extends over larger regions of space. These can be calculated using the Schrödinger equation. But in diffraction experiments, as well, the diffraction patterns are made visible by the fact that every electron incident on a fluorescent screen produces a short pulse of light there. In spite of the wave picture which explains the diffraction patterns, the electrons appear as particles when they are detected. The wave and particle representations of the electrons' motion seem to be contradictory. But upon more precise examination, we see that they are only in conflict with our habits of thought, and not with the experimental facts.

Our habits of thought are based on the assumption that natural effects can be observed continuously. Accordingly, we describe the phenomena of nature either as a continuous movement of particles which are (more or less) points in the sense of Newtonian mechanics, or as a continuous propagation of extended wave fields. The foundation of these representations is the assumption of a continuous space-time, in which these processes take place.

Doubts about the *continuum hypothesis* of classical physics were raised, however, by the atomic (Section 2.1.1) and quantum hypotheses (Section 4.4.4). In agreement with the latter hypotheses, modern experimental and observational techniques prove that all observations take place in a series of discrete and, therefore, countable *elementary events* (Section 6.6.1). An electron current is measured as a series of individually detectable electrons and a light beam, as a series of individual photons, if the experiments aim for the maximum possible precision.

Since all measurements and observations ultimately rely on counting of elementary events, the continuum representations we draw of physical processes can only be treated as aids which we use to discuss these processes and foster the development of new ideas. But these intuitive representations do not correspond directly to a space-time reality.

As long as an electron triggers no observable elementary event in its surroundings, we can treat it as a wave whose motion obeys the Schrödinger equation. This wave motion is, therefore, fundamentally not directly observable. In particular, an electron bound in the hydrogen atom can be described as a standing wave $\psi(\mathbf{r})\exp(-i\omega t)$, without fear of contradiction with any measurement results. That is, provided no quantum jump takes place, the electron actuates no observable event.

The question remains as to whether the wave functions $\psi(\mathbf{r},t)$ of the electrons are associated with any observable elementary events that the electrons might trigger. This question was answered by Max Born in 1926. Thus, the square of the absolute value of the wave function, $|\psi(\mathbf{r},t)|^2$, is the probability density for finding an electron at time $t$ at position $\mathbf{r}$. This is, therefore, one of the statistical relationships between wave functions and the occurrence of elementary events. The wave function, itself, is to be treated as a *probability amplitude*. Like the amplitudes of classical waves, it has a phase as well as a magnitude. When two wave functions overlap, interference structures can develop, as reflected in the spatial or temporal distribution of observed elementary events.

According to Born's interpretation, an electron beam described by the wave function $\psi(\mathbf{r},t) = A\exp(i(\mathbf{k}\cdot\mathbf{r} - \omega t))$ has a particle density of $n = |A|^2$. The momentum $\mathbf{p} = h\mathbf{k}/2\pi$ of the electrons implies that they are moving with velocity $\mathbf{v} = (h/m_e)\mathbf{k}/2\pi$. Thus, the particle flux (intensity) of the electron beam, $I = n|\mathbf{v}|$ is proportional to $k|A|^2$. This relationship, $I \propto k|A|^2$ has already been referred to in our discussion (Section 5.2.2) of the transmission and reflection of an electron wave at a potential barrier.

---

**Problem 5.9**

An electron beam of intensity $I = 1\,\text{mA}/\text{mm}^2$ is to be described as a plane wave. The amplitude $A$ of the wave is to be chosen so that $|A|^2 = n$. Calculate the wave amplitude for the case where the electrons have been accelerated by a voltage of $10\,\text{kV}$.

---

| **Notes** | When an electron beam is described as a plane wave |

$$\psi(\mathbf{r},t) = A\exp(i(\mathbf{k}\cdot\mathbf{r} - \omega t))$$

the square $|A|^2$ of the absolute value of the amplitude is a measure of the particle density $n$ in the beam.

## 5.3 THE ELECTRON CLOUDS OF ATOMS

Rutherford's experiments (Section 5.1.1) showed that neutral atoms consist of a nucleus with a Z-fold positive charge and Z electrons moving around the nucleus. The negatively charged electrons would be attracted by the positive charge $Ze$ of the nucleus, but repelled by one another. The motions of the electrons are, therefore, coupled. It is hardly worthwhile to deal with such a complicated system if this involves describing the fundamental relationships. Therefore, we make the greatly simplifying assumption that the interaction of the electrons among themselves can be neglected. Then the problem of the many-electron atom reduces to the problem of a single electron system, which has already been solved. As for the hydrogen atom, the stationary states of the individual electrons of a many-electron atom can be solved using the Schrödinger equation for the motion of an electron in a Coulomb potential. Many properties of the spectra of many-electron atoms and of the periodic system of the elements (Section 5.3.3) can actually be explained on this basis if it is additionally assumed that the electrons obey a fundamental principle formulated by Wolfgang Pauli in 1924, the *exclusion principle*, which is named after him.

## 5.3.1 The Pauli principle and the shell structure of the electron clouds of atoms

If we disregard Coulomb repulsion, then the electrons move independently of one another in the Coulomb potential $V(r) = -Ze^2/4\pi\varepsilon_0 r$ of the atomic nucleus. In that case, the discrete energy levels and stationary states of the individual electrons can be calculated using the Schrödinger equation. In their lowest energy states the electrons have an energy $E_1 = -13.6Z^2$ eV; that is, an energy $|E_1|$ must be supplied in order to remove electrons in this state from the atom. Thus, for an atom with $Z = 10$, $|E_1| = 1.36$ keV and for $Z = 100$, it would go to 136 keV. The measured ionization energies of the elements are, however, a lot lower. They are all below 25 eV (Figure 5.15). This indicates that not all the electrons are in the lowest energy state.

The values of the ionization energies of the chemical elements are a clear example of the well-known periodicity of chemical properties that led to the creation of the periodic table. Pauli's *exclusion principle* makes it possible to explain both the ionization energies and the periodic system (Section 5.3.3) of the elements.

When the electron spin with its two different spin orientations $m_s = \pm\frac{1}{2}$ is taken into account, the Schrödinger equation implies that a total of $2n^2$ different stationary electron states (*quantum states*) are associated with each energy level $E_n$. As with the hydrogen atom, they can be characterized by the four quantum numbers $n$, $l$, $m$, and $m_s$. These states are indicated by the symbol $|n,l,m,m_s\rangle$. It is with reference to these states that Wolfgang Pauli formulated the fundamental law for many-electron systems that is named after him:

---

### The Pauli Principle

Every quantum state $|n,l,m,m_s\rangle$ can be occupied by, at most, one electron.

---

In this form the Pauli principle provides a conceptual basis for explaining various properties of many-electron atoms. In its generalized form, it will be needed in order to explain theoretically the properties of molecules and solids. Thus, it is fundamental to chemistry, solid state physics, and semiconductor physics.

The Pauli principle implies that the *electron cloud* surrounding a many-electron atom has a *shell structure* (Figure 5.12). This is because, according to the Pauli principle, all the electrons cannot be in the lowest energy state. In a helium atom ($Z = 2$), both electrons can be in the $n = 1$ state. But by the time we get to the lithium atom ($Z = 3$), one of the three electrons has to be in an electron state with $n = 2$. It moves at a greater distance from the nucleus than the two electrons in the ground state with $n = 1$. Since the latter electrons partially screen the nuclear charge, the outer electron essentially moves in a Coulomb potential with an effective charge $Z_{\mathrm{eff}} \approx 1$. Thus, this electron is bound to the atom with a binding energy $E_B \approx \frac{1}{4} \cdot 13.6\,\mathrm{eV} = 3.4\,\mathrm{eV}$. In fact, a somewhat higher energy, about 5 eV, is actually required to ionize a Li atom.

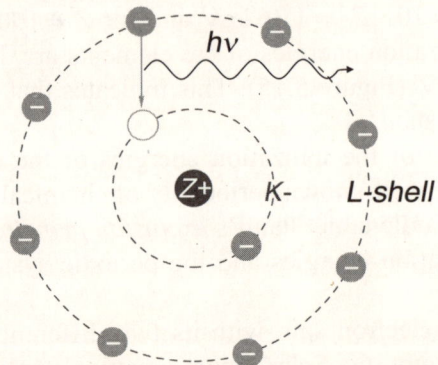

**FIGURE 5.12** The shell structure of an atom and the origin of the characteristic x-ray emission.

Because of the Pauli principle, the electrons of atoms with nuclear charges $Z > 2$ are bound with different binding energies to the atom and are correspondingly arranged in shells. The shells with $n = 1, 2, 3, \ldots$ are referred to as the K-, L-, and M-shells, respectively. The binding energy of an electron in the K-shell is roughly

$$E_B\,(\text{K-shell}) \approx 13.6 Z^2 \;\mathrm{eV}.$$

In the heaviest natural element, uranium with $Z = 92$, therefore, the K-electrons have a binding energy of about 100 keV. The binding energies of the electrons in the outer shells are substantially lower. On one hand, the binding energy falls off as $n^2$. On the other, the screening of the nuclear charge by the electrons in the inner shells must be taken into account. For electrons in the L-shell the effective nuclear charge is $Z_{eff} \approx Z - 2$ and for those in the M-shell, $Z_{eff} \approx Z - 10$. The outermost electron moves essentially in a Coulomb potential with an effective charge of $Z_{eff} = 1$. Thus, in all elements the outermost electron is similar to the electron in a hydrogen atom and is bound with an energy of a few eV.

### Problem 5.10

Make a sketch of the model of an atom for the elements hydrogen ($Z = 1$) and argon ($Z = 18$). How do the radii $r(Z,n)$ of the electron orbits change with the nuclear charge $Z$ and the principal quantum number $n$?

**Notes**    Every stationary electron state $|n,l,m,m_s\rangle$ of a many-electron atom can be occupied by, at most, one electron. The atom is in its ground state if the $Z$ electron states with the lowest energies are occupied. Since all the states with a given principal quantum number $n$ have the same energy, the electron cloud surrounding the atom has a shell structure.

## 5.3.2 Characteristic x-ray spectrum

One piece of evidence for the shell structure of atoms is provided by studies of x-ray spectra. The x-ray spectrum emitted from a molybdenum anode bombarded by electrons with sufficiently high energies consists, in part, of the continuum x-ray bremsstrahlung spectrum (Section 4.3.3). Superimposed on this bremsstrahlung spectrum are two discrete spectral lines which are characteristic of a Mo anode. The photon energies of these spectral lines do not change as the anode voltage $U_{AK}$ of the x-ray tube is varied. These spectral lines are, in fact, a characteristic of the anode material. Only if the anode material is changed will other lines be observed.

The *characteristic x-ray spectrum* appears when an electron is knocked out of the K-shell during bombardment of the anode. The kinetic energy of the incident electron must exceed the binding energy of the K-electron. After ionization of the K-shell, an electron can jump from one of the outer shells into the K-shell and reoccupy the free electron state that was produced there (Figure 5.12). A photon corresponding to the K$\alpha$- or K$\beta$-line is emitted during a quantum jump from the L- or M-shell. The energies of these photons can be estimated on the basis of the shell model for the electron cloud surrounding the atom. For a molybdenum anode ($Z = 42$) the K$\alpha$- and K$\beta$-lines correspond to photon energies of about 18 and 22 keV, respectively.

---

**Problem 5.11**

Calculate the minimum accelerating voltage $U_B$ required to excite the characteristic x-ray lines of molybdenum. Show also that energies of the x-ray lines given above are consistent with the shell structure of the Mo atom.

---

The lines in the characteristic x-ray spectrum can only be emitted if the K-shell has been ionized beforehand. Likewise, an electron in the K-shell can be excited into the L-shell only if an electron state in the L-shell is unoccupied. For this reason the characteristic x-ray lines emitted by Mo atoms cannot be absorbed by other Mo atoms (in the ground state). Absorption of x-rays is possible only if the energy of the x-ray photons is high enough to ionize the K-shell. With increasing energy, therefore, the absorption increases abruptly when the energy of the x-ray photons reaches and exceeds the binding energy of the K-electrons. For zirconium ($Z = 40$) these *absorption edges* lie exactly between the two characteristic x-ray lines of molybdenum. Thus, they can easily be observed experimentally. (Absorption is also possible through ionization of the L- or M-shells. But the absorption coefficients are very small if the energy of the x-rays is much greater than the binding energy of the electrons.)

---

### Experiment 5.5 Absorption of x-rays

For demonstrating the absorption edge of zirconium (Zr) we use the same experimental setup as for studying x-ray bremsstrahlung in Section 4.3.3. The x-rays emitted by an x-ray tube with a molybdenum anode is spectrally decomposed with a crystal spectrometer (Section 5.6.1). Besides the x-ray bremsstrahlung continuum, this spectrum contains two distinct, narrow-intensity maxima at wavelengths $\lambda = 0.06$ and $0.07$ nm (Figure 5.13). These wavelengths correspond to photon energies $E = hc/\lambda$ of 22 and 18 keV, respectively, and are the characteristic x-ray lines of molybdenum.

After taking a measurement without an absorber, a Zr foil is placed in the path of the beam. It strongly absorbs x-ray photons with energies above the binding energy $E_B \approx 13.6 \cdot 40^2 \approx 21.7 \, \text{keV}$ of the K-electrons in the Zr atom. For photon energies $h\nu < E_B$, on the contrary, the x-rays are barely attenuated. That is why the short wavelength end of the x-ray spectrum up to the K$\beta$-line is strongly absorbed at the absorber, but not the K$\alpha$-line at 18 keV.

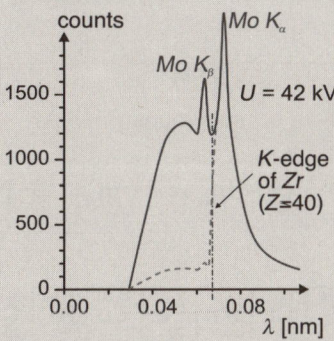

**FIGURE 5.13** The x-ray spectrum of a Mo anode without (smooth curve) and with (dashed curve) a Zr absorber.

## Problem 5.12

Sketch the x-ray spectrum that would be measured with a Mo absorber. In this case, approximately where does the absorption edge lie?

| **Notes** | The lines of the characteristic x-ray spectrum are emitted following ionization of the inner shells of an atom. Electrons from the outer shells then undergo transitions into the unoccupied states in the inner shell. |

Absorption of x-rays is possible only if the photon energy $h\nu$ is high enough to ionize the inner shells, in particular, the K-shell of an atom.

### 5.3.3  The periodic system of the elements

In the periodic system the chemical elements are ordered by their atomic number (nuclear charge) $Z$ and collected into eight (principal) groups and further subgroups (Figure 5.14) based on their chemical and physical properties. The periodicity in the chemical properties of the elements which shows up here is explained by the shell structure of atoms. That is, the tendency of atoms to form molecular bonds with other atoms is primarily determined by the number of *valence electrons*, i.e., the number of electrons in the outermost occupied shell.

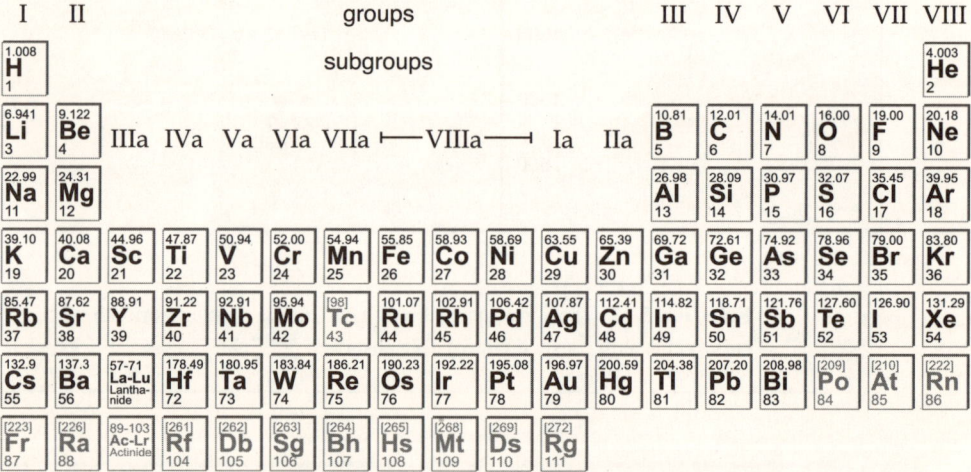

**FIGURE 5.14**  The periodic table of the chemical elements with the atomic masses and nuclear charges (atomic numbers). (The elements in blue have only unstable isotopes.)

In the innermost shell there are only two electrons. Thus, the first period of the periodic table includes only two elements, hydrogen and helium. In the second period the L-shell is filled. First the two states with $n = 2$ and $l = 0$ are filled, and then the six states with $n = 2$ and $l = 1$. This sets the position of the lightest elements in the eight groups of the periodic table. The lithium atom, with only one valence electron, belongs to the alkali metal group. The elements with a closed outer shell stand on the other side of the periodic table. This is the group of chemically inert gases (the noble gases), which begins with helium and neon. Immediately next to the noble gases are the elements in which one electron state in the outermost shell is unoccupied. The elements with one of these *electron hole states* form the halogen group.

After the L-shell, the M-shell ($n = 3$) is filled, with the electron states with $l = 0$ and 1, of course, being filled first. Like the second period, the third contains only eight elements. Because of the screening of the nuclear charge by the electrons in the inner shells, the binding energies of the electron states with $n = 3$ and $l = 2$ are relatively low. Thus, these electron states are only filled along with the $n = 4$ states. In addition, there are the elements in the subgroups. Altogether, the fourth period and, in analogous fashion, the fifth period each contain 18 elements.

The chemical properties of the elements depend essentially on the *ionization energy* and the *electron affinity* of the atoms, that is, respectively, on the energy necessary to remove a valence electron or the energy released when an extra electron is attached. Figure 5.15 shows the ionization energies $I_Z$ of the elements as a function of their nuclear charge $Z$. The ionization energies of the noble gases are the highest, since a closed shell has to be broken up upon ionization. In the case of the alkali metals, on the other hand, the valence electron moves largely outside the closed shells in a Coulomb potential with an effective nuclear charge $Z_{\text{eff}} = 1$. This electron is, therefore, very loosely bound. Hence, the ionization energies of the alkali metals are the lowest.

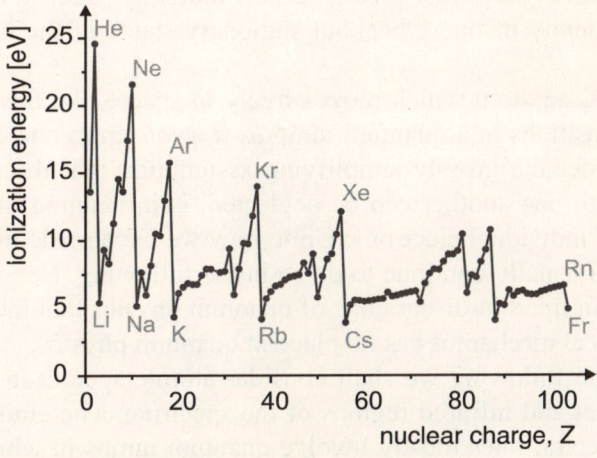

**FIGURE 5.15** Ionization energies of the chemical elements.

**Problem 5.13**

Niels Bohr assumed that electrons move in circular orbits in atoms. In units of $h/2\pi$, the orbital angular momentum of these electrons was given by $l = n$. Draw a sketch of a model of the atom in which the electrons have angular momenta $l < n$. Indicate the orbit with $l = 0$. Using this model, explain how the binding energy of the electrons in a shell should be lower with increasing $l$.

| Notes | The chemical properties of atoms are primarily determined by the binding energies and quantum states of the valence electrons. Because of the shell structure of the cloud of |
|---|---|

electrons surrounding an atom, with increasing $Z$ the valence electrons have configurations that repeat themselves in an approximately periodic fashion. The shell structure of the electron cloud thereby provides an explanation for the periodic system of the elements.

## 5.3.4 Atomic spectra

Bohr's model of the atom relates the spectral lines of the hydrogen atom to quantum jumps of the electron between the discrete energy levels of the atom. The electron clouds of all the other chemical elements are made up of multiple electrons. In this case, in general, there are no stationary states of individual electrons, since they repel each other mutually and, therefore, do not move independently of one other, but stationary states of the *entire* electron cloud do exist.

Thus, an atom which moves freely in space, like the atoms of a gas, undergoes transitions in a quantum jump *as a whole* from one energy level to another. Only under the grossly simplifying assumption that the interaction of the electrons with one another can be neglected, is it meaningful to speak of stationary states of individual electrons. Up to now we have made use of this simplification and shall usually continue to do so in the following. Nevertheless, it is necessary to keep in mind in discussions of quantum physics that the microscopic approach of classical mechanics has no place in quantum physics.

In the following we shall consider atomic spectra in the visible and nearby ultraviolet and infrared regions of the spectrum. The emission and absorption of these spectral lines mostly involve quantum jumps in which the states of the valence electrons change. By analogy with the energy level diagram for the H atom, the energy levels of the stationary states of the atomic states of other elements can be represented graphically. Here the energy levels are usually ordered according to the energy, as well as according to other characteristic features of the stationary states. As a simple example, Figure 5.16 shows an energy level diagram for sodium. As an alkali metal, the Na atom has only one valence electron. In this case the energy levels can be ordered in terms of the orbital angular momentum quantum number $l$ of the (almost) stationary states of the valence electron. Depending on whether $l = 0$, 1, or 2, we speak of s-, p-, or d-electrons, respectively.

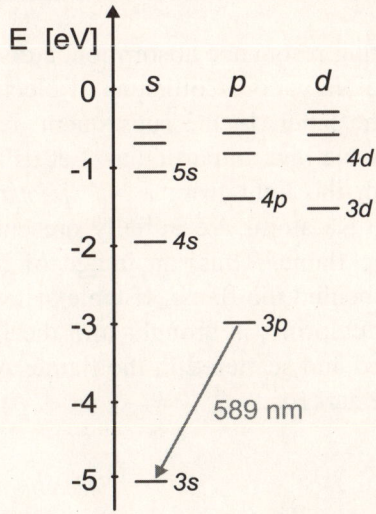

**FIGURE 5.16** Energy level diagram of sodium.

Although in the alkali metals, as in hydrogen, the transitions essentially involve a single electron, the energy level diagrams of the alkali metals differ fundamentally from that of hydrogen. Electronic states with the same principal quantum number $n$, but a different orbital quantum number $l$, have different energies. This difference is a consequence of the fact that the valence electron of an alkali metal not only moves outside the closed shells, but also penetrates into the inner shells. Thus, unlike the electron in a H atom, the valence electron in an alkali metal atom does not move in a pure Coulomb potential with a central charge $Z = 1$.

For this reason, in the sodium atom, an electron in the 3p-state has a binding energy roughly 2 eV lower than that of an electron in the 3s-state. The characteristic yellow line at $\lambda = 589$ nm of the sodium atom is emitted in 3p-3s transitions. Since atoms moving freely in space are usually in the ground state, these atoms can absorb the yellow spectrum line. Then the valence electron jumps from the 3s ground state to the excited 3p-state. Within a time of about $10^{-8}$ s the electron then jumps back into the ground state and thereby emits another photon of the yellow Na line. In this way a photon with the same wavelength but propagating in another direction is produced. Thus, a beam of yellow sodium light will be attenuated as it passes through Na vapor.

---

### Experiment 5.6     Resonance absorption of the yellow sodium line

For demonstrating resonance absorption, a discharge is struck in an electron tube containing Na vapor. Collisions of electrons with the Na atoms cause excitation of the latter. In the subsequent decay to the ground state, they emit the yellow Na line, in particular. Let us illuminate the flame of a Bunsen burner with the light from a Na *spectrum lamp* of this type (Figure 5.17). Since no Na atoms are initially present in the flame, the light is not absorbed in the flame. Thus, an image of the light source appears on a screen located behind the flame. If table salt, the chemical compound NaCl of sodium and chlorine, is brought into the flame, however, then the light will be absorbed and scattered in the flame. At the same time, the image of the light source darkens.

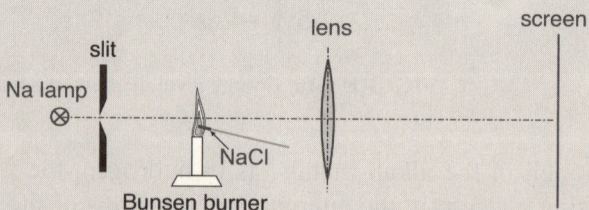

**FIGURE 5.17** Experimental setup for demonstrating resonance absorption.

In this experiment the flame gives off yellow light, even if it is not illuminated. This emission shows that the NaCl molecules in the flame are not only vaporized and dissociated, but are also excited. The number of excited atoms relative to the number of atoms in the ground state is, of course, very small. The excitation probability is essentially determined by the Boltzmann factor $\exp\left(-(E_{3p} - E_{3s})/kT\right)$ (Section 2.1.4). Since the excitation energy

$\left(E_{3p} - E_{3s}\right)$ of the Na atom in the 3p-state is about 2 eV and $kT \approx 0.1$ eV in the flame, the excitation probability is roughly $e^{-20} \approx 2 \times 10^{-9}$. In absolute terms, of course, many atoms are excited in the flame, so the flame emits the sodium line. Nevertheless, the number of excited atoms is only a tiny fraction of all the Na atoms in the flame. Most of the Na atoms are in the ground state, so they are able to absorb the incoming light beam.

**Problem 5.14**

The light from a hydrogen spectrum lamp is directed onto a flame and the intensity of the Balmer-$\alpha$ line is measured behind the flame. How does the intensity of the line change if hydrogen is brought into the flame? Discuss the result of an experiment of this sort. Which hydrogen atoms can absorb the Balmer-$\alpha$ line?

**Notes**   In gaseous discharges the atoms of a chemical element emit a spectrum of discrete spectral lines that is characteristic of that element. The energies of the photons of these lines are calculated from the separations of the energy levels in an energy level diagram. If a gas is in thermal equilibrium at moderate temperatures, most of the atoms will be in the ground state. The atoms of most of the chemical elements have excitation energies of a few eV. Since $kT = 25$ meV $\ll 1$ eV at room temperature, in thermal equilibrium the fraction of excited atoms is usually very low.

## 5.4 THE ATOMIC NUCLEUS

In Rutherford's model of the atom, the atomic nucleus is treated as a point mass compared to the atom as a whole, with a positive charge Ze in which almost all the mass of the atom is concentrated. The quantum mechanical descriptions of the Rutherford model also proceed from the assumption of a structureless nucleus that is almost a point mass. Nevertheless, under suitable experimental conditions atomic nuclei are also revealed as objects with an internal structure.

In our physical description of gases (Section 2.1.2) we have also treated atoms initially as structureless mass points. This assumption was justified in quantum mechanics. The atoms of a gas appear to be structureless units provided their thermal energy $kT$ is small enough compared to the minimum energy required to excite them. Only when the energy changes are greater than the excitation energies of the atoms, as in the Franck-Hertz experiment (Section 5.1.4) or the Rutherford scattering experiments (Section 5.1.1), do atoms manifest their internal structure. Below the excitation threshold the atoms appear to be indivisible units.

Likewise, atomic nuclei can be treated as indivisible units if the energy transfer during a given process is low compared to the excitation threshold for atomic nuclei. This excitation threshold is, however, about six orders of magnitude higher for nuclei than for atoms. The excitation thresholds for nuclei are in the MeV range ($1\,\text{MeV} = 10^6\,\text{eV}$), rather than in the eV range as for atoms.

The first indication of internal structure in atomic nuclei was provided by studies of the *radioactivity* of atoms. Radioactivity was discovered at the end of the 19th century. After the discovery of the atomic nucleus, radioactivity was quickly recognized as a property of the nuclei. But only after the discovery of the *neutron* by Chadwick in 1932 was it possible to develop a model of the nucleus which permitted a quantum physical interpretation of nuclear structure and nuclear processes.

## 5.4.1 The structure of atomic nuclei

A nucleus with *mass number A* and *nuclear charge* (atomic number) $Z$ consists of $Z$ protons and $N = A - Z$ neutrons. The protons have a positive electrical charge and the neutrons are electrically neutral. Therefore, the proton number $Z$ determines the structure of the electron cloud surrounding the atom and, thereby, the chemical properties of the atom. The properties of the nucleus are determined both by the number of protons and by the number of neutrons. Nuclei with equal numbers of protons but different numbers of neutrons are referred to as *isotopes*. The atoms containing nuclei of this sort (i.e., with the same number of protons) have identical chemical behavior.

The proton has a mass $m_p = 1.6 \times 10^{-27}\,\text{kg} = 1836 m_e$. It is, therefore, almost 2000 times heavier than an electron. Its mass corresponds to a rest energy $m_p c^2 = 983.3\,\text{MeV}$. The neutron is about 0.1% heavier than a proton, so its rest energy is 1.3 MeV higher. Because of this energy excess of 1.3 MeV, it can decay into a proton on emitting an electron (rest energy about 0.5 MeV). As a free particle, the neutron is therefore unstable, but it is stable in many atomic nuclei. Protons and neutrons are not only similar in terms of their masses. They have many physical properties in common. Hence, together they are referred to as *nucleons*.

An overview of the nuclei which exist in nature and the many artificially created nuclei is provided by a *chart of the nuclides* (Figure 5.18). On this chart, the abscissa is the number of neutrons and the ordinate is the number of protons. The stable nuclei are indicated in black. Nuclei with mass numbers $A < 40$ have roughly equal numbers of protons and neutrons. By contrast, the heavier nuclei have an increasing neutron excess for higher $A$. In addition, there are many unstable nuclei which decay into adjacent nuclei through the radioactive decay (Section 5.4.2).

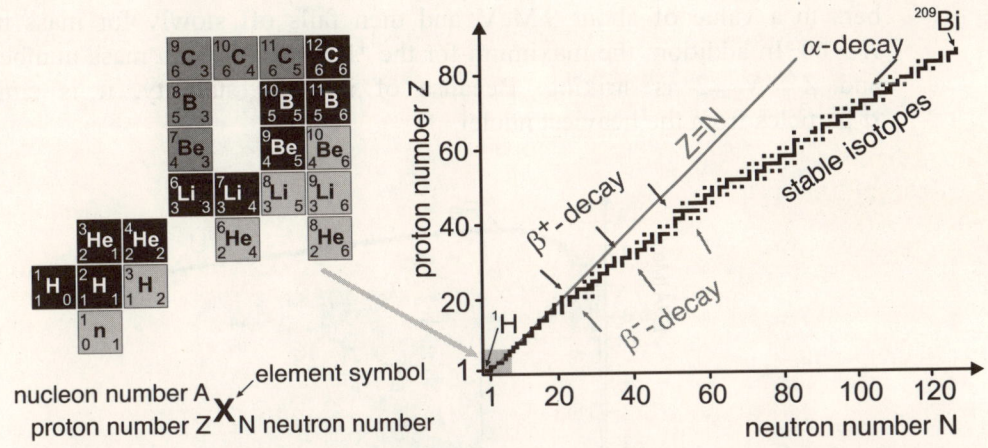

**FIGURE 5.18** Chart of the nuclides with stable (black) and unstable (blue) nuclei (left: $A < 12$; right: $A < 209$).

The stability of the nuclei follows from the binding energy of the nucleons in a nucleus. Only if the binding energy of a nucleus is greater than the binding energy of adjacent nuclei with an equal mass number will the nucleus be stable. The binding energy $E_B$ of the atomic nucleus is derived from its mass. Thus, according to Einstein's special theory of relativity, energy and mass are equivalent (Section 1.2.4):

$$E = mc^2.$$

Because of the equivalence of mass and energy, the mass $M_K$ of a nucleus is smaller than the sum of the masses of its protons and neutrons. The *mass defect* $\Delta M = Zm_p + Nm_n - M_K$ is obtained from the binding energy $E_B$ of all the nucleons,

$$\Delta M = \frac{E_B}{c^2}.$$

The binding energy $E_B$ thus represents the minimum amount of energy which must be supplied in order to break a nucleus up into its $A$ nucleons. This binding energy increases in rough proportion to $A$. In order to graphically highlight some of the interesting structures in the binding energy as a function of the mass number $A$, it is convenient to deal with the average *binding energy per nucleon* $E_B/A$ (Figure 5.19). For the stable nuclei, $E_B/A$ increases initially for low mass numbers to a value of about 9 MeV and then falls off slowly for mass numbers $A > 60$. In addition, the maximum for the $^4$He nucleus with mass number $A = 4$ and $Z = N = 2$ is striking. Because of its high stability, it is emitted as $\alpha$-particles from the heaviest nuclei.

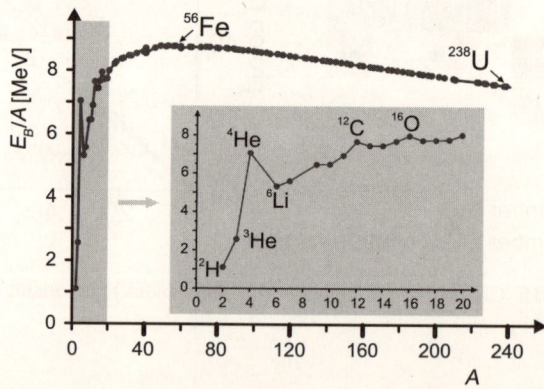

**FIGURE 5.19** Binding energy per nucleon of the stable nuclei with mass numbers $A < 238$ (inset: $A < 20$).

The nucleons are held together in the nucleus by forces which are insignificant outside the nucleus, but bind the nucleons to one another with energies on the order of 10 MeV inside the nucleus. For this reason these forces are known as the *nuclear force*. It operates between the nucleons only over distances $R < 1.4 \times 10^{-15}$ m. The volume of nuclei increases, somewhat like the volume of a water drop, with the number $A$ of particles contained in it. Nuclear radii $R_K$ are therefore on the order of $R_K \approx 1.4 \times 10^{-35}$ m $\cdot A^{1/3}$. The nuclear force is opposed by the repulsive Coulomb force of the positively charged protons. The Coulomb force reduces the binding energy of the protons in a nucleus. Therefore, heavy nuclei will be preferentially made up of neutrons and have a *neutron excess* that increases with $Z$.

**Problem 5.15**

The mass of a carbon atom with mass number $A = 12$ is $m\left({}^{12}\mathrm{C}\right) = 12 \cdot 1.6606 \times 10^{-27}$ kg. Calculate the binding energy $E_B$ of the ${}^{12}\mathrm{C}$ nucleus.

| Notes | Atomic nuclei are made up of protons and neutrons. The two nucleons have roughly the same rest mass. The rest energy $E = mc^2$ of the nucleons is just under $1\,\mathrm{GeV} = 10^9$ eV. |
|---|---|

Protons have a single positive charge. Neutrons are electrically neutral. In a nucleus the nucleons are bound with a binding energy per nucleon of several McV.

## 5.4.2 Radioactivity

Some of the atomic nuclei found in nature and many artificially produced nuclei are unstable. They decay by emitting charged particles or $\gamma$-rays with energies in the MeV range. The *radioactivity* of some heavy elements, e.g., uranium, was discovered in 1896. During radioactive decay an atomic nucleus changes in the same way that an atom changes during a quantum transition; that is, the decay takes place spontaneously and obeys the laws of chance. For an initial number of nuclei $N_0$, the number $N$ of nuclei that have not yet decayed decreases exponentially with time:

$$N(t) = N_0 \exp(-t/\tau).$$

Thus, one can only specify an average lifetime $\tau$ for unstable nuclei. Usually the *half-life* $\tau_{1/2} = (\ln 2)\tau$ is tabulated. It represents the time in which half of a given amount of nuclei will decay. The half-lives of the two naturally occurring *isotopes* ${}^{235}\mathrm{U}$ and ${}^{238}\mathrm{U}$ of uranium (with $A = 235$ and 238, respectively) are so long that not all of the uranium nuclei have decayed since they were created about $10^{10}$ years ago. Their half-lives are on the order of $10^9$ years. Other nuclei decay much more rapidly.

The radioactive isotopes that exist in nature decay in three different ways, namely, $\alpha$-, $\beta$-, and $\gamma$-decay. $\alpha$-decay occurs especially in the heaviest elements. In $\alpha$-*decay* a nucleus loses an $\alpha$-particle, which consists of two protons and two neutrons, so it is equivalent to the nucleus of a helium atom. During the decay of uranium, $\alpha$-particles are ejected with energies somewhat over $4\,\mathrm{MeV}$. This reduces the nuclear charge by two and the mass number by 4 units. Thus, the uranium isotopes ($Z = 92$) yield isotopes of thorium ($Z = 90$). In general, $\alpha$-decay of a nucleus $X$ into a daughter nucleus $Y$ follows the pattern:

$$_Z^A X \rightarrow _{Z-2}^{A-4} Y + _2^4 \alpha.$$

The long lifetime of many $\alpha$-emitters can be explained in terms of the tunnel effect (Section 5.2.2). The $\alpha$-particle and the daughter nucleus interact through both the nuclear forces and Coulomb forces. At close distances the attractive nuclear forces predominate, while at larger distances the repulsive Coulomb forces predominate. Thus, in the neighborhood of the parent nucleus the potential energy of the $\alpha$-particle varies as shown in Figure 5.20(b). An $\alpha$-particle can therefore be bound in the potential well of the nucleus, even if its energy level lies above the separation energy. But it can escape the nucleus if it tunnels through the so-called Coulomb barrier. The higher and wider the barrier is relative to the energy level of the $\alpha$-particle, the longer the lifetime of the nucleus.

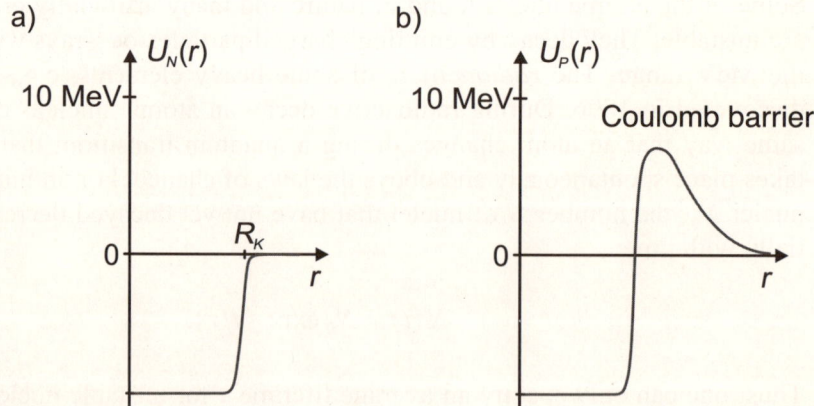

**FIGURE 5.20**  Potential energy of (a) a neutron and (b) an $\alpha$-particle in the neighborhood of a uranium nucleus.

In nature there are three $\alpha$-emitters with half-lives on the order of $10^9$ to $10^{10}$ years, namely the uranium isotopes $^{235}$U and $^{238}$U, and the thorium isotope $^{232}$Th. These are the parent nuclei of the radioactive *decay series* which end in the lead isotopes $^{207}$Pb, $^{206}$Pb, and $^{208}$Pb, respectively. In these series, other $\alpha$-decays take place, as well as $\beta$- and $\gamma$-decays.

Although during $\alpha$-decay the emitted particle is made up of nucleons, i.e., original constituents of the nucleus, during $\beta$-decay, the emitted electrons are only formed during the decay process. Here a neutron changes into a proton.

During $\beta$-decay the nucleon number $A$ does not change, but the high neutron excess of the heavier nuclei is reduced.

Besides the $\beta^-$-decay, which occurs in the natural decay series with the emission of an electron, there is also $\beta^+$-decay in which a proton changes into a neutron and *positrons* are emitted. Positrons have the same mass as electrons, but carry a positive elementary charge. $\beta^+$-decay occurs in the proton-rich nuclei which can be produced in nuclear reactions (Section 5.4.3).

Measurements of the energy balance in $\beta$-decay led to the conclusion that, in addition to a charged particle, a neutral (and almost) massless particle, the neutrino $\nu$ (or antineutrino) is emitted. The $\beta^-$-decay which takes place in the natural decay series thus obeys the following reaction equation:

$$\,_{Z}^{A}X \rightarrow \,_{Z+1}^{A}Y + e^- + \bar{\nu}.$$

The total energy of the electron and neutrino is thus equal to the difference in the rest masses of the parent and daughter nuclei.

Nuclei, like atoms, have excited quantum states that lie above the ground state. Often the daughter nucleus remaining after $\beta$- or $\alpha$-decay is in an excited state. This state then decays with the emission of one or more high-energy photons into the ground state. The energy of these *$\gamma$-decays* follows, as does the energy of electronic transitions in atoms, from Bohr's second postulate (Section 5.1.3). In $\gamma$-decay neither the charge nor the mass number of the nucleus changes.

---

**Problem 5.16**

The radioactive isotope $^{137}$Cs is produced in nuclear reactions and has a half-life of 30 years. After the Chernobyl disaster large areas of land were contaminated by it. Calculate the time for the amount of this Cs isotope to decrease by a factor of 10.

---

**Notes**    Radioactive rays are produced during nuclear transformations. They exist in the form of $\alpha$-, $\beta$-, and $\gamma$-rays. During $\alpha$- and $\beta$-decay charged particles, specifically He nuclei and electrons, respectively, are emitted. In $\gamma$-decay, electromagnetic rays, or photons, are produced. In all radioactive decay processes energies in the keV to MeV ranges are released.

The activity of radioactive samples falls off exponentially. The amounts of radioactive nuclei are reduced by half over the half-life.

### 5.4.3 Nuclear fusion and nuclear fission

The sun seems to be an inexhaustible energy source. For a few billion years it has irradiated the earth with its light. The solar constant (Section 4.4.3), $S = 1.4 \times 10^3$ W/m$^2$, implies that the sun radiates a total power $P \approx 4 \times 10^{26}$ W into space. According to the Einstein relation $E = mc^2$, the gigantic amount of energy radiated away by the sun since its creation almost $10^{10}$ years ago corresponds to a mass equivalent of on the order of 0.1% of the sun's mass $M_S \approx 2 \times 10^{30}$ kg. This energy is produced in the sun's interior by the *fusion* of hydrogen to helium.

---

**Problem 5.17**

Estimate the energy radiated by the sun since its creation assuming that its radiative power has not changed and calculate the mass equivalent to that energy.

---

When four protons and two electrons fuse into an $\alpha$-particle (and two neutrinos), an energy roughly equal to the binding energy of the $\alpha$-particle (Figure 5.19), $E_B = 28$ MeV, is released. The fusion process takes place in a number of steps, in which the heavy isotopes of hydrogen with mass numbers $A = 2$ and 3 are first formed and ultimately fuse to form helium:

$$^3_1\text{H} + {}^2_1\text{H} \rightarrow {}^4_2\text{He} + n + 17.6 \text{ MeV}.$$

Since the hydrogen nuclei are charged particles, fusion can only take place if their kinetic energy is high enough to overcome the Coulomb repulsion, or, at least, if tunneling of the Coulomb barrier is probable enough. Nuclear fusion, therefore, occurs only at high temperatures. In the interior of the sun the average thermal energy $kT \approx 1$ keV. This corresponds to a temperature above $10^7$ K.

Up to now it has not been possible to use this process for commercial energy generation. It is more convenient to obtain nuclear energy through fission of heavy nuclei. Figure 5.19 shows that the binding energy per nucleon, $E_B/A$, has a maximum at $A = 60$. Hence, binding energy is released both in the fusion of light nuclei and in the fission of heavy nuclei. The fission of heavy nuclei can be initiated by neutrons. Since neutrons are uncharged, there is no Coulomb barrier to keep them from penetrating a nucleus (Figure 5.20). They can thus initiate nuclear reactions at ordinary room temperatures. Most nuclei will simply capture the neutrons; that is, the neutrons become part of the nucleus. The binding energy of about 7 MeV released in this way is given up as photons through $\gamma$-decay.

In 1938 Otto Hahn and Fritz Strassmann discovered that neutrons can also cause uranium to undergo fission. When a $^{235}$U nucleus captures a neutron, at first a highly excited nucleus of the isotope $^{236}$U is produced. It vibrates in a way similar to the water drops from a tap and then breaks up into two daughter nuclei with mass numbers between $A = 140$ and $A = 90$ (Figure 5.21). The binding energy per nucleon of these nuclei is about 1 MeV higher than in uranium. Hence, the fission of a uranium nucleus liberates a total binding energy of roughly 200 MeV.

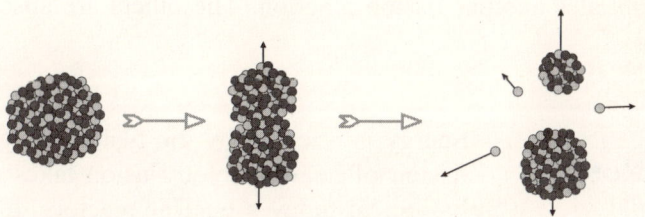

**FIGURE 5.21** Breakup of a uranium nucleus.

This binding energy corresponds quite closely to the potential energy owing to Coulomb repulsion of the two daughter nuclei at the moment of their creation. The highly charged nuclei repel one another and fly apart with a kinetic energy of about 200 MeV.

**Problem 5.18**

Estimate the Coulomb energy of the daughter nuclei during breakup of a uranium nucleus.

When a uranium nucleus breaks up, besides the two daughter nuclei, an average of roughly three neutrons is produced. In this way the large neutron excess, which was favorable for the stability of the uranium nucleus, but not for the formation of stable daughter nuclei, is somewhat reduced. Because of the neutrons generated during fission, a single neutron in uranium can initiate a *chain reaction*. Of course, one neutron is absorbed in every fission event, but at the same time three new ones are produced. If these neutrons again strike fissionable uranium, then a chain reaction can lead to an explosion.

---

### Problem 5.19

Calculate the energy released by the fission of 1 kg of $^{235}$U. How long does a 1000 MW power plant take to produce this much energy?

---

In order to be able to use nuclear energy as an energy source, the chain reaction must be controlled. During operation of a nuclear reactor the neutron flux is controlled. On the average, only one of the three neutrons produced during fission initiates another fission reaction. The others are absorbed in a suitable absorber.

**Notes**   Energy is released by the fusion of light nuclei and by the fission of heavy nuclei. Fusion takes place in the sun and fission, in today's nuclear reactors. In both cases, mass is converted into energy in accordance with the Einstein equation

$$E = mc^2.$$

Fusion processes in the sun yield about 7 MeV per nucleon and nuclear fission, about 1 MeV per nucleon.

## 5.4.4 Absorption of nuclear radiation

Particles or $\gamma$-rays with energies in the keV to MeV range are emitted in all radioactive decay. These energies are many times the ionization energies of atoms (Figure 5.15). Thus, nuclear radiation can ionize many atoms and dissociate many molecules in matter. A single $\alpha$-particle with, for example, an energy of about 4 MeV, can ionize more than $10^5$ atoms as it passes through the matter in a thick foil. The high ionizing ability of nuclear radiation can, on one hand, be used to detect the radiation. On the other hand, for the same reason this radiation presents a great danger if its intensity exceeds certain limits.

In general, high-energy radiation is a part of our environment. It reaches us as cosmic rays from outer space or as radiation from the earth, especially near uranium ore deposits. Some radioactive elements are even present in the bodies of all living things. This natural radiation burden is an important consideration in setting exposure limits for radioactive materials.

## Experiment 5.7    The Wilson cloud chamber

The passage of $\alpha$-particles through matter can be observed in a cloud chamber (Figure 5.22). Supersaturated steam tends to condense on ions and form fog droplets. Thus, $\alpha$-particles leave behind a track of fog droplets in a cloud chamber. The tracks are essentially straight lines, since the high-energy $\alpha$-particles are barely deflected when ionizing the atoms. The length of the tracks follows from the mean free path for ionization ($l \approx 10^{-7}$ m in air) and the average energy loss per ionization event (about 30 eV) for the $\alpha$-particles.

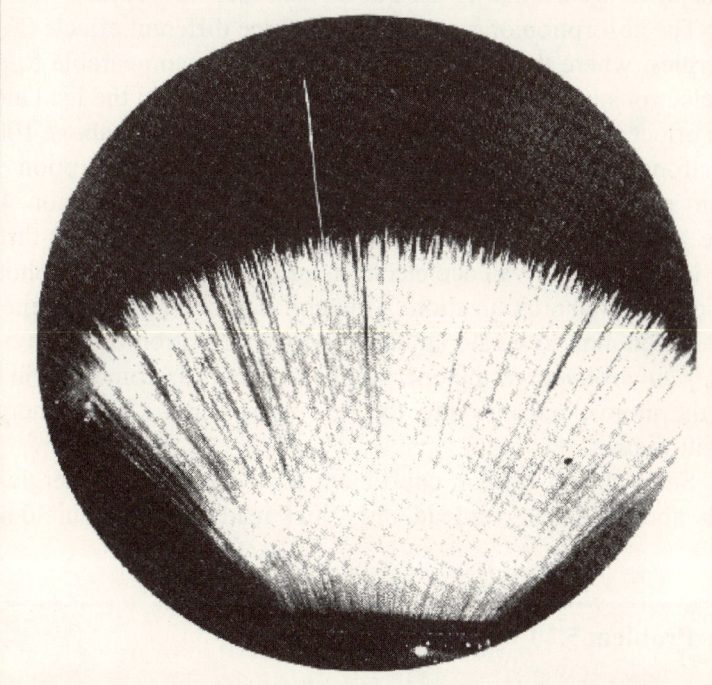

**FIGURE 5.22** Tracks of $\alpha$-particles in a cloud chamber.

In air, $\alpha$-particles with energies of a few MeV lose their energy over a distance of a few cm. In denser matter they are slowed down over correspondingly shorter distances. The product of the range $d$ of $\alpha$-particles and the density $\rho$ of the material is approximately the same for all materials for particles with equal energies. In solids $\alpha$-particles are therefore slowed down over a few times $10^{-5}$ m. $\beta$-rays with energies in the MeV range penetrate solid layers to a few mm. This has already been shown in our measurements of the absorption of electron beams in Al foils (Section 5.1.1).

Like $\alpha$- and $\beta$-rays, all high-energy charged particles are rapidly slowed down in matter. $\gamma$-rays are less efficiently absorbed in matter. The intensity $I$ of $\gamma$-rays falls off exponentially with the thickness $d$ of the absorbing layer; that is,

$$I(d) = I_0 e^{-\mu d}$$

The absorption coefficient $\mu$ depends strongly on the energy of the $\gamma$-rays and the nuclear charge $Z$ of the absorber. Materials with high $Z$ are the best absorbers. Thus, lead is customarily used as an absorber for $\gamma$-rays.

The absorption of $\gamma$-rays relies on three different effects (Figure 5.23). At low energies, where the energy $h\nu$ of the $\gamma$-rays is comparable to the binding energy of electrons in the cloud surrounding the nucleus in the lead atom, the photoelectric effect predominates (Section 4.3.1). For energies above 100 keV, it is mainly electrons in the K-shell that are ionized during the absorption of photons. At medium energies around 1 MeV, the Compton effect (Section 4.3.4) is dominant. The $\gamma$-ray photons give up a large fraction of their energy through scattering on the loosely bound valence electrons of lead. The scattered photons are ultimately absorbed by photoionization. At energies above 1 MeV the energy of the photons is high enough to create an electron and a positron (Section 5.4.2). In this process, *pair production*, the energy of the photon is converted in the Coulomb field of the atomic nucleus into the rest energy and kinetic energy of an electron-positron pair.

Since the absorption coefficient for $\gamma$-radiation at energies of a few MeV is only about 0.4 cm$^{-1}$ for lead, only thick lead plates can shield against $\gamma$-rays.

**Problem 5.20**

How thick must a lead shield be in order to reduce the intensity of MeV $\gamma$-rays by a factor of 1000?

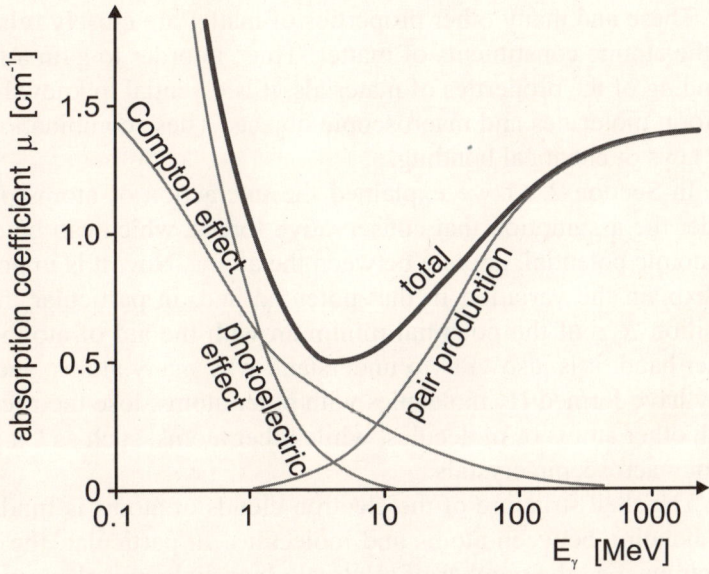

**FIGURE 5.23** The absorption coefficient for $\gamma$-rays in lead.

> **Notes**
>
> All nuclear radiation consists of particles or photons with energies in the keV to MeV range. It can, therefore, ionize or dissociate many thousands of atoms or molecules, respectively, and thereby activate them chemically. Thus, when certain limits are exceeded, radiation causes damage to all living things and can also destroy materials.

## 5.5 CHEMICAL BONDS

Atoms are the building blocks of matter. Thus, the structure of atoms is crucial in determining the properties of macroscopic substances. They can be gaseous, liquid, or solid. Some substances are electrical conductors and others, insulators, while some are transparent and others, opaque. Some substances are permanently magnetic, while others can be magnetized only in a magnetic field. Many substances have amazing properties at very low temperatures. For instance, they can be *superconducting* or *superfluid*; that is, these substances lose their electrical resistance (Section 3.4.4) or viscosity (Section 3.2.1).

These and many other properties of matter are closely related to the structure of the atomic constituents of matter. Thus, in order to gain a fundamental understanding of the properties of materials, it is essential to know how atoms combine to form molecules and macroscopic objects. These combinations are governed by the laws of chemical bonding.

In Section 2.1.1 we explained the interaction of atoms phenomenologically under the assumption that conservative forces, which can be described by an interatomic potential, operate between the atoms. Now it is important, on one hand, to explain the variation in that potential and, in particular, the depth of $\varepsilon_0$ and position $R_{\min}$ of the potential minimum with the aid of atomic structure. On the other hand, it is also vital to understand why many atoms, such as H atoms after they have formed $H_2$ molecules with other atoms, lose their tendency to combine with other atoms or molecules, while other atoms, such as Cu and Fe, combine to form macroscopic crystals.

The shell structure of the electron clouds of atoms is fundamental for chemical bonding between atoms and molecules. In particular, the number of valence electrons, i.e., the number of relatively loosely bound electrons in the outer shell, determines the chemical activity of an element. The noble gases are especially inactive chemically. They are characterized by a closed outer shell, i.e., one that is fully populated by electrons. Atoms with incompletely filled outer shells have a tendency to close that shell by combining with other atoms. There are different types of closure and ways in which closure is obtained. Basically, five types of bonding are distinguished. These will be outlined in the following.

### 5.5.1 Ionic bonds and hydrogen bridges

We begin by describing bonds between dissimilar atoms. An easily comprehensible situation occurs when two atoms with complementary valence shells combine with one another. A typical example is table salt, a compound of sodium and chlorine. As an alkali metal, sodium has a loosely bound valence electron and the Cl atom, as a halogen, is one electron short of filling its valence shell (Figure 5.24). Closed shells result when the valence electron of the Na atom is taken by the Cl atom.

If the two atoms are widely separated from one another, this process requires an energy of 1.4 eV. That is because an ionization energy $I_{Na} = 5.1$ eV is required to ionize the Na atom, while an energy of only $A_{Cl} = 3.7$ eV, the *electron affinity* of chlorine, is given back when an electron is bound to a Cl atom. However, on the whole, energy will be released if the Coulomb attraction of the two resulting ions comes into action. For a distance between the nuclei of $R \sim 2.5 \times 10^{-10}$ m, at which the electron clouds of the two ions come slightly into

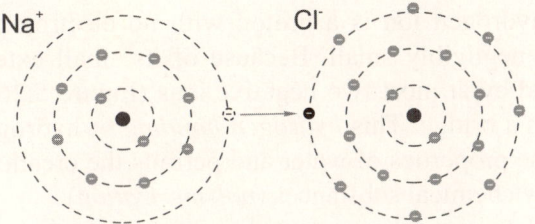

**FIGURE 5.24** Ionic bonding of Na and Cl atoms.

contact, the potential energy is about $\Delta E_{pot} = 6\,\text{eV}$ lower than when the separation is large. Thus, the binding energy $E_B$ of the NaCl molecule is

$$E_B(\text{NaCl}) \approx \Delta E_{pot} + A_{Cl} - I_{Na} \approx 4.6\,\text{eV}.$$

The *heteropolar* molecules formed in this way are electric dipoles. The dipole moment of the $\text{Na}^+\text{Cl}^-$ dipole is given by $d \approx eR$. These dipole moments have the effect that the resulting molecules accumulate next to one another like little permanent magnets. They organize themselves into a crystal lattice in which ions with opposing charges alternate with one another. In this way salt crystals develop from NaCl molecules, with each $\text{Na}^+$ ion surrounded by six chlorine ions and each $\text{Cl}^-$ ion, by six sodium ions.

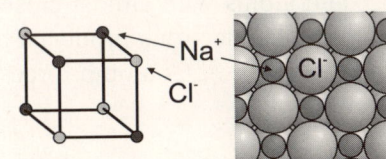

**FIGURE 5.25** A crystal of NaCl (table salt).

Crystal formation causes the binding energy per NaCl molecule to increase by a factor of 1.75 from 4.6 eV to 8 eV. Thus, the creation of 1 mol of table salt with $N_A = 6 \times 10^{23}$ molecules from 1 mol of Na and 1 mol of Cl releases just under 800 kJ of heat.

Because of their relative sizes, $\text{Na}^+$ and $\text{Cl}^-$ ions have a good spatial fit within a crystal. The situation is different when hydrogen ions are involved in the formation of molecules, for example, of the $H_2O$ molecule or of the radicals OH and $NH_2$, which are often encountered in chemical compounds. In the $H_2O$ molecule the oxygen atom attracts both of the electrons from the hydrogen atoms and is able to form a closed valence shell in that way. The remaining positively charged hydrogen ion is a proton with no electron cloud. Its spatial extent is, therefore, negligibly small. Because of its small extent, a given proton can be surrounded by at most two negative ions (Figure 5.26). To some extent the proton acts as a bridge. This *hydrogen bonding* or hydrogen *bridge bonding* is decisive for the properties of water and permits the creation of large chain molecules from many chemical substances (*polymerization*).

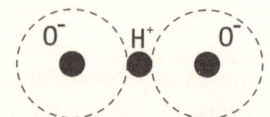

**FIGURE 5.26**  Hydrogen bonding.

---

**Problem 5.21**

Determine the conversion factor from eV per molecule to kJ per mol, which is important for calculating heats of reaction.

---

| **Notes** | Ionic bonds form between atoms with few valence electrons and atoms with almost closed shells. The rearrangement of valence electrons produces positive and negative ions with |

closed shells which, because of Coulomb forces, join together to form a crystal lattice.

## 5.5.2  Covalent bonding

Molecules and crystals can consist not only of dissimilar atoms, but also of similar atoms. Examples of molecules made up of similar atoms include the diatomic molecules $H_2$, $N_2$, and $O_2$ of the gases hydrogen, nitrogen, and oxygen. The well-known metals of the elements Cu, Ag, Au, etc., are examples of crystals made up of similar atoms. Because of the similarity of the atoms, ionic bonding does not apply to these chemical bonds. For chemical bonds between similar atoms, the essential distinction is between two types of bond, covalent and metallic bonding.

We shall explain the fundamental mechanism of *covalent bonding* using the example of the hydrogen molecule $H_2$. Like the He atom, the $H_2$ molecule has two electrons. And as in the He atom, the two electrons form a closed shell in the hydrogen molecule. But, while the electron cloud is positioned concentrically about the doubly charged nucleus, the two electrons of the hydrogen molecule move in the bicentric potential of the two protons (Figure 5.27). Again, as in the He atom, the electron wave functions of the bicentric potential of the $H_2$ molecule have a resonance state with the lowest energy occupied by the two electrons with opposite spins.

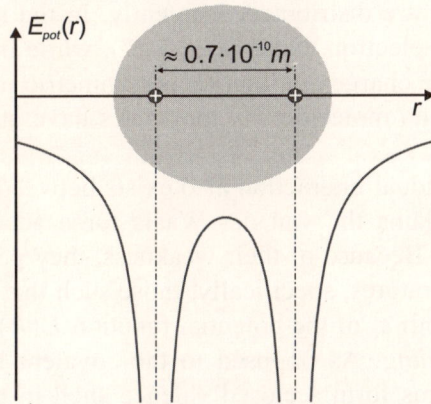

**FIGURE 5.27** The potential and electron cloud of an $H_2$ molecule.

In calculating the binding energy of the molecule, it is important, on one hand, to take account of the binding energy of the electrons in the bicentric potential. It increases as the distance $R$ between the two protons is reduced. At a distance $R = 0$ they would have the same binding energy as the electrons in a He atom. On the other hand, the Coulomb repulsion of the two protons must be kept in mind. It works against bond formation. Thus, the total energy $E(R)$ of the molecule varies with the distance $R$ between the two protons. When the two protons come closer together, it initially decreases, but because of the Coulomb repulsion it again increases at short distances and increases without bound as $R \rightarrow 0$. The function $E(R)$ determines the forces acting between the hydrogen atoms. Similar potential functions also exist for other atoms. An example of such a function is the interatomic potential shown in Section 2.1.1.

For hydrogen atoms, $E(R)$ has a minimum at $R_0 \approx 0.7 \times 10^{-10}$ m with a depth $\varepsilon_0 \approx 4.5$ eV. This value is roughly equal to the binding energy of the $H_2$ molecule. In order to dissociate a hydrogen molecule into two atoms, at least 4.5 eV will be required.

Hydrogen atoms do, indeed, form molecules, but they tend, like helium atoms, not to combine into larger objects. Thus, hydrogen is gaseous at ordinary temperatures and only becomes liquid at temperatures below 27 K (the critical point). This chemical inertia of the molecule is a consequence of the complete filling (closure) of the shell. Indeed, the resonance states of the electron wave functions in the bicentric potential of the two electrons are fundamentally different from the electron states in the potential of the doubly charged He nucleus. But the difference is not severe. The charge distribution of the electrons changes only slightly on going from the He atom to the $H_2$ molecule, since the energy of the $n = 1$ state is far lower than any of the other electron states. Only the positive nuclear charges are distributed differently. In the hydrogen molecule they lie at the edge of the electron cloud (Fig .5.27), while in the He atom they are in the center. Since the charge distribution is symmetric overall with respect to the center of mass of the molecule, $H_2$ molecules have, as expected, no electric dipole moment.

A weak residual interaction also exists between atoms with a closed electron shell. This explains the van der Waals force acting between noble gas atoms (Section 2.2.4). Because of their weakness, they permit a chemical bond only at very low temperatures, specifically, those such that $kT$ is sufficiently small compared to the depth $\varepsilon_0$ of the potential function $E(R)$. In this case we speak of *van der Waals bonding*. As opposed to the covalent bonding of the $H_2$ molecule, where both atoms form a closed valence shell in common, with van der Waals bonding each individual atom has a closed outer electron shell. This van der Waals bonding also makes the condensation of hydrogen possible at low temperatures.

---

### Problem 5.22

Estimate the binding energy of a van der Waals bond between two hydrogen molecules from the critical temperature $T_K = 27$ K of hydrogen. Compare your estimate with the heat of vaporization $Q_{mol} = 454$ kJ/kg of hydrogen at normal pressure.

---

**Notes**  In covalent bonds a full K-shell occupied by two electrons develops between the atomic residues, or cores, of two atoms (for hydrogen atoms the atomic core consists just of the corresponding nucleus, a proton), analogous to the ground state configuration of the He atom. These electrons are highly localized there. Molecules bound in this way have no electric dipole moment. A typical example is the $H_2$ molecule. The binding energy of the $H_2$ molecule is 4.5 eV.

### 5.5.3 Metallic bonding

Metallic bonding, like covalent bonding, makes compounds of similar atoms possible. But, whereas in covalent bonding the two common valence electrons which bind the two atoms together are firmly attached to the pair of atoms, in metallic bonding the valence electrons are much more loosely bound to the individual atoms. Thus, atoms with only one valence electron can also form crystals.

Let us explain metallic bonding using the example of the alkali metals. Like hydrogen, the alkali atoms have one valence electron, but unlike hydrogen they also have filled inner shells. The valence electron is, therefore, in a higher state $n \geq 2$, so there are not just two electron states with approximately equal energies as for $n = 1$, but at least eight. The greater multiplicity of electron states has the consequence that when the distance between the atomic nuclei is varied, the electronic states change more distinctly than in the case of the $H_2$ system. Furthermore, more electrons can exist with roughly equal binding energies within the same region of space.

Because of the larger multiplicity of electronic states, a $Na_2$ dimer made up of two Na atoms tends to bind other sodium atoms to itself, unlike the hydrogen $H_2$ molecule. The charge distribution of the valence electrons in a $Na_2$ dimer extends well beyond the nuclei (Figure 5.28), unlike in the $H_2$ molecule, where it is localized between the nuclei. In this way, other Na atoms can be added to the dimer.

In a sodium crystal consisting of very many Na atoms, the valence electrons are no longer attached to specific atoms. The states of the valence electrons of a metal are preferentially described by wave functions that represent waves extending over the entire crystal (Section 6.3.1). In addition, the charge distribution of the valence electrons therefore extends evenly over the whole crystal.

**FIGURE 5.28** The charge distribution of the valence electrons in a $Na_2$ dimer.

---

### Problem 5.23

The density of the alkali metal sodium (mass number $A = 23$) is $\rho = 0.971 \times 10^3 \text{ kg/m}^3$. Thus, it floats in water. Calculate the distance between neighboring atoms in a sodium crystal and compare the result with distance between the protons in a $H_2$ molecule.

---

| **Notes** | In metallic bonding, as in covalent bonding, the valence electrons are shared by more than one atom. But the electrons that bind the atoms to one another are not localized |

between the atoms. Thus, metallic bonding leads to crystal formation. In a crystal the valence electrons form a gas of conduction electrons that is common to all the atoms.

## 5.5.4 Crystal formation

When a proton and an electron combine, the result is always a hydrogen atom, usually in the ground state. All of the hydrogen atoms created in this way are exactly identical and indistinguishable. In order to change a hydrogen atom, an excitation energy of at least 10 eV must be applied to it.

And when two H atoms combine, a $H_2$ molecule is always the result. In a sufficiently cold environment, this molecule will be in the ground state. In order to set it to rotating, an energy $E_{rot} > (h/2\pi)^2 / (m_H R^2) \approx 8$ meV is required. Thus, rotation can be excited by photons from thermal radiation at ordinary temperatures $T \sim 300$ K.

When a large number of atoms form a crystal, however, very different structures can result, even if the number and type of atoms is exactly specified in advance. A crystal has so many energetically dense quantum states lying close to one another that it is never in a definite quantum state at any time, but constantly switches back and forth between different quantum states because of its interaction with the surroundings. The resulting quantum jumps are so tiny that they appear to be continuous processes when examined macroscopically. Crystallization, therefore, essentially proceeds according to the concepts of classical physics.

While hydrogen atoms and molecules practically form on their own, it takes a great deal of skill to manufacture the high-purity single crystals (monocrystals) required in modern electronic technology. They are often grown under strictly controlled conditions from a melt. In the ideal case, all the atoms in a single crystal are ordered regularly in strictly periodic structures. However, even under optimal experimental conditions, *crystal defects* can never be eliminated. We now demonstrate the origin of crystal defects using a model experiment.

## Experiment 5.8 "Condensation" of steel balls

A large set of small steel balls are dispersed on a framed flat surface. Slightly tilting the surface "condenses" this gas of steel balls onto the lower edge of the frame (Figure 5.29). At first, mostly only small regions with periodically arranged steel balls appear. The steel balls are in a *polycrystalline* ordering. When the holder is slightly agitated (tempered) the balls become more and more ordered, and a *single crystal* with few defects results.

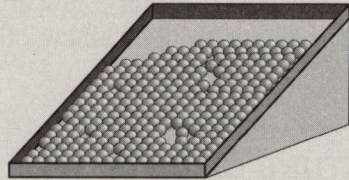

**FIGURE 5.29** A crystallized ensemble of steel balls.

Usually the following defects appear:

## Crystal Defects

■ **Vacancies:** Unoccupied sites in an otherwise strictly periodic ordering.
■ **Interstitial atoms:** Atoms in places which do not match the crystal ordering.
■ **Dislocations:** Crystalline structural defects in the form of a line.
■ **Impurity atoms:** In order to avoid the inclusion of impurity atoms, crystals must be grown in ultraclean laboratories.

Large, flawless, high-purity single crystals are used in modern semiconductor technology. Since the basic structural elements of micro- and nanoelectronics are smaller than 1 μm in size (about 100 of these structural elements would fit into the cross section of a human hair!), the atoms in a crystal wafer on which these elements will be deposited must be in strictly periodic order. In order for electronic components to function faultlessly, the crystalline order must be preserved under thermal load. The manufacture of high purity and thermally stable single crystals is most successful with silicon. It is the starting material for modern electronics.

---

**Problem 5.24**

Calculate how many elementary circuit elements are on a 100 cm$^2$ surface if the individual elements have a size (diameter) of (a) 1 μm or (b) 100 nm $= 10^{-7}$ m. How many atoms are in a circuit element?

---

| Notes | In single crystals the atoms of one or more chemical elements (except for a few crystal defects) are in a strictly periodic ordering in all directions. |
|---|---|

## 5.6 LATTICE STRUCTURE OF CRYSTALS

As ice crystals on a window pane and as polished gems in jewelry crystals fascinate us. Since the invention of the transistor (1947) and of integrated circuits (1958), crystals have been commercially manufactured with increasing perfection for use in modern micro- and nanoelectronics. The regular structure of crystals at an atomic level has enabled the development of highly integrated circuits, gigabit memories, and micromechanical components.

In crystals of different chemical elements the atoms can be arranged in very different ways. They form various types of *crystal lattices*. Just as ice crystals draw our attention by their beautiful symmetric shapes, every crystal lattice has characteristic symmetries, on the basis of which it is classified. Crystals of table salt and the close packing obtained when spheres are piled up are two examples of simple lattice structures. The diamond lattice, which is important in semiconductor physics, is more complicated. We shall use these lattice structures to illustrate the significance of these symmetries in the physics of crystals.

The ordering of the atoms in familiar crystals is easily visualized using models of spheres and rods. It is more difficult to determine the lattice structure of an unknown crystal. The classical procedure for determining the lattice structure is x-ray diffraction. In 1912, Max von Laue proved by diffraction of x-rays both that x-rays are waves and that crystals are regular arrangements of atoms. We shall explain x-ray diffraction in the following section.

## 5.6.1 X-ray diffraction

When a plane light wave is incident on a double slit, an elementary wave (Section 4.2.3) emerges from each of the slits (with width $b \sim \lambda$). Owing to the interference of the two elementary waves, a diffraction pattern develops behind the double slit with intensity maxima and minima. When electromagnetic waves (or other waves that propagate in space) are incident on a spatial lattice structure, a diffraction pattern corresponding to the properties of the lattice develops in the same way. At each lattice point the incident wave is scattered and the scattered waves interfere with one another. Because of the periodic structure of the lattice, intensity maxima develop only in directions such that *all* the scattered waves are in phase. In all other directions, the entire set of scattered waves interfere destructively, so that they cancel each other out.

When a single crystal is irradiated with monochromatic x-rays of wavelength $\lambda$, a pattern of many points shows up as a diffraction pattern on a screen mounted behind the crystal (Figure 5.30). The symmetry of the diffraction pattern is a reflection of the symmetry of the crystal. The angular separations between the points of the intensity maxima are correlated with the distances $d$ of the atoms in the crystal lattice. For $\lambda \ll d$ the angular separations $\Delta\varphi$ of the intensity maxima are obtained in the same way as for the double slit using the equation $\Delta\varphi \approx \lambda/d$. The diffraction pattern in Figure 5.30 was thus obtained with x-rays whose wavelength is small compared to the distances between the atoms in the lattice.

The angular positions of the intensity maxima can be determined geometrically in a simple way following the argument of William L. Bragg. The atoms of a crystal lattice form a series of parallel lattice planes in a number of ways (Figure 5.31). The x-rays scattered on the atoms of a lattice plane are in phase in the direction of reflection and therefore interfere constructively there. In order for the waves scattered on atoms in mutually parallel lattice planes to interfere constructively, the lattice plane separation $d$, the reflection angle $\alpha$, and the wavelength $\lambda$ must satisfy the *Bragg's law* for reflection,

$$2d\cos\alpha = n\lambda. \qquad \text{(Bragg's law)}$$

Thus, an x-ray beam is reflected on a series of parallel lattice planes only if, as with the reflection of light at thin films (Section 4.2.1), the optical path difference of the waves reflected at different planes is an integral multiple of the wavelength $\lambda$.

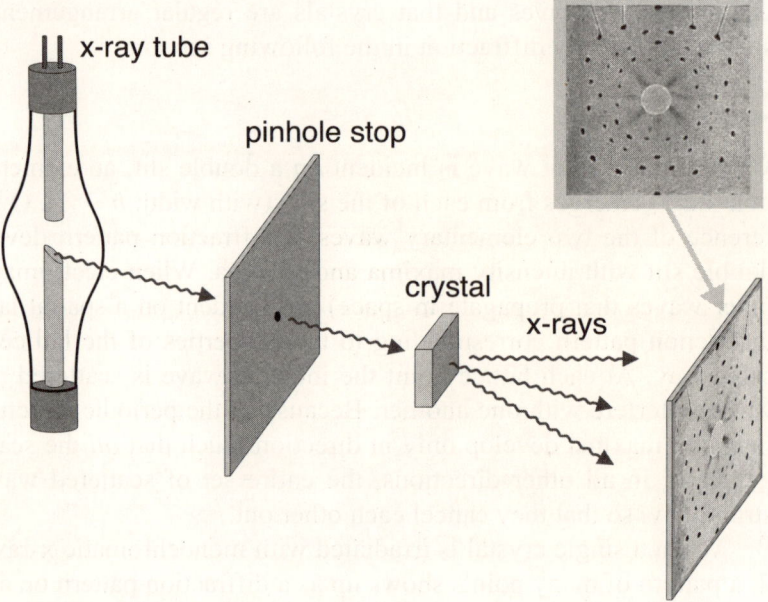

**FIGURE 5.30**  X-ray diffraction by a single crystal and a diffraction pattern (von Laue method).

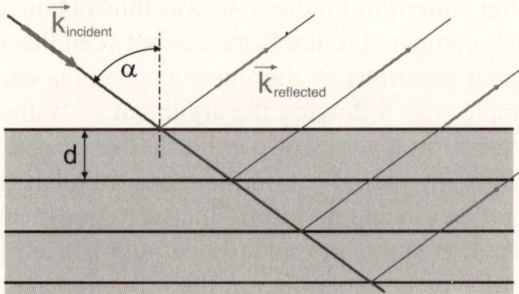

**FIGURE 5.31**  Reflection of x-rays at lattice planes.

Studies are often made of Bragg reflection of x-rays at grazing incidence on a crystal surface. For example, we have made use of this measurement technique (Section 5.3.3) when we measured the spectrum of an x-ray tube with a Mo anode. In measurements at grazing incidence it is convenient to refer to the complementary angle $\alpha' = \pi/2 - \alpha$ of $\alpha$. Then the Bragg law has the form $2d \sin \alpha' = n\lambda$.

Besides x-rays, particle beams can also be used for studying crystal structures. In particular, diffraction experiments have been done with thermal neutrons on crystals. At room temperature thermal neutrons have kinetic energies on the order of $kT = 25$ meV. This implies a wavelength $\lambda = hc/\sqrt{kTm_n c^2}$ on the order of $\lambda \approx 10^{-10}$ m for the neutrons. Thus, the neutron wavelength is in an interesting range for diffraction experiments on crystal lattices.

**Problem 5.25**

A cubic lattice has axes with trigonal and tetragonal symmetry, i.e., it is in congruent positions after rotations of $2\pi/3$ and $\pi/2$, respectively. Take a die and determine its trigonal and tetragonal symmetry axes.

| **Notes** | The arrangement of the atoms in crystal lattices can be studied using x-ray diffraction and neutron diffraction. The symmetry of the diffraction angles corresponds to the symmetry of the crystal structure. |
|---|---|

## 5.6.2 Ionic lattices

Molecules with ionic bonds (Section 5.5.1) form lattices consisting of densely packed ions. The alkali halides form typical ionic lattices. The best known representative is table salt, NaCl. It forms a lattice in which each positively charged $Na^+$ ion has six negatively charged $Cl^-$ ions and each negatively charged $Cl^-$ ion has six positively charged $Na^+$ ions as nearest neighbors (Figure 5.32). With this crystal structure, the space is optimally filled. The small spheres of the $Na^+$ ions lie between the substantially larger spheres of the $Cl^-$ ions. The atomic spheres would be densely packed if the radii $R$ and $r$ of the spheres were in the ratio $1 : \sqrt{2} - 1$.

This example makes it clear that the structure of ionic crystal lattices is essentially determined by the size of the ions. Alkali halides whose ions have a different size ratio form other crystal lattices. In CsCl, the $Cs^+$ ions do not find enough room between six $Cl^-$ ions. Thus, in a CsCl crystal the $Cs^+$ ions have eight $Cl^-$ ions as their nearest neighbors. They lie on the eight corners of a cube surrounding a cesium ion (Figure 5.33).

The symmetry of these crystals follows from the cubic arrangement of the ions. They have different symmetry axes. If a cube is rotated on a body diagonal, then after a rotation by an angle of $2\pi/3$, it has the same cubic body element as in its initial position. Thus, the body diagonal is an axis of *trigonal symmetry* of the crystal. By contrast, cubic axes which are perpendicular to the faces have tetragonal symmetry.

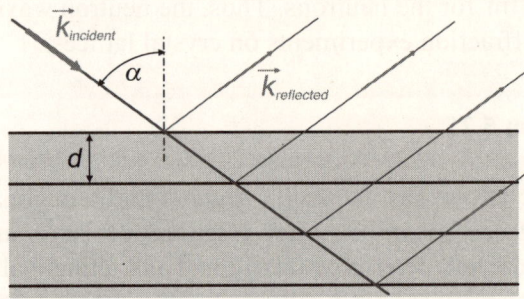

**FIGURE 5.32**  The ionic lattice of a NaCl crystal.

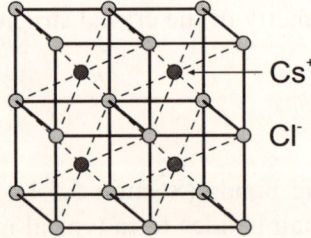

**FIGURE 5.33**  The ionic lattice of a CsCl crystal.

---

## Problem 5.26

Determine the optimum ratio $R : r$ of the ion radii for closest packing of the CsCl lattice.

---

| **Notes** | The lattice structure of typical ionic lattices is essentially determined by the ratio $R : r$ of the ion radii. In a NaCl lattice, the small $Na^+$ ions have six neighboring $Cl^-$ ions. |
|---|---|

### 5.6.3 Closest packing of spheres

*Crystals of a single element* consist only of atomic spheres of the same size. In many of these crystals the atoms are arranged like packed apples of uniform size. The resulting lattice structures are *closest packed*. With closest packing of spheres, each sphere has twelve nearest neighbors. In one plane, a given sphere can be surrounded by six spheres of the same size (Figure 5.34). In both the planes below and above that plane, three other spheres are pressed against this sphere which lies in the middle.

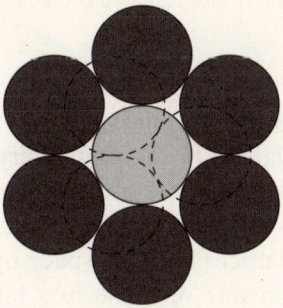

**FIGURE 5.34** Six spheres arranged around a central sphere of the same size.

The *coordination number*, i.e., the number of nearest neighbors (here 12), does not, however, determine the lattice structure uniquely. Spheres can be arrayed strictly periodically in two different ways. That is, there are two possibilities for arranging the spheres relative to one another above and below the original plane. Either they overlay the same three holes between the spheres in the original plane, or they are shifted relative to one another. In the first case, a *hexagonal close packed* (hcp) structure results, and in the second, the spheres form a *face centered cubic* (fcc) lattice.

The face centered cubic lattice grows during the buildup of tetrahedral pyramids of equally sized spheres. First, the spheres of the fourth layer lie vertically above the spheres of the first layer. The face centered cubic lattice is thus characterized by an *A-B-C-A-B-C...* layer sequence. It can, however, also grow out of a cubic structure (Figure 5.35). In the cube, the eight corners and the midpoints of the six faces are occupied by atoms. The tetrahedral pyramid shown in Figure 5.35 demonstrates that this lattice is identical to the lattice with *A-B-C* spherical packing. The name of this lattice configuration is based on the cubic representation.

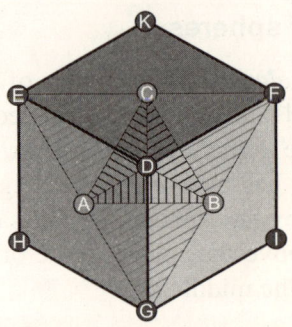

**FIGURE 5.35** A face centered cubic lattice with a tetrahedron indicated.

Hexagonal close packing is associated with a periodic *A-B-A-B*... sequence of layers. It grows through accumulation of a hexagonal pyramid in which the spheres form an equilateral hexagon in every other layer. Thus, the configuration shown in Figure 5.34 would also form out of a layer with three spheres and then one sphere which lies vertically above the center sphere.

The fcc and hcp lattices both fill space optimally with spheres. Many metallic elements crystallize in one of these two lattice configurations. The electrically conducting metals of the first subgroup of the periodic table, Cu, Ag, and Au, form a face centered cubic lattice, while the metals in the second group, such as Be and Mg, form a hexagonal lattice. Crystals with a coordination number $n \neq 12$ are formed from elements in which the atoms are held together by covalent bonds (Section 5.6.4).

## Problem 5.27

Construct pyramids from spheres or apples of the same size and see if you can recognize the cubic structure of the fcc lattice.

**Notes**

There are two closest packing configurations for spheres. An *A-B-C-A-B-C*... layer sequence corresponds to a face centered cubic lattice and an *A-B-A-B*.... layer sequence, to a hexagonal close packed lattice. Both lattices have coordination number $n = 12$.

### 5.6.4  Diamond lattices

The elements in the fourth group of the periodic table, carbon and silicon, are chemically quadrivalent. The binding forces for these elements are directed in four directions like the corners of a tetrahedron. Hence, these elements do not form closely packed crystals, but form lattices with a coordination number $n = 4$. In particular, diamonds and silicon have this type of lattice. Many compound semiconductors have similar crystal lattices (Section 6.5.1). For this reason, we shall examine the structure of the diamond lattice in somewhat more detail.

The diamond lattice originates from a face centered cubic lattice in which an additional four atoms are inserted in the interior of each of these cubes. This doubles the number of atoms per cube. Then, since in the fcc lattice the atoms are positioned either at the corners of the cube or in its faces, only half or 1/8 of these atoms are situated in the interior of the cube whose corners are the centers of the atoms that are there. Thus, overall, the fcc lattice also only has four atoms per cube.

The four additional atoms lie on the four principal diagonals of the cube, that is, at the centers of four equilateral tetrahedra. One of the tetrahedra is accentuated in Figure 5.35. The four corners consist of the atoms $A$, $B$, and $C$ on the three faces with a common corner, plus the corner atom $D$ at that corner. The three other tetrahedra occupied by the additional atoms lie at the corners $H$, $I$, and $K$.

---

**Problem 5.28**

Prove that the atoms $A$, $B$, and $C$ on the sides and the corner atoms $E$, $F$, and $G$ lie in a plane and show that four out of the eight corners of a cube form the corner points of a scalene tetrahedron.

---

The additional atoms, taken alone, again form an fcc lattice. It emerges from the original fcc lattice through a displacement in the direction of one of the principal diagonals $d$ by a distance $d/4$. This translational property of the diamond lattice is of great significance for semiconductor crystals, consisting of equal parts of atoms from the third and fifth groups or, additionally, from the second and sixth groups of the periodic system. In these III-V and II-VI compounds each component forms an fcc lattice by itself. The two components together have the same lattice structure as a diamond lattice.

## Problem 5.29

Construct a diamond lattice from balls of putty and matchsticks to show that a diamond lattice is made up of two fcc lattices shifted relative to one another. Using your model explain how closest packing fills space with more than twice as much material as a diamond lattice. The corresponding *packing fractions* are 74% and 34%.

| Notes | A diamond lattice has a coordination number $n = 4$. Every atom is surrounded tetrahedrally by four nearest neighbors. The atoms of a diamond lattice form two fcc lattices that are |

shifted by a quarter of the principal diagonal of the cube with respect to one another.

# 6 Quantum Gases

## Summary

- Lattice vibrations and phonons
- Lasers
- Energy band model
- Electron mobility in crystals
- Semiconductors
- Elementary events

In many respects, the ideal gas of classical physics serves as a model for the description of physical processes in solids. An ideal gas is an ensemble of very many point masses whose motion in free space follows Newton's law of inertia, but is subject to the laws of chance in collisions. Many processes in solids can be understood in terms of models based on the concept of an ideal gas. For example, in Section 4.3.1 we treated the conduction electrons in metals as an ideal gas.

The model of an ideal gas provides a much deeper understanding of processes in solids if the motion of the individual particles is described quantum mechanically. We then speak of a *quantum gas*. Since all particles also have a wave character in quantum mechanics and all waves have a particle character, the quantum gas model picture can be applied to wide-ranging branches of physics.

Thus, for example, the conduction electrons in a metal form a quantum gas. The photons of electromagnetic wave fields also form a quantum gas. Other wave fields, such as lattice vibrations in a crystal, can also be described as a quantum gas, specifically, in this case, as a quantum gas of *phonons*.

Unlike the atoms of an ideal gas in classical thermodynamics, the particles in quantum gases also have a wavelength, the de Broglie wavelength $\lambda_{dB}$ (Section 5.2.1). The ratio of $\lambda_{dB}$ to the average distance between particles determines whether the gas is to be treated classically or by wave mechanics. For thermal electrons the de Broglie wavelength is on the order of $\lambda_{dB} \sim hc/\sqrt{kTm_e c^2}$. Thus, for $T = 300$ K, $\lambda_{dB} \sim 5$ nm or substantially longer than the diameter of an atom. Since every atom in a metal contributes an electron, the particle density of the gas of conduction electrons is $n > \lambda_{dB}^{-3}$, or

$$\lambda_{dB} > \left( \frac{1}{n} \right)^{1/3}.$$

This relationship makes it clear that the conduction electrons in a metal cannot be described as a gas of classical particles, but must be described as a quantum gas.

Over the last ten years (since 1995) it has been possible to cool atomic gases to such low temperatures that they manifest the typical properties of a quantum gas (Bose-Einstein gas). In this case the de Broglie wavelength of the *atoms* is of major concern. *Bose-Einstein condensation* of a gas occurs when the temperature of the gas is so low that $\lambda_{dB}$ exceeds the average distance $d = n^{-1/3}$ between neighboring atoms.

## 6.1 LATTICE VIBRATIONS AND PHONONS

In an ideal crystal the atoms are absolutely in periodic order to infinity. A real crystal, on the other hand, has a finite temperature $T$, many crystal defects, and a surface. Thus, the periodicity is disturbed in many ways. According to the equipartition theory of classical physics (Section 2.3.3), the atoms move with an average kinetic energy $E_{kin} = \frac{3}{2}kT$. In a crystal they oscillate about their equilibrium positions in the lattice. In terms of classical physics, a crystal can be treated as the three-dimensional variant of the linear chain (Section 3.1.1). Similarly to the point masses in a linear chain coupled by springs, in a crystal the atoms are coupled to their nearest neighbors by the interatomic binding force. If an atom is excited to oscillations about its equilibrium position, waves propagate outward from the excited atom into the crystal, just as in a linear chain. These thermal wave motions of the crystal can be described quantum mechanically in terms of a quantum gas of phonons.

In order to get a look at the behavior of this quantum gas we proceed as with the linear chain. First we consider a "crystal" made up of a single atom that can move back and forth in a parabolic potential well. Like a sphere mounted between two springs, an atom of this sort is a harmonic oscillator (Section 1.6.1). However, this oscillator is now to be described quantum mechanically. Some of the essential properties of the quantum mechanical oscillator follow immediately from Bohr's postulates (Section 5.1.3).

## 6.1.1 The harmonic oscillator

In a three-dimensional parabolic potential an atom experiences a restoring force $\mathbf{F} = -D\mathbf{r}$, which is directed toward the center and has a magnitude which increases with the distance $r$ from the center. The potential energy therefore increases with the square of the distance $r$ and the kinetic energy increases as the square of the velocity $v$ of the atom (of mass $m$):

$$E_{\text{pot}} = \frac{1}{2}Dr^2 \text{ and } E_{\text{kin}} = \frac{1}{2}mv^2.$$

In order to apply the Bohr postulates in an easy way, we shall assume that the atom, like the electron in the Bohr model (Section 5.1.3) of the hydrogen atom, moves in a circular orbit about the center. In this case, the centripetal acceleration $a = \omega^2 r$ follows from the angular velocity $\omega$ and the radius $r$ of the orbit. According to Newton's action principle, we have $Dr = m\omega^2 r$, so that for the linear harmonic oscillator

$$D = m\omega^2.$$

Since $v = \omega r$, the kinetic and potential energies of the orbiting atom are equal. Its total energy is $E_{\text{tot}} = E_{\text{kin}} + E_{\text{pot}} = m\omega^2 r^2$.

That is as far as the arguments of classical mechanics go. In terms of quantum mechanics, particles bound in a potential well have discrete energy levels, which follow from the quantization of angular momentum ($L = nh/2\pi$, Bohr's third postulate) for simple cases. The magnitude $L$ of the angular momentum of the orbiting atom is $L = m\omega r^2$. Thus, $E_{\text{tot}} = \omega L$. Consequently, according to Bohr's theory, the energy levels $E_n$ of the harmonic oscillator are given by

$$E_n = n\hbar\omega.$$

The energy levels of the harmonic oscillator are equally spaced on an energy scale, like the steps on a ladder (Figure 6.1).

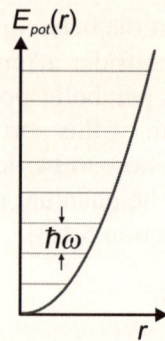

**FIGURE 6.1** A sketch of the energy levels and potential curve for a harmonic oscillator.

According to classical physics, a harmonic oscillator with frequency $\omega$ can radiate waves with that frequency. A charged particle such as an ion would emit electromagnetic waves. A neutral atom in a crystal, on the other hand, emits sound waves. Quantum mechanically the frequency of the emitted waves is given by the separation of the energy levels in accordance with Bohr's second postulate. In a *harmonic* oscillator, transitions only take place between neighboring energy levels. Thus, according to quantum physics, a harmonic oscillator only emits waves at the oscillator frequency $\omega$.

## Problem 6.1

Estimate the oscillator frequency $\nu$ of a Na atom (mass number $A = 23$) in a Na crystal and calculate the corresponding energy of the vibrational quanta, $h\nu$, in eV. For this estimate assume that the potential energy of a Na atom increases by about 1 eV for a deviation of 0.1 nm from its equilibrium position.

| **Notes** | The energy levels of a harmonic oscillator are equidistant, with a separation $\Delta E = h\nu$. Thus, $\nu$ is equal to the eigenfrequency of the oscillator from the classical equation of motion. |
|---|---|

### 6.1.2 Molar heat capacity of crystals

A harmonic oscillator in thermal equilibrium with its surroundings is not at rest. According to the equipartition theorem of classical physics (Section 2.3.3) the oscillator has an average kinetic energy $\langle E_{\text{kin}} \rangle = \frac{1}{2} fkT$. A three-dimensional oscillator has $f = 3$ degrees of freedom and, on the average, exactly as much potential energy as kinetic energy. Thus, according to classical physics, an oscillator in thermal equilibrium with its surroundings has an average energy

$$\langle E_{\text{osc}} \rangle = 3kT.$$

A crystal with $N_A$ atoms per mol would thus have an internal energy $U_{\text{mol}} = 3RT$ if it were in thermal equilibrium. This implies that the molar heat capacity of the crystal is $C_{\text{mol}} = 3R$ (the Dulong-Petit law, Section 2.3.3). In fact, many crystals made up of only one kind of atom do obey the Dulong-Petit law at temperatures around $T = 300\,\text{K}$. Figure 6.2 shows a plot of the molar heat capacities $C_V = dU/dT$ of various crystals consisting of single elements as functions of the temperature $T$. According to the Dulong-Petit law, they should be $C_V = 3R \approx 25\,\text{J} \cdot \text{mol}^{-1} \cdot \text{K}^{-1}$ for all crystals. In fact, the molar heat capacity does reach this value at high temperatures, but falls off to zero at low temperatures.

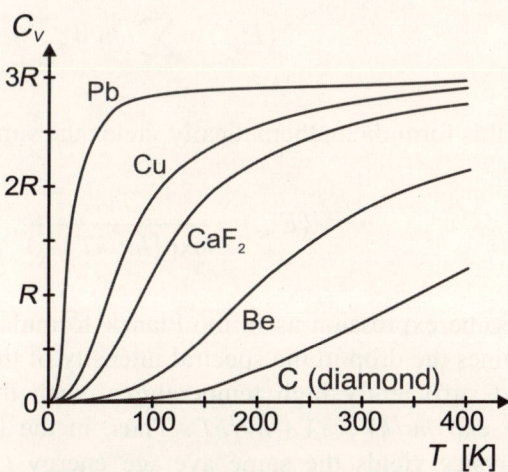

**FIGURE 6.2** The molar heat capacity of crystals.

Quantum mechanically, the *freezing* or *quenching* of the degrees of freedom of atoms in a crystal happens in a way similar to the freezing of the rotational degrees of freedom of the $H_2$ molecule (Section 2.5.4). The average thermal energy of an oscillator is obtained from the probabilities $W_n$ that the energy levels $E_n$ are populated. The probability that a quantum state $|n\rangle$ is occupied thermally is determined by the *Boltzmann factor* (Section 2.1.4):

$$W_n \propto \exp\left(-\frac{E_{th}}{kT}\right). \qquad \text{(thermal population of quantum states)}$$

For a linear oscillator (which can only oscillate in one spatial direction and thus has only one degree of freedom), exactly one quantum state $|n\rangle$ belongs to each energy level $E_n$. For an oscillator of this sort, the probability that an energy level is occupied falls off with $n$ as the terms of a geometric series:

$$W_n = W_0 x^n \quad \text{with} \quad x = \exp\left(-\frac{h\nu}{kT}\right).$$

Here $W_0$ is the probability that the lowest level with $n = 0$ is populated. It is obtained from the condition that the sum of all the probabilities of states being occupied is equal to unity, i.e., $\sum W_n = 1$. The average thermal energy $\langle E_{osc} \rangle$ of a harmonic oscillator is given by the following sum:

$$\langle E_{osc} \rangle = \sum_{n=0}^{\infty} n h \nu W_n.$$

Calculating this formula mathematically yields the simple expression

$$\langle E_{osc} \rangle = \frac{h\nu}{\exp(h\nu/kT) - 1}.$$

This is the same expression as in the Planck formula (Section 4.4.4) where this term determines the drop in the spectral intensity of thermal radiation at high frequencies. At sufficiently high temperatures, such that $kT \gg h\nu$, we have approximately $\exp(h\nu/kT) \approx 1 + h\nu/kT$. Thus, in the limit of high temperatures, quantum physics yields the same average energy $\langle E_{osc} \rangle = kT$ as the classical equipartition theorem for a linear harmonic oscillator. The degree of freedom of the oscillator freezes out when $kT \sim h\nu$, i.e., when the thermal energy $kT$ is comparable to or lower than the excitation energy of the oscillator.

The freezing out of the degrees of freedom of the atomic oscillators in crystals explains the deviations in the molar heat capacity from the Dulong-Petit law. Since the characteristic frequencies $\omega = 2\pi\nu = \sqrt{D/m}$ are determined by the parameter $D$ of the parabolic potential (Section 1.6.1) and the mass $m$ of the atoms, the degrees of freedom of soft crystals with heavy atoms (e.g., lead) are quenched at much lower temperatures than those of hard crystals of light atoms (e.g., diamond). The characteristic quenching temperature $T_D = h\nu/k$ is referred to as the *Debye temperature* of a crystal.

---

**Problem 6.2**

Estimate the Debye temperature of lead and diamond (carbon atoms) and compare your estimates with the experimental curves of Figure 6.2. Rough values for the parameter $D$ of the parabolic potential of the atoms in the crystal lattice can be obtained from the binding energies of the atoms in the crystal and the interatomic distances.

---

| **Notes** | The oscillatory degrees of freedom of the atoms in a crystal are frozen out or quenched if the thermal energy $kT$ is lower than the separation $h\nu$ between the discrete energy levels of the oscillations of an atom in the crystal, i.e., if $kT < h\nu$. |
|---|---|

### 6.1.3 Sound waves in crystals

In a crystal, the motion of any given atom is coupled to the motions of its neighbors through the interatomic force. As a model, a crystal can be represented as a three-dimensional array of point masses which are bound to each other by elastic springs (Figure 6.3), as on a linear chain (Section 3.1.1). As in a linear chain, waves can propagate in the crystal. There are transverse waves, in which the atoms oscillate perpendicular to the direction of propagation of the wave, as well as longitudinal waves, in which the atoms oscillate in the direction of propagation of the wave. Both types of wave contribute to the propagation of sound in crystals. Here we speak in general of *sound waves*.

A linear chain, whose $N$ point masses can oscillate in only *one* direction, has $N$ eigenmodes with frequencies $\nu_n = n(c/2L)$, where $n = 1, 2, ..., N$. Here $L$ is the length of the linear chain (Section 3.1.2). The highest possible frequency corresponds to the eigenmode in which neighboring atoms, separated by a distance $a = L/N$, oscillate in counterphase. The phase velocity $c$ of the waves is approximately constant only for the lowest frequencies, and varies at higher frequencies. Hence, the frequency of the waves does not increase in proportion to the wave number $k$, but satisfies a sinusoidal dispersion relation (Section 3.1.4).

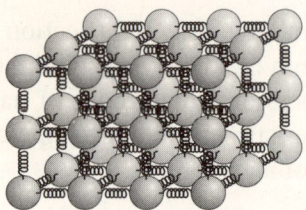

**FIGURE 6.3** A model crystal of elastically coupled spheres.

A crystal with $N$ atoms has $3N$ eigenmodes, since the atoms can oscillate in three directions in space. Here as well, the highest frequencies occur when neighboring atoms oscillate in counterphase. These are frequencies on the order of $\omega = \sqrt{D/m}$, which corresponds to the harmonic oscillations of a single atom (with fixed neighbor atoms) (Section 6.1.1).

The dispersion relations $\omega(\mathbf{k})$ of sound waves in a crystal can be very complicated. They usually depend not only on the magnitude, but also on the propagation direction of the waves. Ionic crystals have a distinctive feature: positive and negative ions alternate in them, so there are two types of waves. The neighboring positive and negative ions oscillate either in phase or out of phase (Figure 6.4). Accordingly, the dispersion relations for these crystals have an *acoustic* and an *optical branch*.

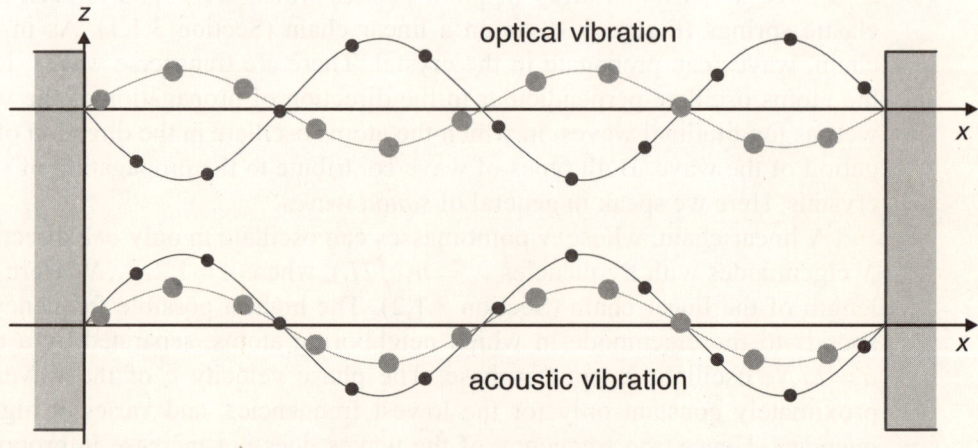

**FIGURE 6.4** Optical and acoustic vibrations of a linear chain with $m_1 \neq m_2$.

For a linear chain with alternating heavy and light point masses, the dispersion relations for the acoustic and optical branches can be calculated (Figure 6.5). Since neighboring point masses always oscillate in counterphase for the waves in the optical branch, the waves with longer wavelengths (i.e., smaller wave numbers) have high frequencies.

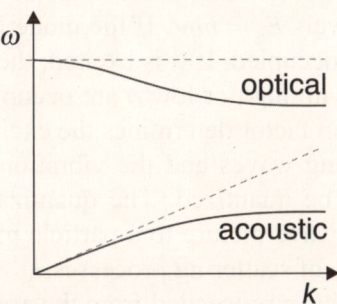

**FIGURE 6.5** The dispersion relation of a linear chain with masses $m_1 \neq m_2$ showing the optical and acoustic branches.

Since the atomic masses in an ionic lattice are charged, so that neighboring masses form electric dipoles, the optical vibrations can emit electromagnetic waves or be excited by them. That is why they are called *optical*.

---

### Problem 6.3

Calculate the dispersion relation $\omega(k)$ of a linear chain consisting of identical atoms.

---

**Notes**

The lattice vibrations of a crystal with $N$ atoms can be represented as a superposition of $3N$ eigenmodes (characteristic vibrations). The dispersion relation $\omega(\mathbf{k})$ between the frequencies $\omega$ and the wave vectors $\mathbf{k}$ of the eigenmodes depends in a complicated way on the crystal structure. Simple dispersion relations can be obtained for linear chains. The dispersion relation of an infinite linear chain of identical atoms (Section 3.1.4) has the form $\omega(k) = 2\omega_0 \sin(ka/2)$. Here $a$ is the distance between neighboring point masses and $\omega_0$ is the natural frequency (eigenfrequency) of a point mass located between two fixed neighbors.

### 6.1.4 Phonons

Every eigenmode of a crystal behaves as a harmonic oscillator. By choosing the corresponding resonance frequency it can be excited to oscillation, just like the standing waves of a linear chain (Section 3.1.2). According to the laws of quantum mechanics, these oscillations are to quantized exactly like the oscillations of single atoms. Thus, each vibrational mode of the crystal with frequency $\nu$ has a series of energy levels $E_n = nh\nu$. If the mode is not excited, then only its ground state with $n = 0$ is occupied. If it is excited, then depending on the strength of the excitation, levels with high or low $n$ are occupied. In the case of thermal equilibrium, the Boltzmann factor determines the excitation probability (Section 6.1.2).

As with standing waves and the vibrational modes of resonators, travelling waves are also to be quantized. The quantization of travelling waves suggests converting from a wave picture to a particle picture. We shall illustrate this conversion for the case of scattering processes.

Up to now we have proceeded from the assumption that sound waves in crystals obey linear differential equations, so that the superposition principle (Section 3.1.3) is satisfied. But this assumption is only satisfied approximately. Thus, one sound wave can be scattered on another. In the course of such scattering, the probabilities of occupying the vibrational modes involved in the scattering will change. In the simplest case, before the scattering event a vibrational mode $a$ is in the first excited quantum state $|1\rangle_a$ and after the scattering event it is in the ground state. Instead, a vibrational mode $b$ is then in the first excited state $|1\rangle_b$.

In the particle picture, this process is described as the scattering of a phonon from vibrational mode $a$ into vibrational mode $b$. Before the scattering process, mode $a$ is occupied by one phonon; however, after the scattering process, mode $b$ is occupied.

Travelling waves are usually described as plane waves with a wave vector $\mathbf{k}$. With respect to travelling sound waves, we say that a phonon with wave vector $\mathbf{k}$ is converted into a phonon with wave vector $\mathbf{k}'$ on scattering.

A vibrational mode can, however, also be in a quantum state with $n > 1$. In that case the vibrational mode is populated by multiple phonons. Since $n$ can be arbitrarily large, an arbitrary number of phonons can exist in a vibrational mode.

The change from a wave picture to a particle picture shows how quantum physical processes can be associated with fundamentally different representations. Rather than considering lattice vibrations as a wave field, we can also treat them as a gas of phonons. These phonons can, like other particles, also have velocities, masses, and momenta ascribed to them. In particular, the momentum $\mathbf{p}$ of a phonon follows from the de Broglie formula as

$$\mathbf{p} = \hbar\mathbf{k}.$$

The energy $E$ of the phonons is $E = h\nu$. In phonon scattering processes, energy and momentum are conserved, just as in elastic collisions.

---

### Problem 6.4

For a linear chain, calculate how the energy of the phonons depends on their momentum. Calculate the (effective) mass of the phonons assuming that the group velocity is equal to the particle velocity.

---

| **Notes** | The propagation of sound in crystals can be described in terms of a wave picture or a particle picture. The energy quanta of the sound waves are known as *phonons*. |
| --- | --- |

## 6.2 LASERS

Starting with the characteristic oscillations (also known as eigenmodes) of a cavity resonator, the electromagnetic field in a cavity can be quantized in the same way as lattice vibrations in a crystal. And just as the quantized lattice vibrations were interpreted as an ensemble of phonons, the quantized eigenmodes of a cavity resonator can be regarded as an ensemble of *photons*. Photons and phonons both form a quantum gas. For the ideal gas of classical physics the average thermal energy of the atoms (and, thereby, the internal energy and the heat capacity, as well) is a consequence of the equipartition theorem:

$$\langle E_{\text{kin}} \rangle = \frac{3}{2} kT.$$

For quantum gases of photons or phonons which are in thermal equilibrium with their surroundings, the internal energy is derived from the average energy $\langle E_{\text{mod}}(\nu) \rangle$ per eigenmode:

$$\langle E_{\text{mod}}(\nu) \rangle = \frac{h\nu}{\exp(h\nu/kT) - 1}.$$

Here $\nu$ is the characteristic frequency (eigenfrequency) of the eigenmode.

A cavity with a small hole emits the black-body spectrum (Section 4.2.2). It is described by Planck's radiation formula (Section 4.4.3). Since the thermal radiation in a cavity can also be interpreted as a quantum gas, Planck's radiation formula can also be derived from the average thermal energy per eigenmode. To do this, all that is needed is to determine the density of the eigenmodes in the cavity.

Ordinary light sources, such as the sun, candles, and light bulbs, are thermal radiators, similar to black bodies. Kirchhoff's radiation law (Section 4.4.2) states that at a given temperature a black body has the highest emittance of any thermal radiator. Few gaseous discharges are even approximately in thermal equilibrium at temperatures of a few thousand K. Hence, even the high spectral emittance of a gaseous discharge near its discrete atomic spectral lines (Section 4.2.2) is lower than the spectral emittance of a black body at the same temperature.

In all thermal light sources a gas or object is initially heated by supplying heat to the light source by a combustion process or electrically. The emission of light is then governed by the laws of thermal radiation. Electrical energy can be directly converted into radiant energy with a *laser* (light amplification by stimulated emission of radiation). Compared to thermal light sources, lasers have a number of fantastic properties.

## 6.2.1 Stimulated emission

According to Bohr's second postulate, quantum jumps take place between the discrete energy levels of atoms and molecules. In a transition from an energetically higher level to a lower energy level a photon will be emitted, and in a transition from a lower to a higher energy level a photon will be absorbed. In both cases, the energy $hv = E_> - E_<$ of the photon is equal to the energy separation between the two levels.

The elementary processes of absorption and emission are illustrated in Figure 6.6. In an absorption process, for example, a photon with energy $hv = E_> - E_<$ strikes an atom in the ground state. After the absorption process the atom is in the excited level $E_>$ and the photon is lost. Absorption processes such as this occur when sodium atoms in the ground state are illuminated with the yellow light of a sodium spectrum lamp in the experiment on resonance absorption of the yellow Na spectral line (Section 5.3.4). The *absorption rate* (that is, the probability per unit time that a sodium atom will absorb a photon) is proportional to the spectral intensity $I(v)$ of the incident light.

In emission, two situations must be distinguished. Either the excited atom will be irradiated by more of the resonance line light or there are no other photons of the resonance line radiation present at the moment. In the latter case the

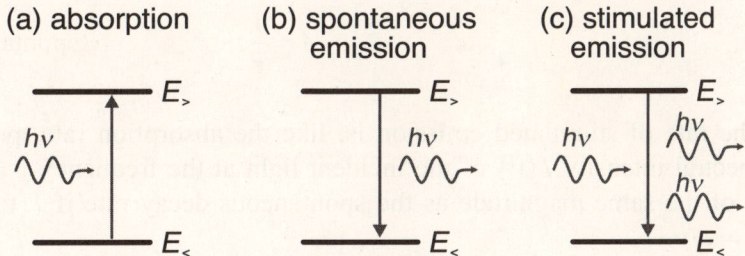

**FIGURE 6.6** The elementary processes of absorption and emission.

atom decays *spontaneously*. As with the radioactive decay of atomic nuclei, spontaneous radiative decay follows the laws of chance. In an ensemble of atoms the number $N(t)$ of excited atoms falls off exponentially with time,

$$N(t) = N_0 \exp(-t/\tau).$$

The time $\tau$ over which the number of excited atoms decreases by a factor of $e \approx 2.7$ is known as the natural lifetime of the excited level. For energy levels which emit photons in the visible range, the lifetime is typically about $\tau \sim 10^{-8}$ s. The direction of emission and the polarization of spontaneously emitted photons are also determined by the laws of chance. Thus, spontaneously emitted photons can be emitted in arbitrary directions and with arbitrary polarizations.

The other situation is *stimulated emission* (induced emission), first considered by Einstein in 1917. When a photon with energy $h\nu = E_> - E_<$ strikes an atom in the energy level $E_>$, it can initiate an emission process. In this case, as with absorption, a photon is present at the start, but here the atom is in an excited energy level. After a stimulated emission process, two photons are present and the atom is in the ground state. The two photons end up in the *same* eigenmode. Thus, they have the same direction of propagation, the same frequency, and the same polarization.

Because of stimulated emission, a light beam incident on an ensemble of excited atoms can be amplified. If an ensemble of Na atoms were to be prepared in such a way that all of them were in the excited state which decays to give the yellow Na line, then the yellow light of the sodium lamp would not be absorbed as it passed through this ensemble of atoms, but amplified.

Stimulated emission competes with spontaneous emission. The probability that an excited atom will emit a photon spontaneously, i.e., the rate $A$ of the spontaneous emission process, follows from the natural lifetime $\tau$.

$$A = \frac{1}{\tau}. \qquad \text{(spontaneous decay rate)}$$

The rate of stimulated emission is, like the absorption rate, proportional to the spectral intensity $I(\nu)$ of the incident light at the frequency $\nu = (E_> - E_<)/h$. It is of the same magnitude as the spontaneous decay rate if $I(\nu) \approx hc/\lambda^3$, where $\lambda = c/\nu$.

---

### Problem 6.5

Estimate the temperature to which a black body must be heated in order for the spectral emittance $E_{BB}(\nu)$ to reach $E_{BB}(\nu) = hc/\lambda^3$. How many photons would occupy the eigenmode with frequency $\nu$ in a cavity at this temperature?

---

**Notes**    If photons of energy $h\nu$ are incident on an excited atom with energy $E_>$, they can induce (stimulate) transitions into a lower energy level $E_<$ if Bohr's frequency condition $h\nu = E_> - E_<$ is satisfied. This raises the number of photons in the eigenmode of the incident (inducing) photon by 1. The induced photon has the same frequency, propagation direction, and polarization as the incident photon.

## 6.2.2 Population inversion

The probabilities $W(E_n)$ of populating the energy levels $E_n$ (i.e., the density of states) of an ensemble of atoms in equilibrium at temperature $T$ for a harmonic oscillator (Section 6.1.2) is determined by the Boltzmann factor,

$$W(E_n) \propto \exp\left(-\frac{E_n}{kT}\right).$$

The probability that a state is occupied falls off exponentially with increasing energy of the atomic levels (Figure 6.7). Thus, for atoms in thermal equilibrium with their surroundings, a higher energy level is always less populated than a lower energy level, i.e.,

$$W(E_>) < W(E_<) \text{ for thermal populations.}$$

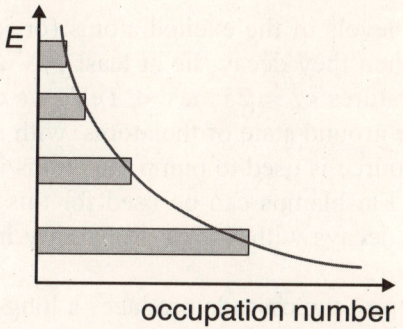

**FIGURE 6.7** The thermal density of states for atomic energy levels.

Thus, light passing through a thermal ensemble of atoms always induces more absorption than emission processes. The light beam will thus be attenuated as it passes through the ensemble.

In order for an ensemble of atoms to act as a light amplifier, evidently an inverted population distribution must be available. That is, an energy level $E_>$ must be occupied with a higher probability than a lower-lying energy level $E_<$, or

$$W(E_>) > W(E_<). \qquad \text{(population inversion)}$$

A population inversion can be obtained in various ways. One common procedure is *optical pumping*. Let us explain this using a four-level system (Figure 6.8) as an example. This method is used, for example, to achieve a population inversion among dopant Nd ions in glass or crystals in neodymium lasers, which emit infrared light with a wavelength $\lambda \approx 1\,\mu\text{m}$.

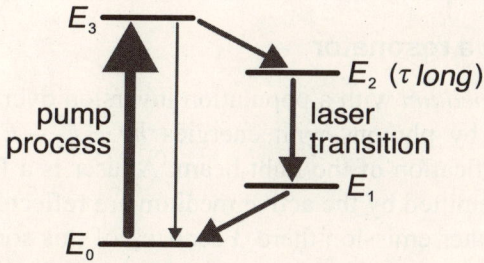

**FIGURE 6.8** Pumping scheme for a four-level system.

The energy levels of the excited atoms (or ions), which emit visible or near infrared light when they decay, lie at least 1 eV above the ground state. Since at ordinary temperatures $kT = 25$ meV $\ll 1$ eV, we can assume that in thermal equilibrium only the ground state of the atoms, with an energy $E_0$, will be populated. A strong light source is used to pump the atoms into a highly excited, short-lived pump level $E_3$. Flashlamps can be used for this purpose for pulsed laser operation. This level decays with a high probability into a three-step cascade back to the ground state.

The first step of the cascade produces a long-lived level $E_2$ with a lifetime $\tau_2$ in the μs range and then a short-lived level $E_1$ with a lifetime $\tau_1$ in the ns range, after which the atom again reaches the ground state. Because of the different lifetimes, there is a *population inversion* between the two intermediate levels. The probabilities $W(E_n)$ that these levels are occupied will be proportional to the lifetimes $\tau_n$, so that

$$\frac{W(E_2)}{W(E_1)} = \frac{\tau_2}{\tau_1}.$$

Here it is assumed that all decay events, in which not only photons, but also phonons (especially in crystals), are produced, are *spontaneous*. If an *active medium* of this sort is irradiated with light that satisfies the Bohr frequency condition $h\nu = E_2 - E_1$, then more photons are produced by stimulated emission than are lost by absorption. In that way, the ensemble acts as a *light amplifier*.

**Notes**   In a population inversion, states with a higher energy $E_>$ in an ensemble of atoms are populated with a higher probability than states with a lower energy $E_<$. Such an ensemble of atoms can amplify light which satisfies the Bohr frequency condition $h\nu = E_> - E_<$.

### 6.2.3 Feedback in a resonator

If an *active medium* with a population inversion over the energy levels $E_1$ and $E_2$ is irradiated by photons with energies $h\nu = E_2 - E_1$, then stimulated emission causes amplification of the light beam. A laser is a light amplifier set up so that the photons emitted by the active medium are reflected back into the medium and stimulate further emission there. Feedback of this sort is achieved by placing the active medium in a resonator or cavity (Figure 6.9).

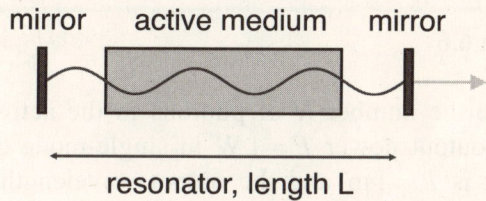

FIGURE 6.9 A sketch of a laser showing the active medium and resonator.

In the simplest case the resonator consists of two plane mirrors with their faces parallel and set a distance $L$ apart, which reflect normally incident light back and forth. During each pass through the active medium, the photons stimulate further emission. But in this way they change the relative population of the laser levels $E_1$ and $E_2$. The medium is active only so long as the population inversion is maintained during operation of the laser, as by optical pumping.

In constructing laser systems, it is important to keep in mind that light also has wave properties. On one hand, this involves ensuring that the reflected waves moving back and forth between the mirrors interfere constructively, so that a standing wave with the maximum possible amplitude can develop. Then the mirrors form a linear resonator (Section 3.1.2). The eigenmodes of the resonator have wavelengths $\lambda_n = 2L/n$. Those eigenmodes with frequencies $\nu_n = c/\lambda_n$ that satisfy the Bohr frequency condition with respect to the laser levels of the active medium will be amplified.

On the other hand, with every reflection the mirrors act as a stop, or diaphragm, at which the light will be diffracted. After diffraction (Section 4.2.3) on a circular mirror aperture of radius $r$, the light beam will have a divergence on the order of $\delta\varphi \sim \lambda_n/r$. The beam of light thus spreads out slightly after each reflection and will only be partially reflected at the next reflection. In order to compensate for these and other losses, the light must be amplified sufficiently strongly as it passes through the active medium. Only when the amplification exceeds a *threshold value* does the cavity generate a laser beam.

In order to let a laser beam out of the cavity, one of the two resonator mirrors should be partially transmitting. For a transmission coefficient of 1% the power emitted from the laser is $P = 0.01 \cdot cE/2L$, or 1% of the energy $E$ stored in the laser resonator divided by the transit time $\Delta t = 2L/c$ required for the light to move back and forth once in the resonator. In single-mode operation of the laser, the energy $E$ is stored in a single eigenmode of the laser resonator.

**Problem 6.6**

Calculate the number $N$ of photons in the active eigenmode of a laser with an output power $P = 1\,\text{W}$ in single-mode operation. The length of the laser is $L = 1\,\text{m}$ and the output wavelength of the emitted light is $\lambda = 1\,\mu\text{m}$. By how many orders of magnitude is the occupation number of the active eigenmode in a laser resonator greater than the occupation number of such an eigenmode for a thermal light source like the sun $(T = 6000\,\text{K})$?

In continuous operation the occupation numbers of the active eigenmodes of a laser are many orders of magnitude higher than the occupation numbers for the eigenmodes at the same frequency in thermal light sources. For this reason, laser radiation manifests many unusual properties:

---

**Properties of Laser Beams**

---

- High spectral intensity $I(\lambda)$ and high total intensity $I$.
- A directed beam with extremely low divergence. (It is ultimately limited by diffraction on the output diaphragm.)
- Large coherence width. (The beam of a single-mode laser is coherent over its entire width.)
- Long coherence length. (The spectral width $\Delta\nu$ of a laser line is often only a few MHz or even kHz. Thus, coherence lengths in the kilometer range are possible.)
- Short light pulses. (Modern lasers can produce light pulses lasting only a few oscillation periods $1/\nu$, i.e., a few times $10^{-15}$ s.)
- Tunability. (Some active media have broadened energy levels. The frequency of lasers operating with such media can be tuned continuously by varying the length of the optical resonator.)

---

**Problem 6.7**

Calculate the density of states of the eignenmodes within the spectral range of the yellow Na line for sunlight.

In laser beams a small number of eigenmodes of the electromagnetic waves are populated by very many photons. The populations are many orders of magnitude greater than for light from thermal light sources.

## 6.2.4 He-Ne lasers

A population inversion also occurs in many gaseous discharges. If a discharge of this sort is created in an optical resonator, then the light emission can be amplified by feedback. In this way the gaseous discharge forms a laser. The He-Ne laser is an example where a discharge is produced in a gaseous mixture of the noble gases helium and neon at a pressure of about $10^2$ Pa. It was one of the first lasers invented, in 1960, and can serve here as an illustration of how lasers function.

Figure 6.10 shows a simplified energy level diagram for the two noble gases. Since closed electron shells must be broken up during excitation, the lowest excited energy levels lie almost 20 eV or 17 eV above the ground states of helium and neon, respectively. Like the ionization energy (Section 5.3.3), the excitation energy of helium is greater than that of neon. The lowest excited states of the noble gases are metastable; that is, in gaseous discharges they do not decay by emission of a photon, but only by collisions with other particles. Thus, in a gaseous discharge, these states have a relatively high probability of being populated.

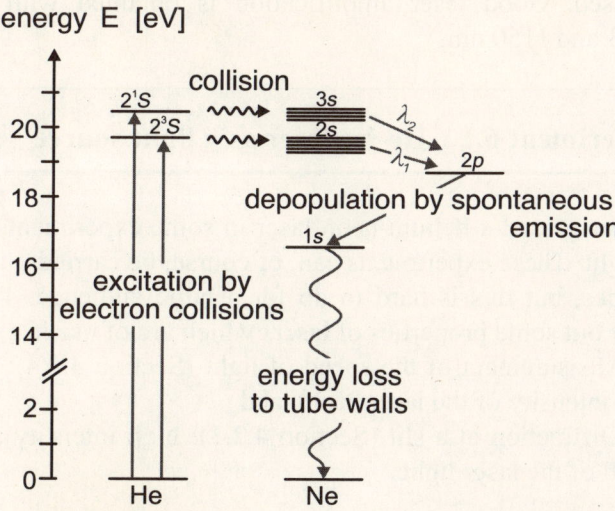

**FIGURE 6.10** Energy level diagrams of helium and neon.

In a gaseous mixture of helium and neon, the metastable helium atoms lose their excitation energy primarily through collisions with neon atoms and excite them to energy levels that have roughly the same excitation energies as the metastable He atoms. Because of this collisional excitation these energy levels of neon are more highly populated than some levels at lower energies, and the former can decay through the emission of visible and near infrared photons. This population inversion is used in a He-Ne laser (Figure 6.11).

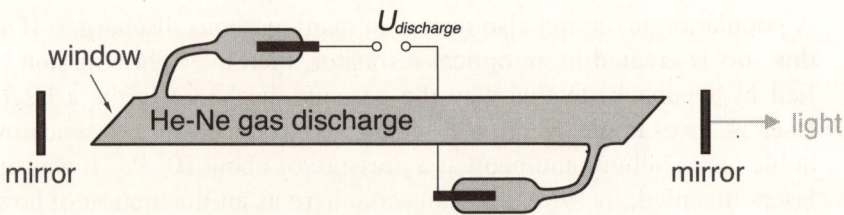

**FIGURE 6.11** Structure of a He-Ne laser.

The discharge tube is placed between the resonator mirrors and is closed off at both ends by oblique glass windows. Placing these windows at an angle keeps them from reflecting the light in one polarization direction as it bounces back and forth between the mirrors. The discharge is formed between electrodes positioned outside the beam path.

He-Ne lasers can emit continuously. Since it can lase on multiple transitions, single-mode operation requires that amplification of undesired spectral lines be suppressed. Good laser amplification is obtained with the spectral lines at $\lambda = 633$ and $1150$ nm.

---

## Experiment 6.1  He-Ne laser as a light source

We have used a helium-neon laser in some experiments on the propagation of light. These experiments can, of course, be carried out with ordinary light sources, but this is hard to do for a large audience. Thus, once again we point out some properties of lasers which are of use in experiments:

Measurement of the speed of light (Section 4.1.4): low divergence and high intensity of the laser beam, and

Diffraction at a slit (Section 4.2.3): high intensity and large coherence width of the laser light.

| **Notes** | In gaseous discharges in a mixture of helium and neon, collisions between metastable He atoms and Ne atoms lead to a population inversion among the Ne atoms. Thus, light in certain spectral lines of the neon atom can be amplified. |
|---|---|

## 6.3 ENERGY BAND MODEL

Like phonons and photons, the conduction electrons in a metal form a quantum gas. Of course, there is an essential difference between these quantum gases. The eigenmodes of a crystal or an optical resonator can accept an arbitrary number of phonons or photons, respectively. Electrons, however, obey the Pauli exclusion principle (Section 5.3.1). Including the electron's spin, every eigenmode of electron waves can be occupied by at most one electron. Accordingly, one distinguishes between *Bosons* and *Fermions*. All particles which exist without limit in an eigenmode are referred to as Bosons (after **Satyendra Nath Bose, 1894–1974**). Particles which obey the Pauli exclusion principle are referred to as Fermions (after **Enrico Fermi, 1901–1954**). Phonons and photons are, therefore, Bosons, while electrons are Fermions.

The different behavior of Bosons and Fermions must be taken into account in calculating probability distributions (Section 2.1.4). At sufficiently low temperatures, the Bosons accumulate in the lowest energy eigenmode. A phase transition, Bose-Einstein condensation, takes place. More than one Fermion cannot, however, occupy a given eigenmode. The $Z$ electrons of an atom fill the $Z$ energetically lowest electronic states of the atom (Section 5.3.1). Thus, at low temperatures $T \to 0$, the $N$ conduction electrons of a metal occupy the lowest-energy eigenmodes for conduction electrons in the crystal.

Besides metals, there are insulators and, especially important for modern electronics, semiconductors. The electrical properties of these crystalline solids are closely related to the Pauli exclusion principle and to the energy levels of the valence electrons of the atoms in the crystal. In order to explain the properties of these solids, we must first determine the energy level diagrams for the electronic states in a crystal and then discuss how electrons populate them.

### 6.3.1 Resonance splitting

While the electronic states in the inner shells of the atomic cloud only change a little when an atom is bound to other atoms or to a molecule, the electronic states of the valence electrons change substantially (Section 5.5). The change in the electronic states is accompanied by a change in the binding energy of these electrons. Here we are primarily interested in the change in the binding energies when atoms or molecules of the same kind are assembled in a crystal.

Since the electronic states in the Coulomb potential of an atomic nucleus can be treated as standing electron waves and, therefore, as the eigenmodes of a resonator, bringing two similar atoms together involves coupling the eigenmodes of these atoms. Thus, here the electronic states behave like two coupled pendulums of equal length (Section 1.6.4).

In their uncoupled state, the pendulums have one and the same eigenfrequency $\omega_0$. When they are coupled, however, the system of coupled pendulums has two different eigenfrequencies. At the lower frequency the two pendulums oscillate in phase and at the higher frequency, in counterphase. When $N$ oscillators are weakly coupled to one another, the combined system (like the linear chain of Section 3.1.1) thus has $N$ different eigenfrequencies, but these lie close to one another. We now demonstrate this behavior with a simulation experiment.

### Experiment 6.2 Coupled oscillator circuits

Five identical electromagnetic oscillator circuits (Section 3.4.4) consisting of coils and capacitors are strung on a plastic rod (Figure 6.12). Oscillations are excited in them by a transmitter with a variable frequency. If only one oscillator circuit is excited to oscillations by the transmitter, the amplitude of the oscillations is varied by tuning the transmitter frequency across the resonance region of the oscillator circuit in the standard way (Section 1.6.3). But, if other oscillator circuits are brought near the transmitter and thereby coupled inductively to the first oscillator circuit, then the resonance curve splits up (Figure 6.13). With each additional oscillator circuit another resonance peak appears.

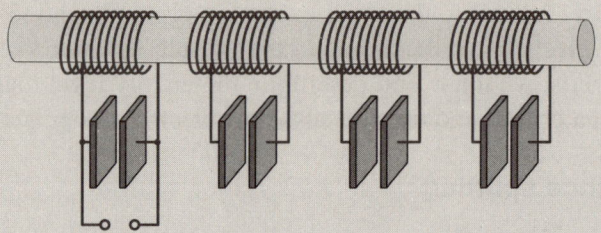

FIGURE 6.12 Coupled electromagnetic oscillator circuits.

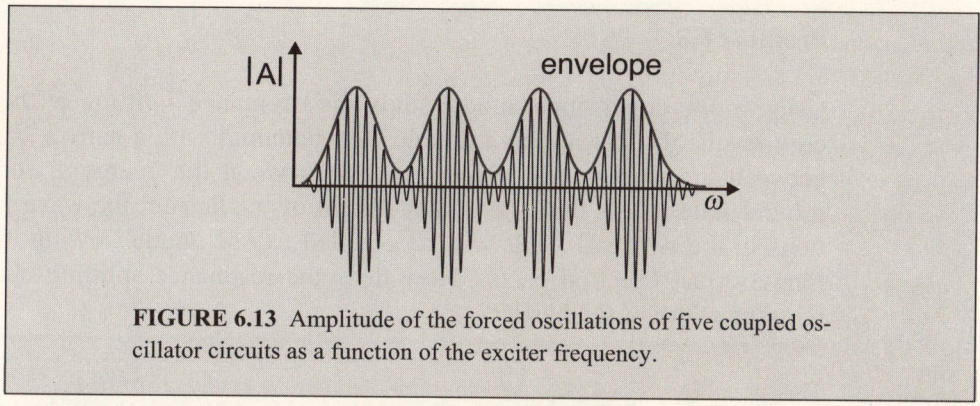

**FIGURE 6.13** Amplitude of the forced oscillations of five coupled oscillator circuits as a function of the exciter frequency.

In similar fashion the energy levels of the electronic states of atoms split when they are brought together into a molecule or a crystal. Thus, the ground state of the electron in a hydrogen atom splits into a bonding (attractive) state and an antibonding (repulsive) state when a (ground state) hydrogen atom is approached by another hydrogen atom (Figure 6.14) or proton (in the $H_2^+$ molecule). Since there are two different spin states, the bonding energy state can accept two electrons with oppositely directed spins. This explains the stability of the covalent bonding of the $H_2$ molecule (Section 5.5.2).

Like the hydrogen molecule, all crystals consist of similar particles (atoms or molecules). Thus, resonance splitting also determines the energies of their electronic states.

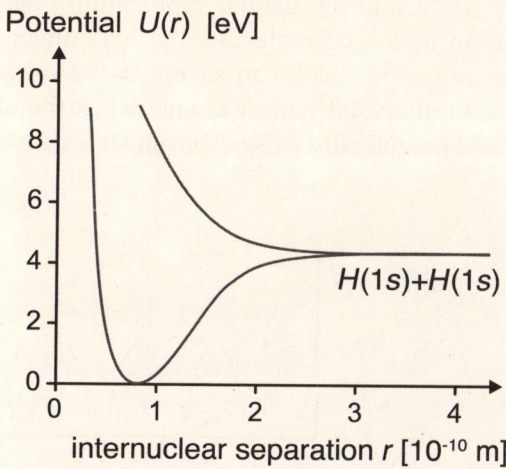

**FIGURE 6.14** Resonance splitting of the ground state of a $H_2$ molecule.

### Problem 6.8

Using a one-dimensional model study the resonance splitting of the energy levels of a particle in a double well potential with a narrow barrier between the two halves of the well. Even when the energy $E_0$ of the ground state is less than the energy height of the barrier, the wave functions in the two half wells will be coupled to one another owing to the tunnel effect (Section 5.2.2). How does the resonance splitting change with the width of the barrier?

| **Notes** | When two atoms of the same kind are brought together, their energy levels split. A bonding and an antibonding molecular state then develop from the ground states of the two |

atoms. This splitting corresponds to resonance splitting of the eigenmodes of coupled oscillators.

## 6.3.2 Energy bands and the Fermi energy

A crystal is made up of a very large number $N$ of atoms or molecules of the same kind. There are $N_A = 6 \times 10^{23}$ atoms per mol. Thus, the energy level of a valence electron in a crystal is split into $N$ close-lying energy levels. Together they form an *energy band*. As an example, let us consider a sodium crystal with $N$ atoms. The 3s ground state of the valence electron in the Na atom has a principal quantum number $n = 3$ and an angular momentum quantum number $l = 0$ (Section 5.3.4). Like the hydrogen molecule, the $Na_2$ dimer has a bonding and an antibonding electron state. Each can accept two electrons with oppositely directed spins. In a sodium crystal with $N$ atoms, when the electron spins are included, a band with $2N$ energetically close electron states develops (Figure 6.15).

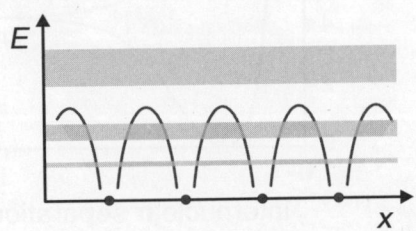

**FIGURE 6.15** The energy bands of the electronic states in a crystal.

The electronic states of the inner shells also broaden into energy bands in a crystal. Of course, the coupling of these electron states is much weaker and the band widths are correspondingly smaller. Other energy bands develop during crystallization from the excited, usually unoccupied electronic states of the atoms. Because of the strong coupling of these states, they are strongly broadened and overlap one another.

The band structure of the electronic states is the same in all crystals. Fundamental differences appear, however, in how the states are populated by electrons. All the energy bands originating from the inner shells are full. They are of secondary significance for the properties of the crystal. The energy band of the valence electrons is important. In a crystal of Na it has $2N$ electron states, but it only has to accept the $N$ valence electrons in the crystal. Thus, it is only half filled.

Partial filling of the valence electron energy band is typical of all metallic conductors (Figure 6.16). This is the origin of the electrical conductivity of metals and is, therefore, referred to as the *conduction band*. In the limit of low temperatures $T \rightarrow 0$, the conduction band is filled up to an upper limit $E_F$, known as the *Fermi energy*.

With insulators, as opposed to metals, the number of valence electrons is equal to the number of electronic states in the valence-electron energy band. The *valence band* of insulators is, therefore, usually full (Figure 6.17). The conduction band lying above it is empty. In order to remove an electron from the valence band of an insulator, so that a site is freed there, or as they say, a *hole (hole state)* is created, an energy of at least $E_G$, the energy gap between the valence band and the next higher energy band, the *conduction band*, must be applied to that electron.

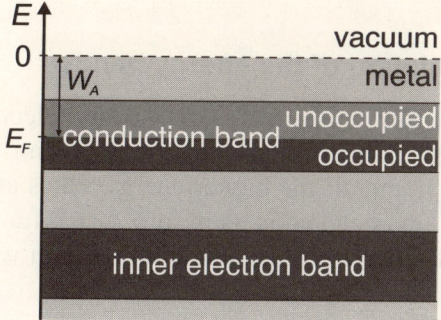

**FIGURE 6.16** Population of energy bands in metals.

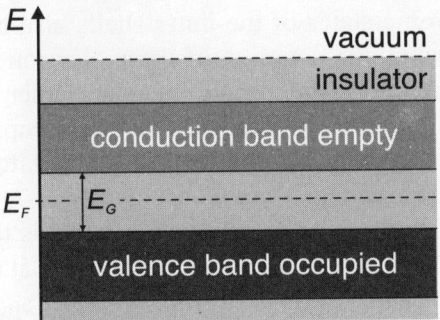

**FIGURE 6.17** Population of energy bands in an insulator.

If the electron gap $E_G \gg 25\,\mathrm{meV}$, then at normal room temperature the valence band will be full and the conduction band, empty. Thus, essentially no electrons can move in the crystal. The crystal is, therefore, an insulator. $E_G > 3\,\mathrm{eV}$ for good insulators.

---

### Problem 6.9

Consider 1 mol of an insulator. Estimate how many electrons are in its conduction band at room temperature $(T = 300\,\mathrm{K})$ if $E_G = 1, 2,$ or $3\,\mathrm{eV}$.

---

| | |
|---|---|
| **Notes** | In metallic conductors the conduction band is partially filled with electrons. But, in insulators the valence band is filled with electrons, while the conduction band is empty. |

There is an energy gap $E_G \gg kT$ between the conduction band and the valence band in insulators.

## 6.3.3 Thermal population of the energy bands

In the limit of low temperatures $T \rightarrow 0$, the electronic states of the conduction band of a metal are filled up to the Fermi energy $E_F$ with probability $W(E < E_F) = 1$, and all the higher-energy states are unoccupied. The probability that the states are occupied is, therefore, a step function (Figure 6.18). For $T \neq 0$, on the other hand, there is a transition region in which this probability decreases from 1 to 0.

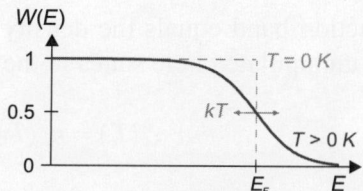

**FIGURE 6.18** Density of states $W(E)$ as a function of the energy $E$ of electronic states for $T = 0$ and $T \neq 0$ (Fermi-Dirac distribution).

Since electrons are Fermions and, therefore, obey the Pauli exclusion principle, the probability distribution function or density of states, $W(E)$, has a different form than that for Bosons:

$$W_F(E) = \frac{1}{\exp((E - E_F)/kT) + 1}. \qquad \text{(Fermi-Dirac distribution)}$$

$W(E) < 1$ for Fermions, in accord with the exclusion principle. For Bosons, on the other hand, it is possible to have $W(E) > 1$. The density of states $W(E)$ for occupation of the eigenmodes of Bosons follows from the average energy $\langle E_{\mathrm{mod}} \rangle$ of the eigenmodes (Section 6.2) and the energy $h\nu$ of the Bosons: $W(E) = \langle E_{\mathrm{mod}} \rangle / h\nu$. Thus, for Bosons

$$W_B(E) = \frac{1}{\exp((E - E_F)/kT) - 1}. \qquad \text{(Bose-Einstein distribution)}$$

**Problem 6.10**

Show that the Fermi-Dirac distribution is a step function with values 1 or 0 in the limit $T \to 0$.

For insulators, as well, the probabilities that the electronic states are populated can be calculated using the Fermi-Dirac distribution function. Of course, a meaningful values for the Fermi energy $E_F$ must first be specified. Since the valence band of insulators is full and the conduction band is empty for $T \to 0$, the Fermi energy lies in the gap between the valence and conduction bands. The exact value of $E_F$ follows from the condition that for $T > 0$ the density $n_{\mathrm{cond}}$ of electrons in the conduction band equals the density $n_v$ of *holes* in the valence band, i.e., the density of unpopulated hole states in the valence band:

$$n_{cond}(T) = n_v(T).$$

Since the energy gap is usually much greater than the thermal energy $kT$, i.e., $E_G \gg kT$, the probability of populating the electronic states in the conduction and valence bands of insulators can be calculated using approximation formulas (Section 6.5.3). In the conduction band, essentially only states at the lower edge $E_{cond}$ of the conduction band are populated. Thus, the probability of populating this state is equal to the Boltzmann factor; that is,

$$W(E_{cond}) = \exp\left(-\frac{E_{cond} - E_F}{kT}\right).$$

In the valence band, however, essentially only those states at the upper edge $E_v$ are unpopulated. The probability that states are unpopulated there is accordingly equal to the Boltzmann factor; that is,

$$W(E_v) = \exp\left(-\frac{E_F - E_v}{kT}\right).$$

Thus, in any insulator there is a certain probability that electrons will be in the conduction band and holes in the valence band. The product $W(E_{cond})W(E_v) = \exp(-E_G/kT)$ is determined by the ratio of the width $E_G = E_{cond} - E_v$ of the energy gap between the conduction and valence bands to the thermal energy $kT$. The quality of an insulator thus depends on the width of its band gap.

**Problem 6.11**

Calculate the product $W(E_{cond})W(E_v)$ for an insulator with $E_G = 3$ eV at room temperature.

> **Notes**
> The probability $W(E)$ that electronic states are occupied in the conduction band of a metallic conductor at temperature $T$ falls from $W(E)=1$ to $W(E)=0$ in the region of the Fermi energy $E_F$. The transition region has a width on the order of $kT$.
>
> In insulators, because $kT \ll E_G$, at temperatures $T > 0$ only a few states at the lower edge of the conduction band are filled with electrons, while an equal number of states at the upper edge of the valence band are empty.

### 6.3.4  Work function and contact potential

In metallic conductors and insulators electrons can not only be excited, but they can also be driven out of the medium. For that to happen, the work function $W_A$ of the material must be overcome. In the photoelectric effect (Section 4.3.1) electrons gain the required energy by absorbing a photon. In thermionic emission (Section 4.3.3) a few electrons have the necessary energy owing to the thermal energy distribution. The potential energy $E_{vac}$ of the electrons at the outside (vacuum) is, therefore, higher by an amount $W_A$ than the energy of the electrons with the highest energy in the material. Hence, for metallic conductors

$$W_A = E_{vac} - E_F.$$

Gold has a high work function ($W_A = 5.1\,\text{eV}$), while cesium has an especially low work function ($W_A = 2.1\,\text{eV}$). Thus, in our experiment on the photoelectric effect (Section 4.3.1) electrons could be knocked out of the Cs cathode by visible photons.

If two metallic conductors with different work functions $W_A(\text{I}) \neq W_A(\text{II})$ are brought into contact, an equilibrium develops such that the Fermi energies of the two conductors are the same (Figure 6.19). Otherwise, electrons would flow from the metal with the higher Fermi energy to the metal with the lower Fermi energy until a potential equilibrium was attained. With equal Fermi energies, however, the two metallic conductors have different potentials outside the respective materials. Thus, a *contact potential* $U_K$ exists between the two conductors:

$$eU_K = W_A(\text{I}) - W_A(\text{II}).$$

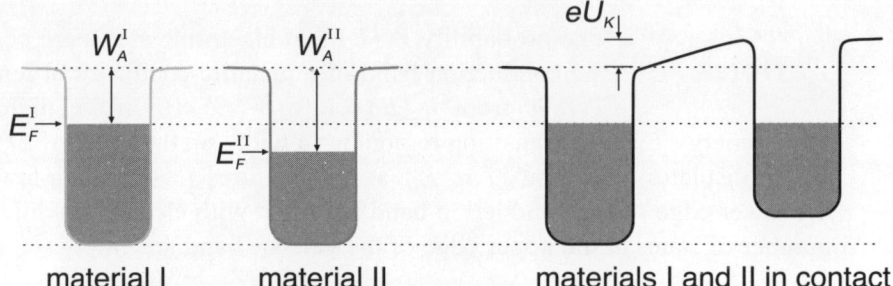

**FIGURE 6.19** Illustrating the interpretation of the contact potential.

## Experiment 6.3   Contact potential

The contact potential between two conducting wires of different materials cannot be measured as usual by a voltmeter. This is because the contacts made when the voltmeter is connected also generate contact potentials. In a closed circuit the contact potentials balance out. Otherwise, a current would flow constantly in the circuit (with its Ohmic resistance leading to dissipation of the current) without an energy source being present.

But the contact potential between two conductors can be measured by detecting the electric field between them. At distances of a few μm a field strength of several hundred V/cm develops between the conductors. These fields are strong enough, for example, to deflect electron beams.

Experimentally it is easier to detect the surface charges which build up at a contact (Figure 6.20). If a pair of copper and aluminum plates are pulled apart suddenly, there is no time for charge equilibration. This leads to the appearance of an electric field **D** between the two plates, whose magnitude $D$ is equal to the charge density on the surfaces of the plates. The voltage $U = dD/\varepsilon_0$ on the charged parallel plate capacitor with a distance $d$ between the plates can be measured with a voltmeter.

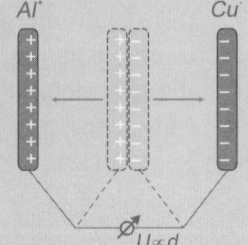

**FIGURE 6.20** Surface charges on conducting plates of Al and Cu that have been suddenly separated.

The contact potentials in an electrical circuit balance out precisely only if the whole circuit is in thermal equilibrium. If two contacts are at different temperatures $T_1 \neq T_2$, then a sensitive voltmeter inserted in the circuit will read a voltage on the order of $\mu V$. This *thermoelectric potential* is consistent with the laws of thermodynamics (Sections 2.3.2 and 2.4.2). They imply that contact potentials depend slightly on the temperature of the contacts. This is because the lattice separation and, thereby, the work function $W_A$ of a conductor change slightly with temperature.

Circuits such as these with two contacts and a sensitive voltmeter are used as thermocouples for measuring temperature differences (Section 2.1.3). For many contacts the thermoelectric potential $U_{th} \propto \Delta T$ is proportional to the temperature difference $\Delta T$ to a good approximation. For Cu-constantan thermocouples the proportionality constant has the value $a = 41\,\mu V/K$.

---

**Problem 6.12**

Consult a technical handbook to find the work functions of metals and determine the contact potentials of various metal pairs.

---

| **Notes** | When two metals are brought into contact, electrons flow from one metal to the other until the Fermi levels of the two metals reach the same height. In this way an electrical dou- |

ble layer develops at the boundary surface. The surface of one metal is negative and that of the other, positive. The resulting contact potential is equal to the difference in the work functions of the two metals.

The contact potentials vary slightly with the temperature of metals. Thermoelectric potentials develop because of the temperature dependence of the contact potentials.

## 6.4 ELECTRON MOBILITY IN CRYSTALS

If a potential $U$ is applied to a metallic conductor, an electrical current $I = U/R$ (Section 3.4.4) flows in the conductor and thereby heats it. The heating power of the current is $P = RI^2$. Here $R$ is the Ohmic resistance of the conductor. These laws of elementary electricity show that the conduction electrons in metallic conductors are not rigidly bound in the electronic states of a crystal, but can also move in the crystal.

In order to describe this motion, let us first treat the conduction electrons as an ensemble of particles which behaves as a classical gas. This model makes it possible to explain a few elementary phenomena. In particular, the Ohmic resistance to an electric current can be attributed to the random thermal motion of the electrons by making an analogy with the viscosity of fluid flows. Since the thermal conductivity of metals is basically also caused by the thermal motion of their electrons, the electron gas model yields a quantitative relationship between the electrical conductivity and the thermal conductivity, the Wiedemann-Franz law. It has been well confirmed experimentally.

The classical model for the motion of conduction electrons in metals also leads, however, to results that are in flagrant conflict with experimental data. A suitable description of the mechanism for conduction in metals and, even more so, in semiconductors is possible only on the basis of quantum physics. Only if the wave character of the electrons and the Pauli exclusion principle are taken into account, i.e., only when the conduction electrons are treated as a quantum gas of Fermions, can electrical phenomena in metals and semiconductors be comprehensively understood.

### 6.4.1 Classical electron gas

According to the simple model picture we have used to interpret the regularities of the photoelectric effect (Section 4.3.1) and thermionic emission (Section 4.3.3), the conduction electrons move like the classical particles of an ideal gas in a potential well whose ground potential $E_0 = -W_A$ is lower than the potential outside the metal, $E_{\mathrm{pot}} = 0$, by an amount equal to the work function $W_A$. Because of their random thermal motion, the conduction electrons have an average kinetic energy $\langle E_{\mathrm{kin}} \rangle = \frac{3}{2}kT$.

If a voltage $U$ is applied to an electrical conductor of length $l$, then the electrons will be accelerated additionally by an electric field $\mathbf{E}$ with $|\mathbf{E}| = U/l$. Hence, they will have a directed motion (toward the positive pole) as well as their undirected thermal motion. The average velocity $\langle \mathbf{v} \rangle$ of this motion depends on the acceleration and also on the average duration $\tau$ of the acceleration.

Just as the directed stream motion in a flowing gas (Section 3.2.2) is continuously converted by interatomic collisions into an undirected random thermal motion, in a conductor the current flow devolves into thermal motion as a result of collisions of the electrons with the crystal lattice. For this reason a conductor heats up when an electric current flows through it.

According to this classical model picture (Figure 6.21), the electrical current density $\mathbf{j}$ (with $|\mathbf{j}| = I/A$) of a current $I$ in an electrical conductor with cross-sectional area $A$ is equal to the product of the particle density $n$ of the conduction electrons, the electronic charge $(-e)$, and the average velocity $\langle \mathbf{v} \rangle$:

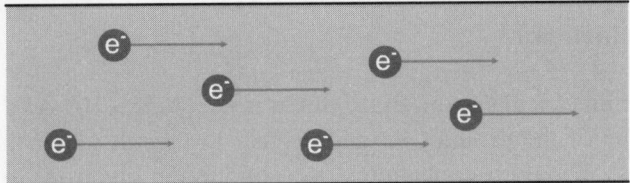

**FIGURE 6.21** An electron current in an electrical conductor.

$$\mathbf{j} = -ne\langle \mathbf{v} \rangle.$$

Here the average velocity $\langle \mathbf{v} \rangle$ of the conduction electrons (with mass $m_e$) is equal to half the final velocity attained by the electrons after the acceleration time $\tau$.

$$\langle \mathbf{v} \rangle = \frac{e}{2m_e} \tau \mathbf{E}.$$

The average velocity of the conduction electrons is therefore proportional to the electric field strength $\mathbf{E}$. The proportionality factor $\mu = (e/2m_e)\tau$ is the *mobility* of the conduction electrons. Along with the particle density $n$ of the conduction electrons, it determines the specific electrical conductivity $\sigma$ of the conductor. This is defined as the proportionality factor in the relationship between the current density $\mathbf{j}$ and the electric field strength $\mathbf{E}$:

$$\mathbf{j} = \sigma \mathbf{E}.$$

Thus, we have

$$\sigma = ne\mu.$$

Substituting the above value of $\mu$ in this equation, we obtain

$$\sigma = \frac{e^2}{2m_e} n\tau.$$

For quantitative estimates of $\sigma$ some assumptions must be made about the values of $n$ and $\tau$. The classical ideal gas model is not appropriate here. But, the product $n\tau$ also determines the thermal conductivity of metals. Thus, the idea that the conduction electrons behave like the atoms of an ideal gas has yielded a fundamental relationship between the electrical and thermal conductivities of metals.

## Problem 6.13

The electrical conductivity of Cu is $\sigma = 0.58 \times 10^8$ A/V·m. Calculate the value of the product $n\tau$ and discuss the result under the assumption that every Cu atom contributes one conduction electron. In that case what is the mean free path of the conduction electrons?

| **Notes** | The random motion of the conduction electrons in a crystal lattice is determined by collisions of the electrons with the lattice atoms. These collisions lead to heating of the crystal |

when an electrical current flows through it.

The electrical conductivity $\sigma$ of metallic conductors is proportional to the product $n\tau$ of the particle density $n$ and mean free time between collisions, $\tau$, of the conduction electrons.

### 6.4.2 The Wiedemann-Franz law

Temperature equilibration by heat conduction was discussed in Section 3.2.3. The formula for the thermal conductivity $\lambda$ of a gas was given there:

$$\lambda \approx c\rho D.$$

For an atomic gas the product $c\rho$ of the specific heat $c$ and density $\rho$ is given by

$$c\rho = \frac{3}{2} kn.$$

The random motion of the atoms (Section 3.2.2) yielded $D = l v_{\mathrm{th}}/3$ for the diffusion constant. On replacing the mean free path $l = v_{\mathrm{th}}\tau$ by the mean time $\tau$ between collisions of the electrons, and given that $m_e v_{\mathrm{th}}^2/2 = \frac{3}{2}kT$, we obtain the following for the diffusion constant of an electron gas:

$$D = \frac{kT}{m_e}\tau.$$

For the thermal conductivity of the conduction electrons this yields

$$\lambda \approx \frac{3k^2 T}{2m_e} n\tau.$$

Like the electrical conductivity $\sigma$, the thermal conductivity $\lambda$ of metals is determined by the product of the density $n$ and mean free time $\tau$ of the conduction electrons in the metal. Thus the ratio $\lambda/\sigma$ obeys the following relation, which is independent of $n$ and $\tau$.

$$\frac{\lambda}{\sigma} \approx \frac{3k^2}{e^2} T. \qquad \text{(Wiedemann-Franz law)}$$

The quotient $\lambda/\sigma$ is, therefore, proportional to the absolute temperature $T$ of the metal. The proportionality factor $L$ is a universal constant, the Lorentz constant. A more exact quantum mechanical theory gives

$$L = \frac{\pi^2 k^2}{3e^2} = 2.4 \times 10^{-8} \text{ V}^2 \cdot \text{K}^{-2}.$$

This relationship between the conductivities $\lambda$ and $\sigma$ is quite well satisfied in all metals. For example, for copper at $T = 293$ K, $\lambda = 393$ W $\cdot$ m$^{-1} \cdot$ K$^{-1}$ and $\sigma = 0.58 \times 10^8$ A/V $\cdot$ m. This gives $L_{Cu} = 2.3 \times 10^{-8}$ V$^2 \cdot$ K$^{-2}$, in good agreement with the theoretical value.

---

## Experiment 6.4  Measurement of the quotient $\lambda/\sigma$ for copper

The setup for the measurement apparatus is shown in Figure 6.22. The Cu rod with a cross-sectional area $A$ is, on one hand, an electrical conductor. In order to determine the electrical conductivity $\sigma$, a current $I$ is passed through the rod and the voltage drop $U$ over a length $s$ is measured. This gives $I/A = \sigma(U/s)$. On the other hand, the Cu rod is a heat conductor. In order to determine the thermal conductivity $\lambda$, one end of the rod is heated electrically and the other end is cooled by a large copper block. In this case, the heating power $P$ is measured and the temperature drop $\Delta T$ along the length $l$ is determined with a thermocouple. This gives $P/A = \lambda(\Delta T/l)$. The two measurements yield a value for the Lorentz constant.

*continued*

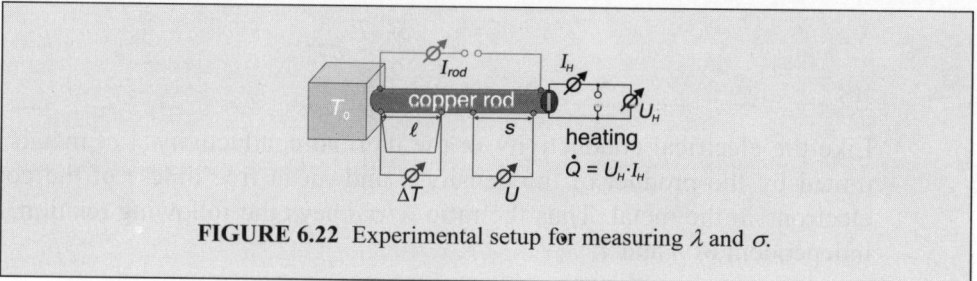

**FIGURE 6.22** Experimental setup for measuring $\lambda$ and $\sigma$.

## Problem 6.14

Check how well the Wiedemann-Franz law is satisfied for the metals iron and aluminum. You can find the electrical and thermal conductivities of these metals in a technical handbook.

---

**Notes**

In metals, the mobility of the conduction electrons determines the electrical conductivity $\sigma$, as well as the thermal conductivity $\lambda$. For that reason the quotient $\lambda/\sigma T = \pi^2 k^2/3e^2$ is a universal constant.

### 6.4.3 Conduction electrons as a Fermi gas

The mobility of the conduction electrons in a crystal is, indeed, decisive for the relatively good thermal conductivity of metals, but the conduction electrons scarcely contribute to the heat capacity of the metal. This is because, according to the Dulong-Petit law (Section 6.1.2), all metals have a molar heat capacity $C = 3R$ at ordinary temperatures. This value follows only from the thermal motion of the atoms about their equilibrium positions. Although the conduction electrons do participate in the thermal motion, they barely make a noticeable contribution to the heat capacity of a metal.

This contradiction is resolved if the conduction electrons are treated as a Fermi gas of Fermions that satisfy the Pauli exclusion principle, rather than as a classical electron gas. Of the electrons in the conduction band of the metal (Section 6.3.2) only a small fraction is actually freely mobile. That is because all the electrons lying in states sufficiently below the Fermi energy, that is, not only all the electrons in fully populated bands, but also most of the electrons in the conduction band, cannot change their quantum states at all, because all the neighboring electronic states are occupied. Only those conduction electrons in quantum states with energies $E$ near the Fermi energy $E_F$ can change the state they are in. Thus, the only electrons that are freely mobile have energies $E$ within a narrow region above and below the Fermi limit $E_F$:

$$|E - E_F| < kT.$$

The propagation of these electrons in the crystal is to be described using wave mechanics.

We now illustrate the wave mechanical concept for describing the conduction electrons in metals with a simple one-dimensional model. It is based on the assumption that the conduction electrons move in a shallow potential well. Starting with this model, we interpreted the Einstein formula for the photoelectric effect in Section 4.3.2. In a potential well of this sort with a length $L$ along the $x$-axis and a work function $W_A$, which is large compared to the kinetic energy $E$ of the electrons, the electron waves have a countable sequence of eigenstates (Section 3.1.2). These are the standing waves

$$\psi(x,t) = \sin(k_n x)\exp(-i\omega_n t) \quad \text{with} \quad k_n = \frac{p_n}{L}.$$

Since $E = p^2 / 2m_e$, they satisfy the dispersion relation

$$\hbar\omega = \frac{(\hbar k)^2}{2m_e} \quad \text{with} \quad \hbar = \frac{h}{2\pi}. \qquad \text{(energy parabola)}$$

These electronic states can each be populated by two electrons with different spin quantum numbers $m_s$. The wave number $k_F$ of the electronic state at the Fermi limit is thus $k_F = (\pi/L)(N/2)$. Here $N = L/a$ is the number of conduction electrons in the linear crystal if $a$ is the distance between neighboring atoms. This yields the Fermi energy $E_F$ for this simple model:

$$E_F = \frac{(\hbar k_F)^2}{2m_e} = \frac{(h/4a)^2}{2m_e}.$$

For the simple model under consideration here, the Fermi energy is on the order of 1 eV. Hence, $E_F \gg kT$. The electrical and thermal conductivities of the metals thus depend on the mobility of only a small fraction of the electrons in the conduction band, and the contribution of these electrons to the heat capacity of the metal can be neglected in the first approximation.

In a shallow potential well the dispersion relation for the electron waves has a parabolic form. The energy $E = h\nu$ of the electronic states can thus be arbitrarily high. But, a crystal with a periodically varying potential has energy bands with gaps between them. The conduction band is up to half filled and, therefore, has a finite width $\Delta E_{\text{cond}}$ on the order of $2E_F$. We shall discuss the dispersion relation of the electron waves in a periodically varying crystal potential in the next section.

---

### Problem 6.15

Estimate the width of the conduction band of a linear crystal where the neighboring atoms are separated by a distance $a = 0.2$ nm. Assume that the Fermi level lies in the middle of the conduction band.

---

| | |
|---|---|
| **Notes** | The conduction electrons in metals can be described as a quantum gas of electrons which obey the Pauli exclusion principle. The electronic states are filled up to the Fermi |

energy $E_F$. $E_F$ is large compared to the thermal energy $kT$.

## 6.4.4 Dispersion relations of electron waves in crystals

The electrons in a crystal move in a periodically structured potential. Only if $\lambda \gg 2a$, i.e., the wavelength of the electron waves is much greater than the distance between neighboring atoms, or $k \ll \pi/a$, is taking the average of the potential, along with the assumption that the conduction electrons move approximately in a shallow potential well, justified. When $k \sim \pi/a$ or the wave number is on the order of the reciprocal of the lattice separation, however, the effect of the periodic potential structure on the electron wave functions must be taken into account. In particular, the periodicity of the crystal potential affects the dispersion relation $\omega(\mathbf{k})$ of the electron waves in the crystal. When the periodicity of the crystal potential is included, the simple energy parabola (Section 6.4.3) breaks up into multiple segments (Figure 6.23). The segments correspond to the energy bands of the electron states (Section 6.3.2).

In order to clarify the transition of the energy parabola of a free electron gas to the energy bands of the electrons in a crystal field, let us again consider a

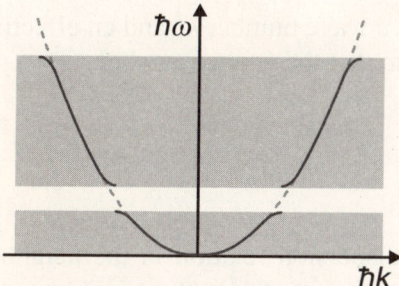

**FIGURE 6.23** The dispersion relation of electron waves in a linear crystal.

linear crystal, and assume that it is of infinite extent. In the periodic crystal field, the wave number $k = \pi/a$ corresponds to two standing waves with different energies (Figure 6.24). If the nodes of the standing wave lie between the atoms, then the electrons are mainly to be found in the potential craters in the neighborhoods of the nuclei of the atoms. Accordingly, the energy $E_<$ of this electron state is comparatively low. If, however, the nodes of the standing waves lie at the nuclei, then the electrons are mainly to be found in the potential peaks between the atoms. The energy $E_>$ of this electron state is, therefore, substantially higher. There is an *energy gap* $E_G = E_> - E_<$ between the energies of the two standing waves within which no electron states exist in the crystal field.

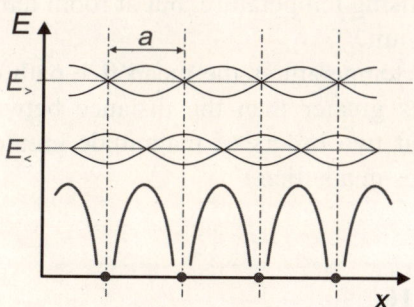

**FIGURE 6.24** Standing electron waves with $k = \pi/a$ in a crystal field.

By contrast, standing waves with wave numbers $k = \pi/a \pm k'$, where $k' \ll \pi/a$, have energies that are somewhat lower than $E_<$ or somewhat higher than $E_>$. The dispersion relations for the electron waves in the crystal thus have a parabolic shape, not only for $k = 0$ but also for $k = \pi/a$ (and all multiples of $\pi/a$) (Figure 6.23). All of these energy parabolas can be interpreted as the waves of particles with a wave number $k'$ and an effective mass $m_{\text{eff}}$. The effective mass is to be chosen so that the energy parabola

$$E'(k') = \pm \frac{(\hbar k')^2}{2 m_{\text{eff}}}$$

is equal to the dispersion relation in the neighborhood of $k' = 0$. If the energy parabola has a minimum at $k' = 0$, so it represents the dispersion relation for the lower edge of an energy band, then the corresponding electronic states describe the motion of electrons. But if the energy parabola has a maximum at $k' = 0$, so the dispersion relation applies to the electronic states at the upper edge of an energy band, then the corresponding electronic states describe the motion of holes (Section 6.3.3). The dispersion relations near the upper and lower edges of energy bands are of special importance for electron mobility in semiconductors. They will be needed to explain the properties of semiconductors (Section 6.5.3).

The wave mechanical description of the motion of electrons in crystals leads to another important consequence. In a strictly periodic crystal travelling waves can travel unimpeded. The electron waves are scattered only where the periodicity of the crystal is disturbed. On one hand, disturbances of this sort occur at crystal defects or imperfections (Section 5.5.4). On the other hand, the thermal motion of the atoms also causes a departure from the periodicity of the crystal structure. Thus, the electron waves in crystals have a finite mean free path $l$. It decreases with rising temperature, but at room temperature, $T \sim 300$ K, is still on the order of 100 nm.

At ordinary temperatures the mean free path of electron waves in crystals is about 100 times greater than the distance between neighboring lattice atoms. Thus, it is about two orders of magnitude greater than the mean free path for classical particles in a lattice.

### Problem 6.16

The electrical and thermal conductivities of metals depend on the product $n\tau$ of the particle density $n$ and the mean free time of flight $\tau$ of the conduction electrons. Examine how the product $n\tau$ changes when the decryption of the conduction electrons is altered from a classical gas model to a quantum gas model.

**Notes**

In the neighborhood of the band edges the electron waves obey parabolic dispersion relations which can be written either as relations between the variables $E$ and $\mathbf{p}$ of the particles, or the variables $\omega$ and $\mathbf{k}$ of the waves:

$$E = \frac{p^2}{2m_{\text{eff}}} \text{ or } \hbar\omega = \frac{(\hbar k)^2}{2m_{\text{eff}}}.$$

## 6.5 SEMICONDUCTORS

At sufficiently low temperatures, crystals whose electronic states can be described in terms of energy band models either become electrical conductors or insulators. All electrical conductors have a partially filled conduction band. In insulators, however, the energy bands are either fully populated or empty. But when the energy gap between the fully populated valence band and the empty conduction band is not too large compared to the thermal energy $kT$, some electrons can reach the conduction band from the valence band at ordinary temperatures. The density of states for the electrons at the lower edge of the conduction band and the probability of hole states at the upper edge of the valence band are given by the Fermi distribution (Section 6.3.3).

Nonmetallic crystals with band gaps in the range of 1 eV cannot, however, serve as insulators. But they are of great technical interest as *semiconductors*. Semiconductor technology is based on ultrahigh purity crystals of elements in the fourth group of the periodic table, especially silicon, whose band gap $E_G = 1.2$ eV, and the III-V and II-VI compounds. They crystallize in diamondlike lattices (Section 5.6.4). The electronic band structure of these crystals arises from sp$^3$ hybrid bonding. This is decisive for many properties of the semiconductors that are of technical interest. We begin by examining sp$^3$ hybrid bonding in somewhat more detail.

### 6.5.1 sp$^3$ hybrid bonding

sp$^3$ hybrid bonding is similar in many ways to the covalent bonding of the $H_2$ molecule. It determines the crystal structure of a diamond (Section 5.6.4). As opposed to the H atom with its single valence electron, carbon atoms have four valence electrons. Thus, four other C atoms can be bonded to a given C atom, so that crystals develop in which each atom has four nearest neighbors.

The four valence electrons of the C atom have a principal quantum number $n = 2$. The $n = 2$ shell is thus half filled. The electronic states in the $n = 2$ shell are approximate solutions of the following (time-independent) Schrödinger equation (Section 5.2.3):

$$-\frac{\hbar^2}{2m_e}\nabla^2\psi(\mathbf{r},t) - \frac{Z_{\text{eff}}e^2}{4\pi\varepsilon_0 r}\psi(\mathbf{r},t) = E\psi(\mathbf{r},t).$$

Here it is assumed, for simplicity, that the valence electrons move independently of one another in a Coulomb potential with the effective charge $Z_{\text{eff}}$ of the $C^{4+}$ ion. In this approximation all the electronic states with $n = 2$ (as in the H atom) have the same binding energy. To determine how the four valence electrons are distributed among the eight electronic states of the $n = 2$ shell requires a more exact theory and depends on the external circumstances.

In free (chemically unbound) C atoms, the two electronic states with $l = 0$ and two states with $l = 1$ are populated; that is, two s- and two p-states with oppositely directed spins in each case. In a chemical bond, such as a $CH_4$ molecule (methane) or a diamond, however, one s- and three p-states are occupied. Thus, we speak of an $sp^3$ bond.

Here we speak of a *hybrid bond*, because the four electrons in the $sp^3$ configuration can be regarded as electrons with equivalent wave functions that differ only in their spatial orientation (Figure 6.25). This deviation from the $|n,l,m,m_s\rangle$ classification is possible since the Schrödinger equation is a linear differential equation, so that all superpositions (Section 3.1.3) of its eigenstates with equal eigenvalues are also eigenstates of the equation. Electronic states with a tetrahedral charge distribution, as in a $CH_4$ molecule, can be formed by a superposition of s- and p-states. By pairing with the valence electrons of H atoms, the four valence electrons of a C atom can bind four H atoms. Like the bond in a $H_2$ molecule, each bond in the $CH_4$ molecule is formed by two electrons with oppositely directed spins. Thus, a stable electron configuration resembling the closed shell of a He atom develops between the C atom and each of the H atoms.

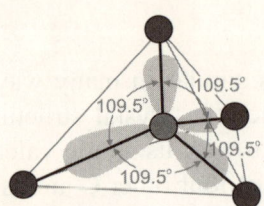

**FIGURE 6.25** The tetrahedral structure of the $CH_4$ molecule.

In a way similar to what happens with the $CH_4$ molecule, a helium-like electron configuration exists between every pair of neighboring C atoms in a diamond. The stability of this configuration explains the hardness of diamonds.

The elements silicon and germanium also belong to group IV of the periodic table and, therefore, have four valence electrons. They also form crystals with a diamond lattice structure. While diamond is a good insulator, silicon and germanium are typical semiconductors. Compared to diamond with its large energy gap between the valence and conduction bands (5.3 eV), Si and Ge crystals have relatively small band gaps, $E_G = 1.12$ and $0.67$ eV, respectively.

As noted in Section 6.3.1, the band structure of the electronic states in a crystal follows from the resonance splitting of the atomic energy levels owing to crystal formation. In the elements of the fourth group of the periodic table the valence electrons occupy the states with $l = 0$ and $l = 1$. In free atoms, that is, when the distance between the atoms is large, the s-states have somewhat lower energies than the p-states (Figure 6.26). As the distance between neighboring atoms decreases, the s- and p-states initially split into bands. But with splitting, the electronic states themselves also change.

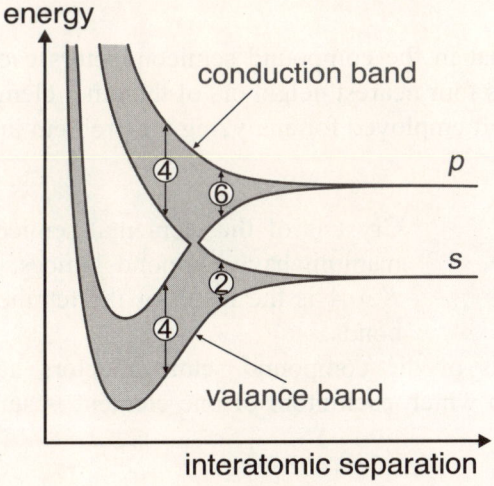

**FIGURE 6.26** Band splitting in a diamond lattice as a function of interatomic separation.

For large interatomic separations the energetically lower s-band can only accept two electrons (with opposite spins) and the energetically higher p-band can, thus, accept six electrons per atom, while for short separations one bonding and one antibonding energy band are formed, each of which can accept four electrons per atom. The two energy bands correspond to the bonding (attractive) and antibonding (repulsive) energy states of the $H_2$ ion or the $H_2^+$ molecular ion (Section 6.3.1). In a crystal (at sufficiently low temperatures) the bonding band is fully populated and the antibonding band is empty. They form the electronic valence and conduction bands, respectively, in a diamond lattice.

The energy gap between the valence and conduction bands depends on the distance between neighboring atoms in the crystal lattice. C atoms, with only one closed inner shell, are separated by only 0.154 nm in a diamond lattice. In Si crystals, however, the atoms (which have two closed inner shells) are separated by 0.235 nm. Here, in accord with the larger separation, the energy gap between the valence and conduction bands is considerably smaller than in diamond.

As in the case of the *elemental semiconductors* Si and Ge, energy bands are formed in crystals of the *compound semiconductors*, in which one atom of group III or II of the periodic table is surrounded tetrahedrally by four atoms from group V or VI (and vice versa). The band gaps $E_G$ of the compound semiconductors differ substantially. For some II-VI compounds, such as ZnS, $E_G > 3\,\text{eV}$, while for some of the III-V compounds, such as InAs, $E_G < 1\,\text{eV}$.

---

**Problem 6.17**

Show that in the compound semiconductors each atom of a given element has four nearest neighbors of the other element. Use a lattice model of the sort employed for analyzing the problem in Section 5.6.4.

---

**Notes**   Crystals of the elemental semiconductors silicon and germanium have diamond lattices. The coordination number $K = 4$ is the result of the tetrahedrally directed $sp^3$ hybrid bonds.

Crystals of the compound semiconductors have a diamondlike lattice structure in which each atom of one element is surrounded by four atoms of the other.

## 6.5.2 Doped semiconductors

The elementary and compound semiconductors considered up to now are so-called *intrinsic* semiconductors. In these intrinsic semiconductors the density $n_{cond}$ of the electrons in the conduction band is always equal to the density $n_v$ of holes in the valence band. In Si crystals with a band gap $E_G = 1.12\,\text{eV}$, at ordinary temperatures $n_{cond} = n_v \approx 1.5 \times 10^{10}$ cm$^{-3}$. *Doped* semiconductors have higher charge carrier densities and different densities of electrons and holes.

Doping is done by incorporating donor and acceptor atoms into a semiconductor crystal. In Si crystals group V elements such as P, As, and Sb act as *donors* and group III elements such as Al, Ga, and In act as *acceptors*. At the lattice site of a donor atom, there is an excess electron which is only weakly bound to the donor atom. The Bohr orbits of these electrons have radii of several nm. Thus, these orbits cover many lattice sites. For that reason, the binding energy with which these electrons are attached to the donor atom is only on the order of 10 meV. In the energy band model these electrons occupy energy levels that lie just below the lower edge of the conduction band (Figure 6.27). The electrons can easily reach the conduction band through thermal excitation from these states. Semiconductors doped with donors thus have a high electron density $n_{cond}$ in the conduction band with a nearly fully occupied valence band ($n_v = 0$). Since the charge carriers are negatively charged, they are referred to as *n-type semiconductors*.

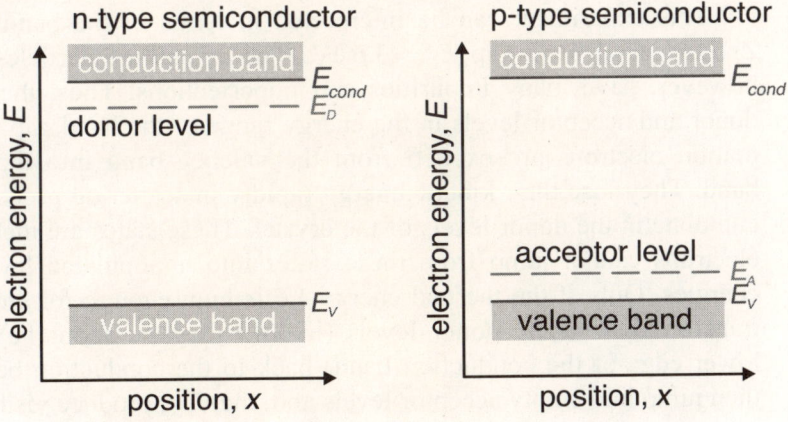

**FIGURE 6.27** Donor and acceptor levels in doped semiconductors.

Semiconductors doped with acceptors have one less electron than otherwise at the lattice sites of acceptor atoms. Hence, an unfilled electronic state exists there, which can accept an electron from the valence band of the crystal. Since the acceptor levels lie only a few meV above the valence band, at ordinary temperatures most of them are filled by electrons from the valence band. Thus, semiconductors doped with acceptors have a high density $n_v$ of holes in the valence band while the conduction band is empty. Since holes behave as positive charge carriers, in this case we speak of *p-type semiconductors*.

### Problem 6.18

Calculate the density of states of the donor and acceptor levels of a doped semiconductor at $T = 300$ K for the case in which the energy separation from the corresponding band edge is $\Delta E = 10$ meV.

---

### Experiment 6.5    Thermoluminescence

A powder of the compound semiconductor ZnS at room temperature is irradiated by UV light ($h\nu > 3.6$ eV). This causes it to emit light and it continues to emit light for a short time after the irradiation ends. The ZnS powder behaves differently if it is first cooled to the temperature of liquid nitrogen and then irradiated. Initially it does not emit light. It begins to emit light, however, if it is heated by a hairdryer, for example, after it has been irradiated.

This experiment can be interpreted in terms of the band structure of ZnS. It has a band gap $E_G = 3.6$ eV. The crystalline particles of powder, however, have many impurities and imperfections. Thus, there are many donor and acceptor levels in the energy range of the band gap. During irradiation electrons are excited from the valence band into the conduction band. They lose their kinetic energy rapidly in scattering processes and accumulate in the donor levels of the crystal. These states are metastable. The electrons cannot jump from these states into unpopulated levels at lower energies. Only if the thermal energy $kT$ is high enough for electrons to be transferred from the donor levels (in ZnS they lie about 1 eV below the lower edge of the conduction band) back to the conduction band can they then jump into empty acceptor levels and, thereby, produce visible photons.

---

**Notes**

If a Si crystal is doped with group V elements, an n-type semiconductor is produced. The donor levels lie just below the lower edge of the conduction band. On the other hand, if a Si crystal is doped with group III elements, a p-type semiconductor results. The acceptor levels lie just above the upper edge of the valence band.

### 6.5.3  Charge carrier density in semiconductors

In intrinsic semiconductors the density $n_{\text{cond}}$ of the conduction electrons is every-where equal to the density $n_{\text{v}}$ of holes. The charge carrier densities $n_{\text{cond}} = n_{\text{v}}$ can be calculated, on one hand, by determining the respective densities of states $W(E_{\text{cond}})$ and $W(E_{\text{v}})$, which give the probabilities of populating the states at the lower edge $E_{\text{cond}}$ of the conduction band with electrons or the states at the upper edge $E_{\text{v}}$ of the valence band with holes. These follow from the Fermi distribution function (Section 6.3.3):

$$W(E_{\text{cond}}) = \exp\left(-\frac{E_{\text{cond}} - E_F}{kT}\right) \quad \text{or} \quad W(E_{\text{v}}) = \exp\left(-\frac{E_F - E_{\text{v}}}{kT}\right)$$

The product of these probabilities is $W(E_{\text{cond}})W(E_{\text{v}}) = \exp(-E_G/kT)$. In order to calculate the probabilities $W(E_{\text{cond}})$ and $W(E_{\text{v}})$ separately, the Fermi energy must first be determined. For intrinsic semiconductors this follows from the condition that here the density $n_{\text{cond}}$ of the conduction electrons equals the density $n_{\text{v}}$ of holes.

On the other hand, these charge carrier densities depend on the densities $N_{\text{eff}}(E_{\text{cond}})$ and $N_{\text{eff}}(E_{\text{v}})$ of the electronic states (number per volume per energy interval) which at temperature $T$ become effective for population by electrons or population by holes, respectively. These *effective densities of states* follow from the effective masses $m_{\text{n}}$ and $m_{\text{p}}$ of the corresponding charge carriers:

$$N_{\text{eff}}(E_{\text{cond}}) = 2\left(\frac{2\pi m_{\text{n}} kT}{h^2}\right)^{3/2} \quad \text{or} \quad N_{\text{eff}}(E_{\text{v}}) = 2\left(\frac{2\pi m_{\text{p}} kT}{h^2}\right)^{3/2}.$$

For the product $n_{\text{cond}} n_{\text{v}}$ of the charge carrier densities, the so-called *law of mass action* yields the following for the charge carriers in semiconductors:

$$n_{\text{cond}} n_{\text{v}} = 4\left(\frac{2\pi kT \sqrt{m_{\text{n}} m_{\text{p}}}}{h^2}\right)^3 \exp\left(-\frac{E_G}{kT}\right). \qquad \text{(law of mass action)}$$

Since $n_{\text{cond}} = n_{\text{v}}$ for intrinsic semiconductors, it is possible to calculate the charge carrier densities. For doped semiconductors the charge carrier distributions also depend on the densities $n_D$ and $n_A$ of the donor and acceptor atoms in the crystal. At ordinary temperatures almost all donor levels have given up their electrons to the conduction band. Almost all acceptor levels, on the other hand, are filled with electrons from the valence band. Thus, for doped crystals we have

$$n_{\text{cond}} - n_D \approx n_v - n_A.$$

Since the product $n_{\text{cond}}n_v$ is independent of the doping of the semiconductor according to the law of mass action, this implies that a heavily n-doped semiconductor $(n_D \gg n_v)$ is almost exclusively n-conducting; i.e., $n_{\text{cond}} \approx n_D \gg n_v$. Correspondingly, a heavily p-doped semiconductor $(n_A \gg n_{\text{cond}})$ is almost exclusively p-conducting.

---

**Problem 6.19**

By what factor does the product of the charge carrier densities change if the temperature is raised from 0 to 20°C?

---

**Notes**

According to the law of mass action, the product $n_{\text{cond}}n_v$ of the charged particle densities in the conduction and valence bands in a semiconductor is independent of the doping of the semiconductor. The value of this product varies with the temperature of the semiconductor.

## 6.5.4 p-n junctions

Modern electronics relies substantially on semiconductor components in which p-doped and n-doped regions of a semiconductor are in contact. Hence, within the boundary region there are conduction electrons in the conduction band of the n-type semiconductor and holes in the valence band of the p-type semiconductor. The charge carrier densities in such p-n junctions determine the properties of many semiconductors.

First we consider a p-n junction in thermal equilibrium (Figure 6.28). To start with, let us assume that the p- and n-regions are spatially separate and that an equilibrium population exists in both regions, independently of one another. Thus, for temperatures that are not too high, the conduction band of the p-type semiconductor is practically empty and the valence band of the n-type semiconductor is fully populated. The Fermi energy of the p-type semiconductor, therefore, lies between the energies $E_A$ of the acceptor levels and $E_v$ of the upper edge of the valence band and the Fermi energy of the n-type semiconductor lies between the energies $E_D$ of the donor levels and $E_{\text{cond}}$ of the lower edge of the conduction band.

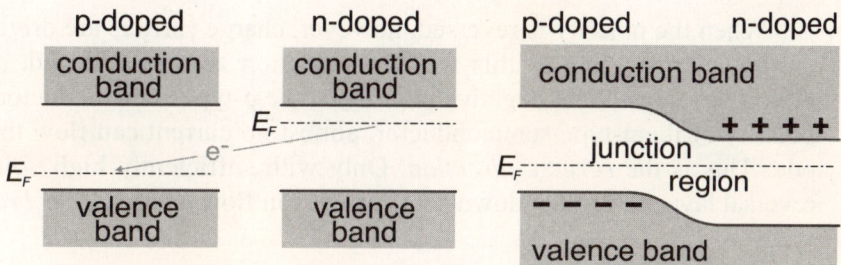

**FIGURE 6.28** Band structure and charge density at a p-n junction.

Second, let us consider the band structure of a semiconductor of which half is p-doped while the other half is n-doped. Both halves together form a *diode*. As when two different metals come into contact (Section 6.3.4), here an equilibration process takes place until the Fermi energies of the two halves are at the same height. In the junction region some of the conduction electrons of the n-type semiconductor migrate to the p-type semiconductor and occupy hole states in the valence band (or free acceptor levels) there. This causes formation of an electrical double layer in the boundary region (as when two metals come into contact; Section 6.3.4), which leads in turn to bending of the band structure. In thermal equilibrium both halves have the same Fermi energy. For this reason, however, in the junction region the charged particle density is lower than in the rest of the crystal. There is a deficit of conduction electrons in the n-type semiconductor and of holes in the p-type semiconductor.

The thermal equilibrium is disturbed if a potential $U$ is applied to the p-n junction (Figure 6.29). If the p-type semiconductor is connected to the positive pole of the supply and the n-type to the negative pole, then the *depletion zone* of the junction is again filled with charge carriers. Both conduction electrons from the n-type semiconductor and holes from the p-type are drawn toward the boundary region. With this polarity, the diode is electrically conducting. This is the *forward* (or conducting) *direction* of the diode.

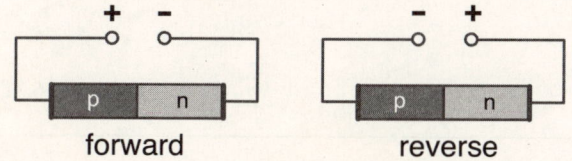

**FIGURE 6.29** Forward and reverse polarities of a p-n junction.

When the polarity is reversed, however, charge carriers are driven away from the boundary region. In this way the depletion zone is widened. An insulating region develops. With negative polarity of the p-type semiconductor and positive polarity of the n-type semiconductor, almost no current can flow through the diode. This is the *reverse direction*. Only with sufficiently high voltages can the reversal layer be broken down so a current can flow again (*Zener breakdown*).

---

### Experiment 6.6 Laser diode

In the boundary layer of a p-n junction a population inversion develops when a potential is applied in the forward direction (Section 6.2.2). The conduction electrons drawn to the p-type semiconductor initially remain in the conduction band, until they *recombine* with holes, that is, until they jump from the conduction band into an unoccupied state in the valence band and thereby give up their excitation energy as phonons or photons (Figure 6.30). In semiconductors with sufficiently high band gaps ($E_G > 1.7\,\text{eV}$), the emitted light can be seen. (This is a light-emitting diode or LED.)

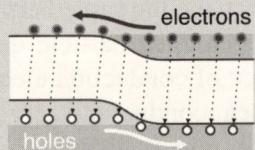

**FIGURE 6.30** Population inversion in the boundary layer of a forward biased diode.

The population inversion in the boundary layer of a forward biased diode is the basis of laser diodes. When the diode current is raised above a threshold value, in suitably constructed diodes (Figure 6.31), the light is amplified by stimulated emission (Section 6.2.3). Here the surfaces of the crystal operate as resonator mirrors.

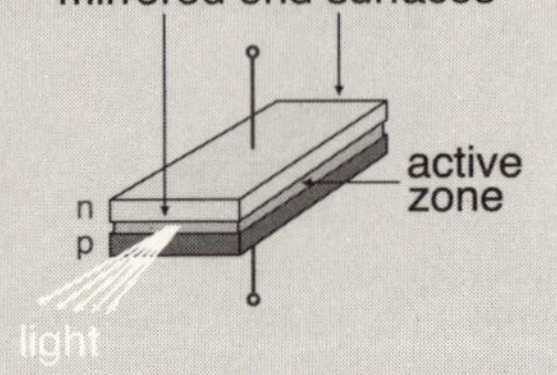

mirrored end surfaces

active zone

n
p

light

**FIGURE 6.31** Structure of a laser diode.

## Experiment 6.7    Solar cell

Electrical power can be converted into light with a diode. Conversely, suitably constructed diodes can convert light into electricity (Figure 6.32). The diode then functions as a solar cell. A sufficiently thin p-conducting layer is irradiated with light. Near the boundary layer, electrons will be lifted out of the valence band and into the conduction band, provided the energy $h\nu$ of the photons is greater than the band gap energy $E_G$. The electrons excited into the conduction band can flow into the n-conducting region until the electrical double layer in the junction region decays (Figure 6.28) and the band bending is balanced out. Here a photoelectric potential $U = E_G/e$ develops between the p- and n-conducting layers of the solar cell, as it did between the cathode and anode in the experiment on the photoelectric effect (Section 4.3.1).

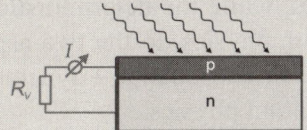

**FIGURE 6.32**  Structure of a solar cell.

## Problem 6.20

How and why do the Fermi energies of doped semiconductors shift when they are heated?

**Notes**    At ordinary temperatures the Fermi energies of n-type semiconductors lie near the lower edge of the conduction band, and the Fermi energies of p-type semiconductors lie near the upper edge of the valence band. Thus, contact potentials develop at the boundary surfaces of n- and p-type semiconductors.

## 6.6 ELEMENTARY EVENTS

The atomic hypothesis (Section 2.1.1), the discovery of the elementary charge (Section 3.4.1), and the quantum hypothesis (Section 4.4.4) are fundamental to the modern world view of physics. Newtonian mechanics and Maxwell's electrodynamics were based on the idea that matter, electrical charge, and electromagnetic fields vary *continuously* in space and time. This continuum picture had to be relinquished step by step. At first the discrete structure of matter, charge, and fields could only be deduced indirectly and, thus, formulated as a set of hypotheses. Modern precision-measurement techniques make it possible, however, to count atoms, electrons, ions, and photons piece by piece, so that they can be detected directly.

The discrete structure of matter and field is the basis of statistical physics. Temperature and heat can only be interpreted under the assumption that natural processes are determined both by the dynamic laws of mechanics and electrodynamics and by the laws of chance. The formulation of the laws of chance is based on discrete structures, while the deterministic laws presuppose a continuum picture. Only in quantum physics do the two aspects of classical physics come to a synthesis. Determinism and chance—dynamic evolution and quantum jumps—both show up in quantum physics.

This synthesis was only possible at the price of intuitive clarity. In terms of the world picture of classical physics it is inconceivable that particles (electrons, as well as atoms) propagate as waves. It was equally difficult to imagine that waves could be detected as particles, specifically photons or phonons. But the concepts of classical physics rely on the assumption, apparently well justified by daily experience, that we can observe the motion of objects and waves *continuously*. A continuous observation is, however, fundamentally impossible, because, ultimately, all measurements involve the *counting of elementary events*.

On one hand, the discrete structure of the measurement process is of fundamental interest and forces us to rethink our world view of the natural sciences. On the other, it is also of great practical interest, because it determines the limits of measurement precision. Even with the greatest care, all measurements are burdened with uncertainties in which certain limits cannot be undercut. Thus, we conclude by taking another look at the measurement process.

### 6.6.1 The measurement process

Precise and reproducible measurements are the foundation of the exact sciences. We pointed this out in at the beginning of this introductory physics course. In the six chapters of this textbook we have tried to present the basic concepts of the physical description of nature. By building on these concepts it is possible to explain, reproduce, and make technical use of many natural processes. Neverthe-less, the measurement process itself is, in many respects, an unsolved problem of modern physics.

Except for the consequences of its discrete structure, the measurement process will not be examined further here. This discrete structure shows up in all measurements in which physical processes are studied with high spatial and temporal resolution. Rather than a continuous process, the measurement apparatus records a discrete sequence of elementary events, which cannot be further broken down or temporally resolved. We now illustrate these observable consequences of elementary events for precision measurements with the aid of a few examples.

Single events are evident and perceptible as *clicks* during detection of radio-active rays (Section 5.4.2). Transient discharges are produced when radioactive rays pass through a *counter tube* (Figure 6.33), invented by Hans Geiger in 1908, and these are measured as current pulses; they can be made audible with a suitable amplifier and loud speaker. A counter tube of this sort consists of a cylindrical gas-filled tube with a thin wire mounted along its axis. A high voltage is applied between the wall of the tube and the wire, but not high enough to ignite a self-sustained discharge. Only when a high-energy charged particle passes through the counter tube and leaves a trail of ions and electrons, will a short discharge pulse be produced by the electrons which are accelerated toward the positively charged wire. The current pulse flowing through the resistor R can be detected with a sensitive electrometer.

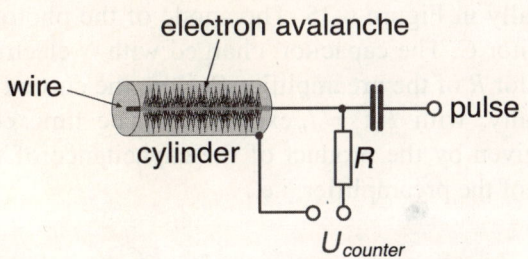

**FIGURE 6.33** A Geiger-Müller counter tube.

The Geiger counter was historically the first apparatus with which elementary events could be counted. Today, photons in the visible portion of the spectrum can be counted with a photomultiplier tube (photomultiplier or PMT, Figure 6.34) in order to determine the intensity of weak light sources. As in our demonstration of the photoelectric effect (Section 4.3.1), electrons can be released from the photocathode of a photomultiplier by photons (with energies $h\nu > W_A$). In a photomultiplier each photoelectron releases more electrons in the course of a cascade. The photocathode is kept at a negative voltage in the kV range relative to the anode. A voltage divider is used to supply an electrical potential to a series of intermediate *dynodes*. The photoelectrons are accelerated toward the first dynode and, on striking it, release approximately 3 secondary electrons, which are, in turn, accelerated toward the next dynode. This process proceeds from dynode to dynode, so that a current pulse of about $10^5$ electrons arrives at the anode. For luminous intensities that are not too high, these current pulses can be counted (a technique known as photon counting).

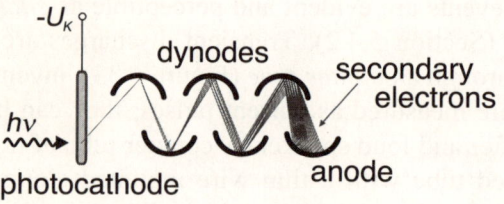

**FIGURE 6.34** A photomultiplier.

Nevertheless, the current pulses initiated by a photoelectron are substantially weaker than the current pulses produced in a Geiger tube following the passage of an ionizing particle. Thus, the current pulse must first be amplified by a preamplifier before it can be counted in a counter. An amplifying circuit is shown schematically in Figure 6.35. The anode of the photomultiplier is indicated here as a capacitor $C$. The capacitor, charged with $N$ electrons, discharges through the input resistor $R$ of the preamplifier $P$. Here the current flowing through $R$ falls off exponentially, with $I(t) = I_0 \exp(-t/\tau)$. The time constant of the exponential decay is given by the product of the capacitance of the capacitor and the input resistance of the preamplifier, i.e.,

$$\tau = RC.$$

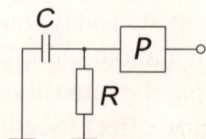

**FIGURE 6.35** A circuit for amplifying the current pulses from a photomultiplier.

Whether a current pulse can be identified uniquely as a pulse depends substantially on the energy $E$ delivered as the capacitor discharges into the input resistance of the preamplifier. Thus, we need to calculate this energy. The charge input $\int I(t)\,dt = I_0\tau$ is equal to the charge Ne at the anode of the photomultiplier and the electrical energy expended in the input resistor during the current pulse is given by $E = \int RI^2(t)\,dt = RI_0^2\tau/2$. As expected, it turns out that $E$ is equal to the energy stored in the charged capacitor $C$ (Section 3.4.4), i.e.,

$$E = \frac{(Ne)^2}{2C}.$$

This *signal* of a photoelectron is to be compared with the thermal *noise* (Section 6.3.3) of the current in the input resistor of the preamplifier. Only when the signal stands out from the noise, that is, the signal-to-noise ratio exceeds unity, can the current pulses be uniquely identified as such.

| | |
|---|---|
| **Notes** | At high temporal and spatial resolution all measurements turn out to be discrete sequences of elementary events. |

## 6.6.2 Statistical noise

The clicking of a Geiger counter does not follow a set rhythm like the ticking of a clock, but is purely random. This randomness, which is characteristic of all elementary events, will become clear in a simple series of measurements:

---

### Experiment 6.8    Measurement of a count rate

A Geiger counter tube is placed near a radioactive sample (Figure 6.36). The counter clicks at irregular time intervals, specifically whenever a particle of the radiation produces a discharge in the detector. The number of discharges per second is to be measured. Since the sequence of discharges is irregular, sometimes more and sometimes fewer events per second will be counted. A good clock, on the other hand, ticks exactly once per second. An

*continued*

essential indication of the randomness of the events is thus the fluctuation in the measured numbers $N_i$ $(i = 1, 2, \ldots, n)$ of events, counted over, say 10 s, in a series of $n$ measurements.

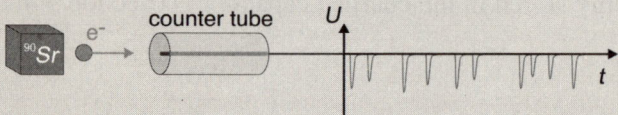

**FIGURE 6.36** Experimental arrangement of a Geiger tube and a random sequence of signals from it.

The measured counts first yield the average

$$\bar{N} = \frac{1}{n}\sum_{i=1}^{n} N_i .$$

The magnitude of the fluctuations about the average is expressed in terms of the *standard deviation s*, given by

$$s = \sqrt{\frac{1}{n-1}\sum_{i=1}^{n}(N_i - \bar{N})^2} .$$

For a strictly rhythmic sequence of events, such as the ticking of a clock, $s = 0$. For a purely random sequence, in which the events are in no way correlated with one another, so they are independent of one another, it can be shown mathematically that in the limit of very many measurements $(n \to \infty)$ $s$ is equal to the square root of the average number of events, that is

$$s = \sqrt{\bar{N}} .$$

These fluctuations in the measurement result, which result from the randomness of the elementary events, is known as *statistical noise* or *shot noise*. It is a fundamental consequence of the physics of the measurement process and cannot, therefore, be eliminated or reduced by even the most ingenious measurement techniques. The only possibility for reducing the measurement errors available to the experimenter is to extend the measurement time (or to make sure that more events take place). Then the number of events increases linearly with time, while the standard deviation only increases as the square root of the number of events. The (relative) accuracy of a measurement is given by the ratio of the standard deviation $s$ to the number of detected events,

$$\frac{s}{N} = \frac{1}{\sqrt{N}}.$$

The relative uncertainty of the measured value thus falls off as the square root of the number of measured elementary events.

---

**Problem 6.21**

How many events must be measured, at the least, for the count rate to be determined with an accuracy of 0.1%? By what factor must the measurement time be increased in order to reduce the uncertainty of the measurement by a factor of 10?

---

| **Notes** | The accuracy of a measurement increases as the square root of the number of measured events. |
|---|---|

## 6.6.3 Thermal noise

In addition to the randomness of elementary events, the thermal motion of atoms and the thermal fluctuations of fields detract from the accuracy of all measurements. For all measurements in which a signal is delivered to the input resistance of a preamplifier (Section 6.6.1), the thermal motion of the conduction electrons in the input resistor must be taken into account. The current owing to the thermal motion of the conduction electrons will also be amplified by the preamplifier and can thus interfere with the measured signals, so that they cannot be identified. Like elementary events, these thermal currents in the input resistor obey the laws of chance, so they are referred to as *thermal noise*.

The thermal noise of a resistor $R$ follows directly from the fact that a purely Ohmic resistance operates the same way as a black body (Section 4.4.3). Like a black body it absorbs all the electromagnetic waves flowing into it along its leads. And just like a black body, it radiates electromagnetic waves. In calculating the spectral emittance of a resistor, however, it should be kept in mind that the electromagnetic waves can only propagate in one direction, that is, along the leads.

Accordingly, we shall consider the radiative power emitted by a resistance $R$ per lead. It originates in the thermal excitation of electromagnetic oscillations on a conducting wire (Section 6.1.1). The radiative power $P$ which a resistor emits within a frequency interval $dv$ is the product $P = P(v)dv$ of the spectral emittance $P(v)$ and the frequency interval $dv$. The spectral emittance has the dimensions of energy and is given by the following well-known expression derived from the Planck radiation formula (Section 4.4.3):

$$P(v) = \frac{hv}{\exp(hv/kT) - 1}.$$

For the frequencies of interest in low- and high-frequency technology, usually $hv \ll kT$. Thus, in this frequency range the following approximation can be used:

$$P(v) = kT. \qquad \text{(Nyquist formula)}$$

In thermal equilibrium the resistance absorbs exactly as much radiative power as it emits. The total power from both leads is thus $4P(v)$.

The energy $E_{\text{th}}$ emitted as thermal radiation over a time $\tau$ is roughly equal to the radiative power emitted over the frequency range $0 < 2\pi v < 1/\tau$ times the time interval $\tau$. Thus, $E_{\text{th}} \approx kT$. The current pulse produced by the discharge of a capacitor with a capacitance $C$ is then only detectable if the energy $E = (Ne)^2/2C$ is significantly greater than the thermal energy $E_{\text{th}} \approx kT$.

---

### Problem 6.22

Estimate the gain a photomultiplier must have if a single photoelectron is to produce a detectable current pulse in the input resistor of a preamplifier.

---

**Notes**

The spectral radiative power $P(v)$ of the electromagnetic waves with $hv \ll kT$ emitted by Ohmic resistors along a lead wire is given by the Nyquist formula:

$$P(v) = kT.$$

### 6.6.4 Determinism and chance

In principle, the results of a measurement can never be exact because of statistical noise, referred to so vividly as *shot noise,* in reference to the granularity of objects and fields, and thermal noise, which originates in the thermal unrest of the elementary grains. All measurement results are fundamentally limited by a measurement uncertainty.

Based on this fundamental insight, let us now look back over the different, often seemingly contradictory concepts of physics. The foundations of the exact sciences are experiment on the experimental side and the logical structure of mathematics on the theoretical side. Both sides are equally important in physics. Mechanics (Chapter 1) is based on the assumption that exact relationships of unrestricted validity can be established among such different quantities as *force*, *mass*, and *acceleration.* This assumption contrasts with experiment, which can only provide measurement values that are burdened with measurement uncertainties.

In fact, the laws of mechanics are valid only to a limited extent. Only for objects that are sufficiently large do the measurement uncertainties owing to statistical and thermal noise become meaningless, so that the laws of mechanics are essentially satisfied exactly. But the smaller these objects become, the more important the noise will be and the more the concepts of mechanics will come into question. Modern micro- and nanotechnologies can no longer be explained on the basis of this world view alone. And these concepts fail completely when describing atomic processes.

For this reason, the atomic hypothesis (Section 2.1.1) initiated the evolution of physics from the world view of mechanics to today's modern concepts of physics. For the first time, the atomic hypothesis offered the possibility of combining the laws of chance with the deterministic laws of mechanics and, thereby, of explaining the phenomena of heat and irreversibility in natural processes.

The successes of thermodynamics and statistical mechanics were, however, bought at the cost of various obvious contradictions in the physical world view: on one side, the continuum picture and the determinism of classical mechanics, according to which all processes are ultimately reversible; and, on the other, chance and the discrete structure of matter, which are necessary to explain irreversibility.

This apparent contradiction is not, however, in conflict with the experimental basis of physics. A Maxwell demon (Section 2.4.1) could only be imagined in the framework of classical mechanics. Given the experimental fact that random elementary events are the basis of all measurements, the notion of a demon that follows the motion of atoms loses any experimental validity. Only with continuous observations could the completely similar atoms of a gas be distinguished and the trajectories of individual atoms tracked.

The contrast between continuum and discrete structure and, thereby, the contrast between determinism and chance, again became a topic of contemporary interest when James Clerk Maxwell formulated the basic laws of the electromagnetic field (Section 3.5.4). This is because, just like mechanics, electromagnetic theory is based on a continuum picture and ultimately leads, as does mechanics, to the conclusion that all processes are reversible. This conclusion is inconsistent with temperature equilibration by means of thermal radiation. The conflict was eliminated, in turn, by discretizing the continuum picture. In this case, Max Planck made the decisive step when he formulated the quantum hypothesis (Section 4.4.4).

With the atom and quantum hypotheses, the foundations of modern physics were laid. After the development of quantum mechanics, which could only be discussed in this book in a rudimentary fashion, starting with Bohr's postulates (Section 5.1.3) and the Schrödinger wave equation (Section 5.2.3), it became possible for the first time to explain atomic spectra (Sections 5.3.2 and 5.3.4) and the radioactivity of atomic nuclei (Section 5.4.2). Some difficulties arise in the intuitive interpretation of quantum mechanics. The duality according to which waves behave as particles and particles as waves, and which is expressed in the commonly used term *wave mechanics*, cannot be explained in terms of the concepts of Newtonian mechanics that have been imprinted on us. But Newtonian mechanics is not the measure of physics, experiment is. And experiment does not argue against the wave-particle duality.

Since the measurement process is based on elementary events, we cannot observe either waves or particles continuously. Nevertheless, we use these pictures in order to make the mathematically formulated laws of quantum mechanics *intuitive*. Only when we treat these intuitive pictures as a space-time reality does quantum mechanics seem contradictory to us, for quantum mechanics is substantially consistent and free of contradictions in its inner mathematical structure and is in agreement with its experimental foundations.

# Appendix

| Physical quantity | Symbol | SI unit |
|---|---|---|
| **Chapter 1** | | |
| Length | $l$ | m |
| Time | $t$ | s |
| Velocity | $v$ | m/s |
| Angular velocity | $\omega$ | $s^{-1}$ |
| Mass | $m$ | kg |
| Force | $F$ | $N = kg \cdot m/s^2$ |
| Work | $W_A$ | $J = kg \cdot m^2/s^2$ |
| Energy | $E$ | $J = kg \cdot m^2/s^2$ |
| Momentum | $p$ | $N \cdot s = kg \cdot m/s$ |
| Angular momentum | $L$ | $J \cdot s = kg \cdot m^2/s$ |
| Torque | $T$ | $N \cdot m = kg \cdot m^2/s^2$ |
| Moment of inertia | $J$ | $kg \cdot m^2$ |
| Frequency | $\nu$ | $Hz = s^{-1}$ |
| **Chapter 2** | | |
| Pressure | $P$ | $Pa = kg \cdot m^{-1} \cdot s^{-2}$ |
| Particle (number) density | $n$ | $m^{-3}$ |
| Temperature | $T$ | K |
| Amount of heat | $Q$ | $J = kg \cdot m^2/s^2$ |
| Amount of a substance | | mol |

| Heat capacity | $C$ | J/K |
|---|---|---|
| Molar heat capacity | $C_V$ | $J/(K \cdot mol)$ |
| Specific heat | $c$ | $J/(K \cdot kg)$ |

**Chapter 3**

| Energy density | $u$ | $J/m^3$ |
|---|---|---|
| Diffusion coefficient | $D$ | $m^2/s$ |
| (Coefficient of) viscosity | $\eta$ | $Pa \cdot s = kg \cdot m^{-1} \cdot s^{-1}$ |
| (Specific) thermal conductivity | $\lambda$ | $J \cdot m^{-1} \cdot s^{-1} \cdot K^{-1}$ |
| Intensity (energy flux) | $I$ | $J \cdot m^{-2} \cdot s^{-1}$ |
| Power | $P$ | $W = J/s$ |
| Electrical charge | $q$ | $A \cdot s$ |
| Electrical current | $I$ | A |
| Current density | $j$ | $A/m^2$ |
| (Specific) electrical conductivity | $\sigma$ | $A \cdot m^{-1} \cdot V^{-1}$ |
| Electric field strength | $E$ | V/m |
| Electric potential | $U$ | $V = J \cdot A^{-1} \cdot s^{-1}$ |
| Magnetic induction | $B$ | $V \cdot s/m^2$ |
| Capacitance | $C$ | $A \cdot s/V$ |
| Inductance | $L$ | $V \cdot s/A$ |
| Resistance | $R$ | $\Omega = V/A$ |
| Displacement polarization | $P$ | $A \cdot s/m^2$ |
| (Electrical) displacement | $D$ | $A \cdot s/m^2$ |
| Magnetization | $M$ | A/m |
| Magnetic intensity | $H$ | A/m |

**Chapter 4**

| Emittance | $E(T)$ | $W/m^2$ |
|---|---|---|
| Spectral emittance | $E(\nu, T)$ | $J/m^3$ |
| | $E(\lambda, T)$ | $W/m^3$ |

**Table B** Physical Constants.

| Physical constant | Symbol | Value in SI units | Value in eV units |
|---|---|---|---|
| Speed of light | $c$ | $2.997\,792\,458 \times 10^8$ m/s | |
| Gravitational constant | $G$ | $0.667 \times 10^{-10}$ m$^3 \cdot$ kg$^{-1} \cdot$ s$^{-2}$ | |
| Boltzmann constant | $k$ | $1.38 \times 10^{-23}$ J/K | $0.862 \times 10^{-4}$ eV/K |
| Avogadro number | $N_A$ | $6.022 \times 10^{23}$ mol$^{-1}$ | |
| Universal gas constant | $R = N_A k$ | $8.314$ J$\cdot$ mol$^{-1} \cdot$ K$^{-1}$ | |
| Elementary charge | $e$ | $1.602 \times 10^{-19}$ A$\cdot$ s (C) | |
| Permeability of free space | $\mu_0$ | $4\pi \times 10^{-7}$ V$\cdot$ s$\cdot$ A$^{-1} \cdot$ m$^{-1}$ | |
| Permittivity of free space | $\varepsilon_0$ | $8.854 \times 10^{-12}$ A$\cdot$ s$\cdot$ V$^{-1} \cdot$ m$^{-1}$ | |
| Coulomb interaction coefficient | $e^2/4\pi\varepsilon_0 = \dfrac{hc}{2\pi \cdot 137.036}$ | $0.23 \times 10^{-27}$ J$\cdot$ m | $1.44 \times 10^{-9}$ eV$\cdot$ m |
| Planck constant (quantum of action) | $h$ $h/2\pi$ $hc$ | $6.626 \times 10^{-34}$ J$\cdot$ s $1.055 \times 10^{-34}$ J$\cdot$ s $1.986 \times 10^{-26}$ J$\cdot$ m | $4.136 \times 10^{-15}$ eV$\cdot$ s $0.658 \times 10^{-15}$ eV$\cdot$ s $1.24 \times 10^{-6}$ eV$\cdot$ m |
| Electron mass | $m_e$ | $0.911 \times 10^{-30}$ kg | $0.511 \times 10^6$ eV |
| Proton mass | $m_p$ | $1.673 \times 10^{-27}$ kg | $0.938 \times 10^9$ eV |
| Proton/electron mass ratio | $m_p/m_e$ | $1836.153$ | |

# Index